Shape Memory Alloys for Seismic Resilience

Cheng Fang · Wei Wang

Shape Memory Alloys for Seismic Resilience

Springer

Cheng Fang
Tongji University
Shanghai, China

Wei Wang
Tongji University
Shanghai, China

ISBN 978-981-13-7039-7 ISBN 978-981-13-7040-3 (eBook)
https://doi.org/10.1007/978-981-13-7040-3

This Springer imprint is published by the registered company Springer Nature Singapore Pte Ltd.
The registered company address is: 152 Beach Road, #21-01/04 Gateway East, Singapore 189721, Singapore

Preface

A resilient city is one, which is able to rebound quickly from a disaster (e.g. earthquake) that has caused structural collapse, casualties, infrastructure disruption and business interruption. Although nowadays, most code-compliant structures have satisfactory collapse resistance under design-level earthquakes, the permanent structural and non-structural damages can be difficult and time consuming to repair, thus posing a great challenge to the affected communities for restoring essential services. After the Christchurch earthquake in 2011, a large number of buildings in the affected zone have been pulled down and it was estimated that the total cost of the rebuild would take up 20% of New Zealand's annual GDP. An independent study claimed that in a developed region such as California, a magnitude 7.8 scenario earthquake would have resulted in an estimated USD113 billion in damages to buildings and lifelines, and nearly USD70 billion in business interruption.

Lessons drawn from past major earthquakes urged a pressing need of a fundamental shift in the design targets for buildings in seismic zones. This provides an impetus for the development of low-damage and high-performance alternatives to conventional structural systems. One of the most promising strategies is to introduce self-centring capability into a structure. The core intention of the self-centring structural design is to minimise the post-earthquake structural damage and to eliminate permanent inter-storey drift. One feasible way is to utilise a unique class of smart metals, namely shape-memory alloys (SMAs), to achieve seismic resiliency. This class of material is capable of recovering large strains either spontaneously or by heating, depending on the thermal–mechanical state. Since the early development in the 1960s, SMAs have been successfully applied in the medical, aerospace, robotic and automobile industries. The consideration of SMA as emerging materials for seismic protection started in the 1990s, and great research progress has been made since then. However, the practical application of SMA to the construction industry has not been common, partially due to the lack of effective knowledge exchange between the communities of material scientists and civil engineers.

This book is intended to make a comprehensive summary of the up-to-date research achievements promoting the use of SMA for civil engineering. It helps to remove the knowledge barriers across disciplines, and sheds considerable light on the opportunity of commercialising SMA products against the seismic hazard. The results from the analysis are demonstrated with religious experimental verifications supplemented by numerical and analytical investigations. The book is mainly written for senior undergraduates, graduate students, academic researchers and practising engineers in the disciplines of civil and seismic engineering. The cutting-edge research introduced in this work aims to provide technical incentives to encourage design professionals, contractors and building officials to use high-performance smart materials in structural design, allowing them to remain at the forefront of construction technology. While this book is also of scientific interest to the mechanical and material science community, much emphasis is placed on making the relevant subjects easier to learn by civil engineers. Therefore, in most cases, the presentation has been structured following the civil engineering custom and terminology.

With the knowledge provided in this work, the readers are expected to get acquainted with the fundamental material properties of SMA, and to gain an in-depth understanding of the detailed working mechanisms and applications of a series of SMA-based structural elements, devices and members. In particular, the state-of-the-art research on SMA-based self-centring connections, braces and dampers is presented in detail, and the dynamic responses of the self-centring structural systems under various types of earthquake excitations are revealed. In addition, the book attempts to provide member-level and system-level design recommendations, which are compatible with the existing seismic design provisions. The economic seismic loss is also evaluated, which allows the reader to have a more direct recognition of the competitiveness of the SMA technology from a financial point of view. It is hoped that this work can also enlighten the researchers and practitioners engaged in other relevant areas such as wind engineering, blast engineering, vibration control, etc. where the SMA technology could be utilised.

The book is mostly from the research findings published over the past decade by the research team at Tongji University. Investigations by many other independent research groups, complemented by the authors' interpretations, are also included in this book with appreciation. The financial supports from the National Natural Science Foundation of China (NSFC), the State Key Laboratory of Disaster Reduction in Civil Engineering (SLDRCE) of China, International Joint Research Laboratory of Earthquake Engineering (ILEE) and Central University Fund for Interdisciplinary Research, Tongji University are gratefully acknowledged, and in addition, the authors would like to express their sincere gratitude to the following individuals: Prof. Yiyi Chen, Prof. Michael Yam, Prof. James Ricles, Prof. Richard Sause, Prof. Roberto Leon, Prof. Yunfeng Zhang and Prof. Bing Qu, for their long-term collaboration and assistance; Prof. Canxing Qiu for kindly sharing the original data for Chaps. 2 and 5; Dr. Yue Zheng for providing technical support for the manufacturing of the SMA cables; graduate students Hongliang Shao, Ce He, Jia Liu, Xiao Yang, Ao Zhang, Qiuming Zhong, Weikang Feng, Junbai Chen,

Linting Huang, Dong Liang and Shuling Hu, for their excellent work in this area over the years; and Zong Cao for assistance with the artwork during the preparation of the manuscript.

Shanghai, China Cheng Fang & Wei Wang

Endorsement

Improving community resilience requires development of more robust structural and non-structural systems and materials for infrastructure, both for new and existing buildings and facilities. In particular, it requires more sustainable and durable construction materials so that our society can respond to natural and man-made disasters in a more equitable way. This book, by providing the first detailed and comprehensive treatment of the use of shape memory alloys in seismic applications, is a very timely and welcome addition to the literature for this class of materials. The book focuses in developments in the last decade and emphasizes the synergies between the material properties and its applications. Its treatment of the topic follows a logical path, starting from basic properties and heat treatments, to their implementation in energy dissipative and recentering devices, and finally to the response of entire structures. It concludes with a very thorough application example that clearly demonstrates the potential for this technology. This clearly written and well-organized book will be an invaluable resource to anyone interested in innovative solutions to seismic design problems.

—Roberto T. Leon, DM ASCE,
D.H. Burrows Professor of Construction Engineering
Virginia Tech (Blacksburg, VA)

Contents

Notation

All symbols used in this book are defined where they first appear. For the reader's convenience, the principal meanings of the commonly used notations are contained in the list below. The reader is cautioned that some symbols may denote more than one quantity; in such cases, the meaning is clarified when read in context.

Abbreviations

2D, 3D	2-dimension, 3-dimension
BRB	Buckling-restrained braces
BRBF	Buckling-restrained braced frame
CoF	Coefficient of friction
CoV	Coefficient of variation
DBE	Design-based earthquake
DCH	Ductility class high
DCM	Ductility class medium
DIC	Digital image correlation
DM	Damage measure
DS	Damage state
DSC	Differential scanning calorimeter
DV	Decision variable
EDM	Electrical discharge machining
EDP	Engineering demand parameter
EVD	Equivalent viscous damping
FE	Finite element
FEMA	Federal Emergency Management Agency
HSS	High-strength steel
IDA	Incremental dynamic analysis
IDR	Inelastic displacement ratio
IM	Intensity measure

IMF	Intermediate moment frame
MCE	Maximum considered earthquake
MID	Maximum inter-storey drift
MRF	Moment-resisting frame
NiTi	Nickel–titanium
OWSME	One-way shapememory effect
PFA	Peak absolute floor acceleration
PFV	Peak floor velocity
PAR	Plasma arc remelting
PBSD	Performance-based seismic design
PEER	Pacific Earthquake Engineering Research
PGA	Peak ground acceleration
PGV	Peak ground velocity
PT	Post-tensioned/posttensioning
RDR	Residual displacement ratio
RID	Residual inter-storey drift
SC	Self-centring
SCBF	Self-centring braced frame
SDOF	Single degree of freedom
SEM	Scanning election microscope
SE-P	Superelastic–plastic
SMA	Shape-memory alloy
SME	Shape-memory effect
SMRF/SMF	Special moment-resisting frame
SE	Superelastic effect/superelasticity
TEM	Transmission electron microscopy
THA	Time-history analysis
TWSME	Two-way shape-memory effect
TIF	Transformation-induced fatigue
UTM	Universal test machine
VIM	Vacuum induction melting
VAR	Vacuum arc remelting

Roman Symbols

a_{max}	Peak absolute acceleration
a_t	Total acceleration
A	Cross section area
A_b	Shank area of bolt
A_{BRB}	Cross section area of the BRB steel core
A_f	Austenite finish temperature
A_s	Austenite start temperature
c	Damping coefficient

C	Spring index of helical springs
C_A	Change of critical stresses that induce reverse transformation with changing temperatures
C_M	Change of critical stresses that induce forward transformation with changing temperatures
C^{AS}, C^{SA}	Clausius–Clapeyron constants for the phase transformations
C_F	Coefficient of friction
d	Wire diameter of helical springs
d_h	Bolt hole diameter
D	Diameter
D_e	External diameter of Belleville washers or ring springs
D_i	Internal diameter of Belleville washers or ring springs
E	Young's modulus
E_A	Young's modulus of austenite SMA
E_M	Young's modulus of martensite SMA
E_s	Young's modulus of single-variant martensite
f_d	Damping force
f_{el}	Maximum force of linear elastic system
f_{max}	Maximum force of inelastic system
f_y	Yield strength/resistance
f_{yp}	Yield strength of the end-plate
$F_{b,pre}$	Preload applied to the SMA bolts
$F_{b,y}$	Yield resistance of SMA bolts/tendons
$F_{BRB,y}$	Yield resistance of the BRB
F_e	Yield force of helical springs
$F_{r,pre}$	Preload applied to the SMA ring springs
F_{rst}	Restoring force
$F_{r,y}$	Yield resistance of SMA ring spring set
$F_{Rd,y}$	Design yield resistance
$F_{w,pre}$	Preload applied to the SMA washers
$F_{w,y}$	Yield resistance of SMA Belleville washers
F_y	Yield strength/resistance (general)
g	Gravitational acceleration
g_1, g_2	Dimensions of steel angles
G	Shear modulus of material
h_i	Distance of the ith bolt row to the rotation centre
h_j	Distance of the jth SMA washer to the rotation centre
H^{AS}, H^{SA}	Scalar quantities in Auricchio model
H	Height of Belleville washers or ring springs
I	Second moment of area
I_a	Second moment of area of angle section
k	Stiffness (general)
k_i	Effective stiffness of SMA bolt with the consideration of preload
k_r	Stiffness of each single SMA ring spring
$k_{r\text{-}SMA}$	Reverse transformation stiffness of SMA material

k_w	Stiffness of SMA washer group
$K_{b,ini}$	Design/estimated initial stiffness of brace
$K_{c,ini}$	Design/estimated initial stiffness of connections
$K_{d,ini}$	Design/estimated initial stiffness of device
K_e	Elastic stiffness of subassembly system
K_o, K_p, K_g	Three-stage stiffness of steel angles under tension
$K_{r\text{-}ang}$	Reverse stiffness of single steel angle
$K_{r\text{-}rot}$	Reverse rotational stiffness of self-centring connection
$K_{r\text{-}SMA}$	Reverse stiffness of single SMA bolt
L_{arm}	Beam arm length
L_b	Beam length/span
$L_{e,BRB}$	Effective working length (yielding region) of BRB
L_p	Parallel/working length of the SMA bolts/tendons
m	Mass
M_{ang}	Moment contributed by steel angles
$M_{b,el}$	Design elastic moment resistance of connected beam
$M_{b,pl}$	Design plastic moment resistance of connected beam
M_f	Martensitic finish temperature
$M_{j,Rd}$	Design yield moment resistance of connections
M_{max}	Moment capacity of connections
M_{np}	Bolt moment strength
M_{pa}	Plastic moment capacity of angle section
M_{rev}	Reverse moment
M_{rst}	Restoring moment
M_s	Martensitic start temperature
M_{SMA}	Moment contributed by SMA bolts
M_{ya}	Yield moment capacity of angle section
n	Total number of SMA outer rings
N	Number of active coils in a helical spring
N_{ya}	Yield capacity of angle section
P	Applied load
P_{ang}	Maximum force of steel angle at the peak deformation
P_{rst}	Restoring bolt force of SMA
P_{rev}	Reverse force of steel angle
P_t	Bolt tensile strength
P_y, P_s, P_u	Three-stage load resistance of steel angle
R	Radius or strength ratio, as specified in the text
R_f	Rhombohedral phase finish temperature
R_s	Rhombohedral phase start temperature
s	Compressive deformation of Belleville washers
S	Overall cross-sectional tensile force of SMA outer ring
S_a	Spectral response acceleration
t	Thickness
$t_{pReq'd}$	Required end-plate thickness for SMA-bolted connections
T	Temperature or natural period of vibration, as specified in the text

T_0	Reference temperature
T_1	Fundamental period of vibration
T_p	Pulse period
$T_{p\text{-}v}$	Equivalent pulse period
u	Displacement
u_{el}	Maximum displacement of linear elastic system
u_{max}	Maximum displacement of inelastic system
u_{res}	Residual displacement
u_y	Yield displacement
V	Base shear
w	Uniformly distributed load along the perimeter of SMA outer ring
W	Seismic weight
W_D	Energy loss per cycle
W_E	Strain energy stored in a linear system that has the same maximum load and deformation as the nonlinear system
y_e	Yield displacement of helical springs
Y_p	Yield line mechanism parameter

Greek Symbols

α	Taper angle of SMA ring springs or post-yield stiffness ratio, as specified in the text
α_{is}	Empirical factor for initial stiffness of device with SMA ring springs
α_r	Parameter for reduction of yield resistance of SMA components
α_{sc}	Parameter for design of SMA preload in SMA-BRB hybrid braces
α^{AS}, α^{SA}, α_m	Material constants in Auricchio model
β	Energy dissipation factor or dispersion
β_c	Ratio of the maximum compression force to the maximum tension force of BRBs
β_m	Internal variable that describes martensite reorientation in Auricchio model
χ	Reduction factor for SMA performance
δ	Deformation
δ_u	Ultimate deformation
Δ	Displacement of each half-piece of SMA outer ring
Δ_b	Beam tip displacement
$\Delta\sigma$	Incremental Kirchhoff stress
$\Delta\varepsilon$	Incremental Kirchhoff strain
ε_{Af}	Reverse transformation finish strain
ε_{As}	Reverse transformation start strain
ε^e	Elastic strain in Auricchio model
ε_L	Transformation strain
ε_{Ms}	Forward transformation start strain

ε_{Mf}	Forward transformation finish strain
ε_{peak}	Peak strain
ε_{pre}	Prestrain
ε_{r}	Transformation-induced residual strain
ε_{res}	Residual strain
ε_{u}	Ultimate tensile strain
θ	Pitch or core angle of helical springs or Belleville washers
θ_{c}	Maximum inter-storey drift associated with the performance objective of collapse prevention
θ_{el}	Elastic inter-storey drift
θ_{p}	Plastic rotation
θ_{res}	Residual rotation
$\theta_{r,max}$	Maximum residual inter-storey drift among all stories
θ_{t}	Strain tensor related to thermal coefficient of expansion
$\theta_{t,max}$	Maximum transient inter-storey drift among all stories
κ	Training parameter in Auricchio model
ν_{A}	Poisson's ratio of austenite phase
ν_{M}	Poisson's ratio of martensite phase
π^{AS}, π^{SA}	Material constants in Auricchio model
ρ^{AS}, ρ^{SA}	Activation factors in Auricchio model
σ	Stress (general)
σ_{As}	Reverse transformation start stress
σ_{Af}	Reverse transformation finish stress
σ_{Ms}	Forward transformation start stress
σ_{Mf}	Forward transformation finish stress
σ_{max}	Maximum stress of SMA
σ_{OM}	Stress at the reference location OM of Belleville washers
σ_{pre}	Bolt prestress
σ_{s}^{cr}	Twinned martensite to detwinned martensite transformation start stress
σ_{f}^{cr}	Twinned martensite to detwinned martensite transformation finish stress
σ_{p}	Stress at the onset of plastic deformation of SMA
σ_{Rs}	Stress when reverse transformation is initiated for the case of $\sigma_{max} < \sigma_{Mf}$
σ_{rst}	Restoring SMA bolt stress
ΣF	Overall frictional force along the perimeter of SMA outer ring
ΣF_{V}	Vertical component of frictional force along the perimeter of SMA outer ring
ΣP	The force applied at one taper face of SMA outer ring
ΣP_{H}	Horizontal component of the force applied at one taper face of SMA outer ring
ΣP_{V}	Vertical component of the force applied at one taper face of SMA outer ring
υ	Velocity exponent

Ω	Phase transformation strain tensor
ω	Strain hardening adjustment factor for BRBs
ξ_{eq}	Equivalent viscous damping (EVD)
ξ_{mv}	Martensite volume fraction
ξ_R	Residual part of martensite volume fraction
ξ_s	Stress-induced martensite volume fraction
ξ_T	Temperature-induced martensite volume fraction
ξ_V	Reversible part of martensite volume fraction
ψ	Coefficient of thermal expansion
ζ	Viscous damping ratio

Chapter 1
Introduction to Shape-Memory Alloys

Abstract The background information on shape-memory alloys (SMAs) is presented in this chapter. A brief history of the development and application of SMAs is first introduced, which is followed by a detailed discussion on two fundamental properties of SMAs, i.e. superelastic effect (SE) and shape-memory effect (SME). The potential of the two unique properties for civil engineering application is then elaborated together with a demonstration of several existing projects that successfully adopted the SMA technology. The necessary manufacturing procedures ensuring satisfactory thermal–mechanical properties of SMA products are highlighted, and the civil engineering-oriented experimental characterization methods are introduced. Finally, the available constitutive models for SMAs are introduced with a focus on phenomenological modelling approaches, which are more suited to the civil engineering application. Although the information provided in this chapter would suffice for practical civil/structural engineers who are interested in this topic, the reader should recognise that a wealth of knowledge exists on the relevant subjects, which are available in a vast body of literature.

1.1 Shape-Memory Alloys—A Brief History and Future Trends

The first discovery of shape-memory alloys (SMAs) can be traced back to 1932 (Ölander 1932), although the term 'shape memory' was first introduced almost 10 years later (Vernon and Vernon 1941). Researchers then confirmed that the phenomenon of shape memory was attributed to reversible martensitic transformation. This metallurgical characteristic received significant attention among the community of material scientists at that time. Prior to the 1950s, however, only a small number of alloys, such as CdAu, CuZn and CuAl, were found to exhibit such transformation phenomenon, and these materials remained unutilised, until Buehler and his co-workers discovered the shape-memory effect (SME) in a nickel–titanium (NiTi) alloy in 1963 (Buehler et al. 1963). It was later found that the addition of a third element such as Fe and Nb could change the material properties (e.g. transformation temperatures) dramatically, and it was also realised that the microstructure,

C. Fang and W. Wang, *Shape Memory Alloys for Seismic Resilience*, https://doi.org/10.1007/978-981-13-7040-3_1

manufacturing procedure and heat treatment/annealing could significantly affect the behaviour of the SMA (Frick et al. 2005). The extensive laboratory work on SMAs also revealed that the strain of the material can be automatically recovered at room temperature, a phenomenon known as superelasticity (SE). These findings promoted follow-up investigations on more diverse SMA types and classes, and eventually led to practical applications of the SMAs in various industries. In honour of the Naval Ordnance Laboratory where the material was first discovered, NiTi SMA is also widely known as nitinol, an ingenious combination of the material composition and the place of discovery. Until the present date, NiTi system is still one of the most successful classes of SMA in practice.

The first commercial success for application of NiTi SMA was the pipe coupler (CryoFit™) in F-14 fighter aircraft developed in 1969 (Kauffman and Mayo 1997). Continued research led to the application of NiTi SMA in the extended military and civil engineering fields such as repairs for battle damage and nuclear reactors (Lagoudas 2008). Since the 1970s, the widespread use of NiTi SMA started to emerge in the biomedical sector, but the true commercial breakthrough was made in the 1990s when NiTi SMA stents were developed and widely accepted by the public. Concurrently, SMAs had found additional applications in electrical, automotive, aerospace, robotics and oil industries. These applications often require that the SMAs exhibit SME or SE with a wide temperature/load hysteresis and stable material behaviour, and in this regard, NiTi SMA seems to be superior to other classes of SMA, as revealed by the extensive laboratory work at that time (Huang 2002). Nevertheless, continuous developments in alloy composition inspiration and manufacturing technology revealed the great potential of copper-based or iron-based SMAs, which could exhibit comparable behaviour to NiTi SMA, or acceptably compromised performance with much-reduced cost (Hodgson et al. 1990; Cladera et al. 2014). The booming development of various SMA classes also attracted great attention in the civil engineering community in the late 1990s, where a particular effort was devoted to the mitigation of seismic effects on engineering structures (Song et al. 2006).

Since the twenty-first century, continuous research efforts have been made in the development and application of SMAs. A survey conducted using the SCOPUS and USPTO search engines (Figs. 1.1 and 1.2) indicated that from 1990 to 2013, more than 19,000 US patents have been issued on SMAs and relevant products, where those related to the biomedical industry take up more than 60% of the total patent number (Jani et al. 2014). Over the same period, nearly 20,000 articles related to SMA have been published, where the topics cover engineering, physics, astronomy, material science, medicine, etc. By the end of 2019, more than 25,000 US patents will have been authorised, and more than 30,000 articles will have been published (Jani et al. 2014). In particular, the application of SMA in civil engineering has received great attention. Using the Scopus search engine with searching keywords of ‘shape memory alloy’ or ‘nitinol’ and ‘civil engineering’ or ‘seismic’, more than 800 relevant journal and conference papers have been published from 1990 to 2017.

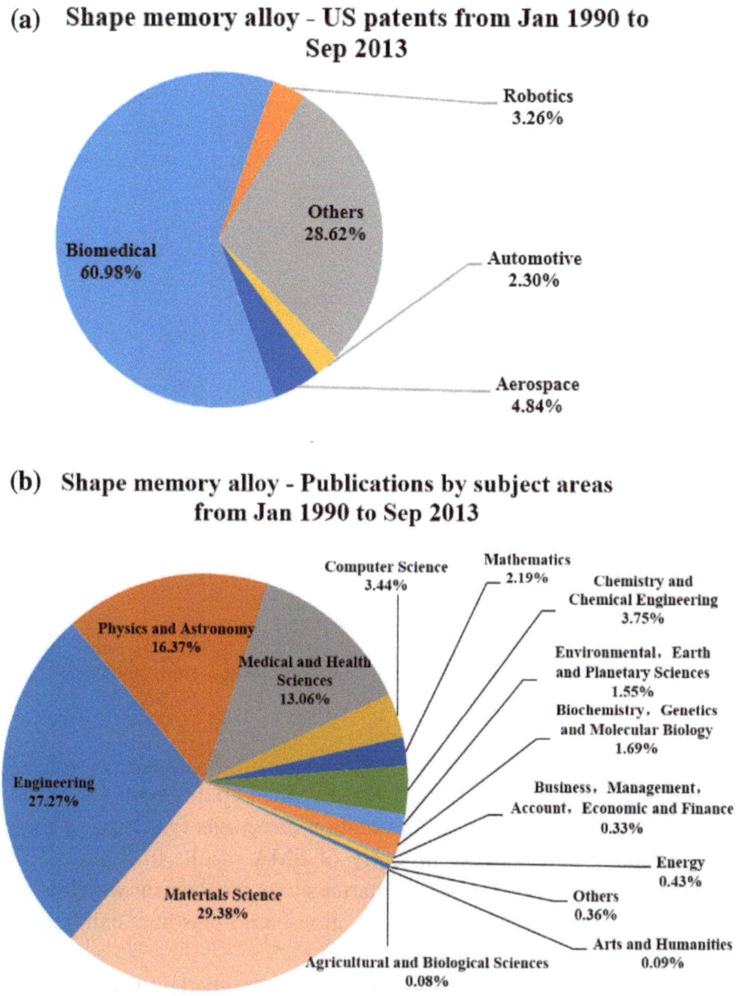

Fig. 1.1 Results by subjects and sectors from the survey using SCOPUS and USPTO search engines: **a** US patents, **b** publications (Jani et al. 2014)

1.2 Basic Properties of Shape-Memory Alloys

SMAs are a unique class of metals capable of recovering large strains either spontaneously or by heating, depending on their thermal–mechanical state. The microstructure and crystallography of SMAs have been extensively investigated by material scientists and engineers (Otsuka and Wayman 1998; Gall et al 1999a, b; Bellouard 2008; Cisse et al. 2016). The material has two prevailing phases, i.e. martensite and austenite, with three different crystal structures, i.e. twinned martensite, detwinned martensite and austenite. The austenite is stable at higher temperatures and the martensite

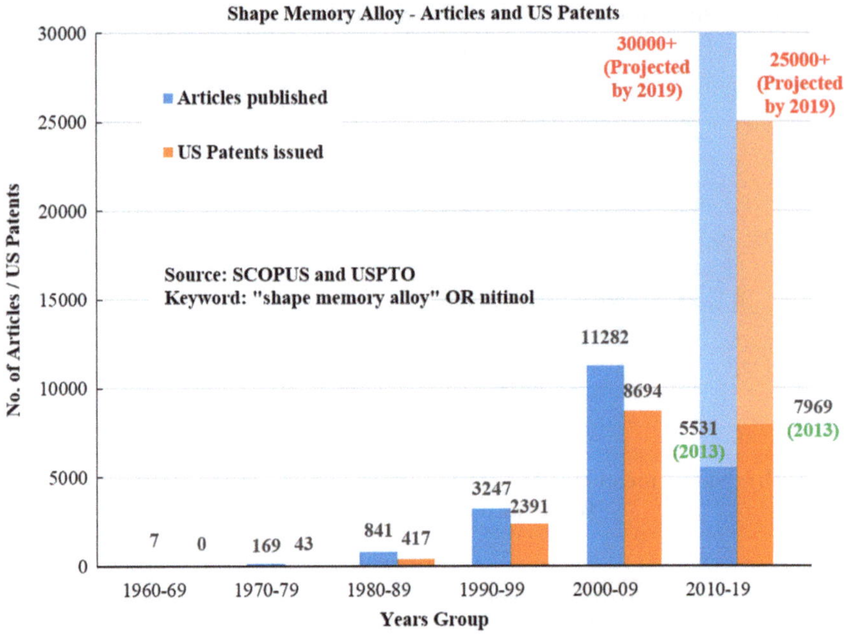

Fig. 1.2 Number of publications and US patents related to SMA—from 1990 to 2013 (Jani et al. 2014)

is stable at lower temperatures. The existence of the two different martensite crystal structures is related to the different orientation directions (variants) of the martensitic crystal. The strain recovery property of SMA results from reversible phase transformations between austenite and martensite, a mechanism which is attributed to shear lattice distortion rather than dislocation-based plasticity exhibited by other constructional metals such as carbon steel.

The phase transition from austenite to martensite is called forward transformation, and that from martensite to austenite is called reverse transformation. There are four characteristic temperatures that are associated with the transformation processes, namely martensitic start temperature M_s, martensitic finish temperature M_f, austenite start temperature A_s and austenite finish temperature A_f. In the absence of an applied load (stress), SMA is transformable between austenite and twinned martensite with changing temperatures, as illustrated in Fig. 1.3. Upon cooling, SMA changes from austenite to twinned martensite. This process starts when the temperature decreases to M_s, and finishes at M_f. The temperature gap between M_s and M_f varies with different types of SMA material. Similarly, when the temperature increases, the material is transformed from the twinned martensite phase to the austenite phase. The transformation initiates at A_s and is completed at A_f. Under certain circumstances, a rhombohedral phase (R-phase), also called a 'premartensitic' phase, can be induced

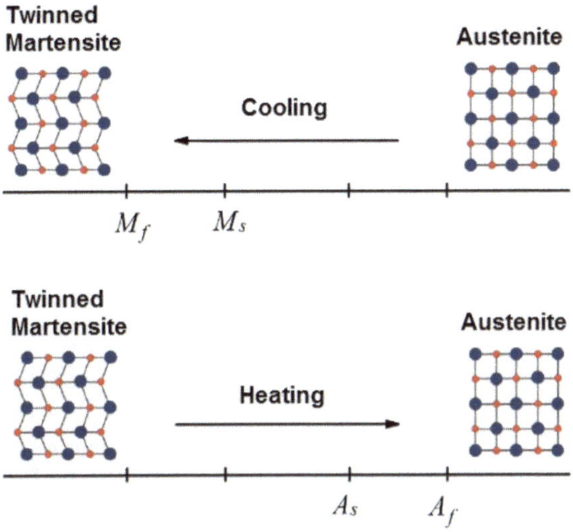

Fig. 1.3 Phase transformations due to changing temperature (Lagoudas 2008)

as a result of certain thermal processing, in which case a two-step transformation with the R-phase existing between the martensite and austenite phases may occur.

The transformation becomes more complicated when stresses are involved in the transformation process. Depending on the external conditions (e.g. environmental temperature), SMAs can exhibit either shape-memory effect (SME) or superelastic effect (SE). When the material is loaded at a temperature below M_f, as shown in Fig. 1.4, a transformation from the twinned martensite to detwinned martensite occurs. This process starts when the stress achieves $\sigma_s{}^{cr}$ and finishes at the end of the loading plateau at $\sigma_f{}^{cr}$. With a further increase in stress, the elastic behaviour of the detwinned martensite is exhibited, until the permanent yield stress of the martensite is reached. If the material is unloaded at a stress level below the permanent yield stress, transformation-induced residual strain ε_r remains, but heating the deformed SMA above A_f can promote strain recovery, i.e. SME. The maximum recoverable strain due to the transformation from detwinned martensite to austenite is termed as transformation strain ε_L. A subsequent cooling process causes a transformation from austenite to twinned martensite with no further change of the shape. The above complete SME cycle is also called one-way shape-memory effect (OWSME). It should be noted that certain SMAs can have a two-way shape-memory effect (TWSME) where the recovered shape changes back during the final cooling stage (Ma et al. 2010). However, TWSME is commercially less applied due to more demanding training requirements and unsatisfactory cyclic stability (Perkins and Hodgson 1990; Huang and Toh 2000).

On the other hand, when the SMA is loaded at a temperature above A_f, i.e. at its austenite phase, the material accommodates the strain by transforming from austenite into detwinned martensite. Upon unloading, up to 8–10% strain can be recovered spontaneously. This phenomenon is termed as SE. As shown in Fig. 1.5, there are four

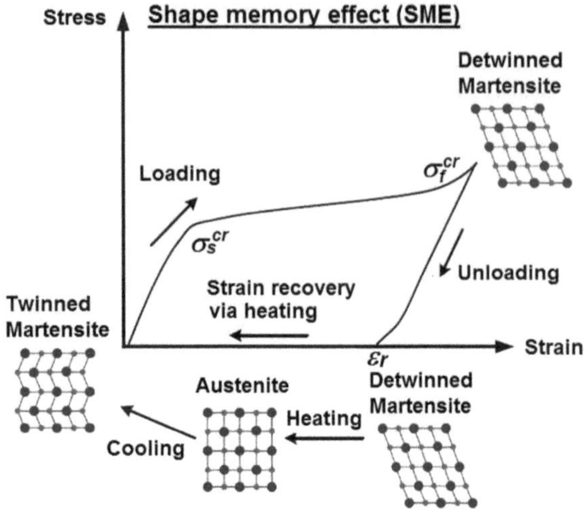

Fig. 1.4 Illustration of a typical SME cycle

characteristic stresses during a complete stress–strain cycle of SE, namely forward transformation start stress σ_{Ms}, forward transformation finish stress σ_{Mf}, reverse transformation start stress σ_{As} and reverse transformation finish stress σ_{Af}. At the beginning of loading, the material behaves linearly with the slope of the stress–strain curve being equal to the elastic modulus of the austenite. Forward transformation starts when the stress achieves σ_{Ms}, and then the stress–strain curve advances into the forward transformation plateau. The shape of the stress–strain curve during the forward transformation stage is quite similar to that reflecting the typical yielding behaviour of normal steel, but noting that the latter leads to permanent unrecoverable strain. The forward transformation finishes when the stress reaches σ_{Mf}, where the material is completely transformed into detwinned martensite. The stress can further increase elastically until reaching the permanent yield stress. Upon unloading, the stress decreases linearly towards σ_{As}, and the onset of reverse transformation is indicated by an evident change of the unloading slope on the stress–strain curve. The reverse transformation plateau, accompanied by strain recovery, initiates at σ_{As} and ends at σ_{Af}. The unloading path tends to intersect with the initial elastic loading path, and upon the completion of the reverse transformation, the stress decreases linearly to zero with, ideally, no residual strain.

As mentioned in Sect. 1.1, among all the available classes of SMA, NiTi SMA has been considered as a preferred choice for applications in various engineering fields because of its remarkable strain recovery, stable thermal/loading hysteresis, satisfactory fatigue and corrosion resistance and well-understood thermal processing techniques. The typical properties of NiTi wires, in either the austenite or martensite phases, are summarised and compared with structural steel in Table 1.1 (Penar 2005; Fugazza 2005; Janke et al. 2005). The physical properties of the alloy are generally comparable with those of structural steel. The yield strength, ultimate strength

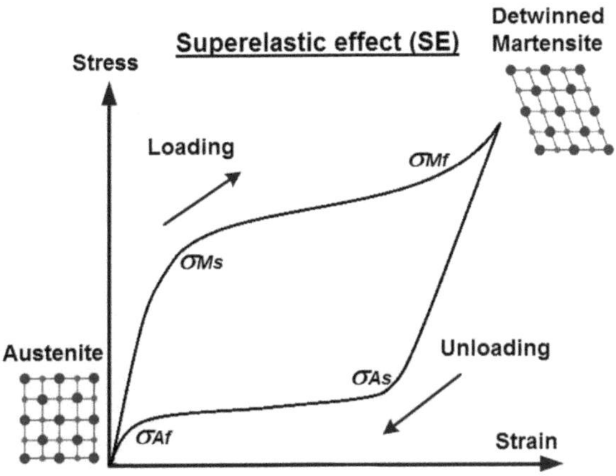

Fig. 1.5 Illustration of a typical SE cycle

Table 1.1 Typical physical and mechanical properties of NiTi SMA wires

Material properties	Austenite NiTi	Martensite NiTi	Structural Steel
Melting point	1240–1310 °C	1240–1310 °C	1500 °C
Density	6.45 g/cm^3	6.45 g/cm^3	7.85 g/cm^3
Thermal conductivity	0.28 W/cm °C	0.14 W/cm °C	0.65 W/cm °C
Thermal expansion	11.3e−6/°C	6.6e−6/°C	11.7e−6/°C
Recoverable elongation	8–10%	8–10%	0.2%
Young's modulus	30–83 GPa	21–41 GPa	205 GPa
Yield strength	195–690 MPa	70–140 MPa	235–460 MPa
Ultimate tensile strength	895–1900 MPa	895–1900 MPa	390–540 MPa
Elongation at failure	15–25%	20–60%	20%
Poisson's ratio	0.33	0.33	0.30
Hot workability	Quite good	Quite good	Good
Cold workability	Not easy	Not easy	Good
Machinability	Not easy	Not easy	Good
Weldability	Quite good	Quite good	Very good
Corrosion performance	Excellent	Excellent	Fair

and ductility are within desirable ranges for engineering applications, although the Young's modulus is lower than that of steel.

When SMA is intended to be used as a superelastic material at room temperature (i.e. to serve at its austenite phase), special attention should be paid to its temperature dependency characteristic. This can be demonstrated by a phase diagram over the stress and temperature domain, as shown in Fig. 1.6. At different temperatures, the characteristic stresses relevant to the forward and reverse phase transformations are marked in the stress–strain curves. These critical stresses coincide with the ordinates of the intersection points at the inclined phase boundary lines. With changing tem-

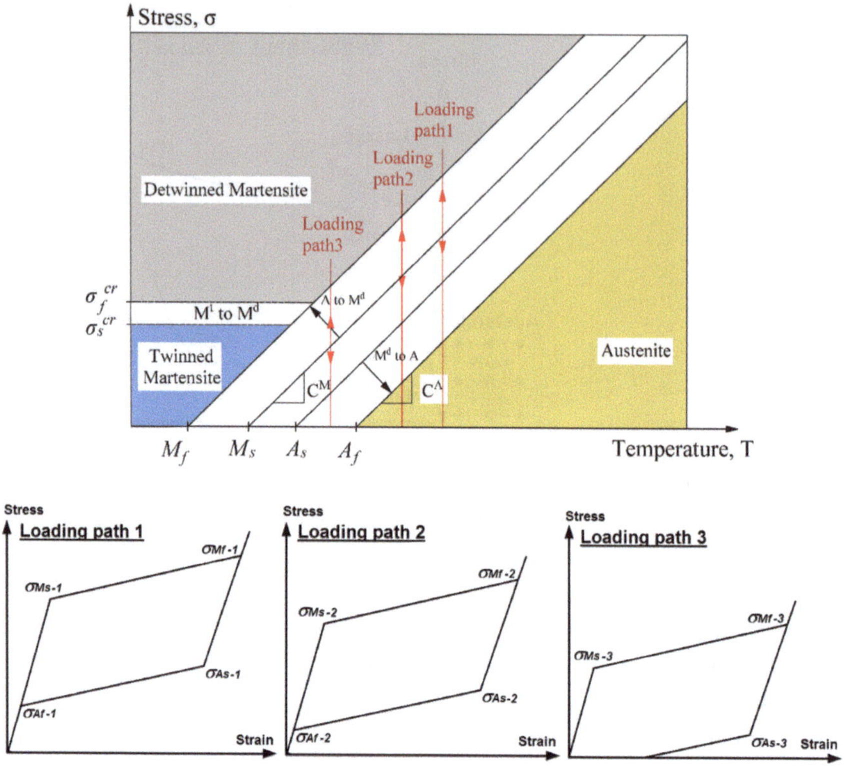

Fig. 1.6 Phase diagram of SMA

peratures, the critical stresses move upwards or downwards with slopes of C_M and C_A. Such temperature-related slopes can vary with different SMA classes. For NiTi SMA, in particular, the typical value of the slopes ranges from 6 to 7 MPa/°C. In other words, if we consider a reference environmental temperature of 20 °C, and expect that the working temperature ranges from, say, 0 to 40 °C, the maximum variation of the 'yield' (forward transformation) plateau can be ±130 MPa. Smaller stress–temperature slopes are observed in Cu–Al–Be SMA, where C_M and C_A values of 1.5–2.0 MPa/°C were reported (Qiu and Zhu 2014). Such uncertainty should be considered in the practical design, may be in the form of reliability-related factors. Figure 1.6 also demonstrates that when the material is loaded at a temperature below A_f, e.g. loading path 3, the unloading plateau decreases and a residual strain is induced. This residual strain is recoverable via subsequent heating to a temperature above A_f.

It is worth mentioning that the uniaxial behaviour of SMA wires can be different from the behaviour of other forms of SMA products (e.g. bars) due to the variation of the necessary annealing and forming procedures (Tyber et al. 2007; McCormick et al. 2007). The ductility of superelastic SMA bars is often lower than their wire

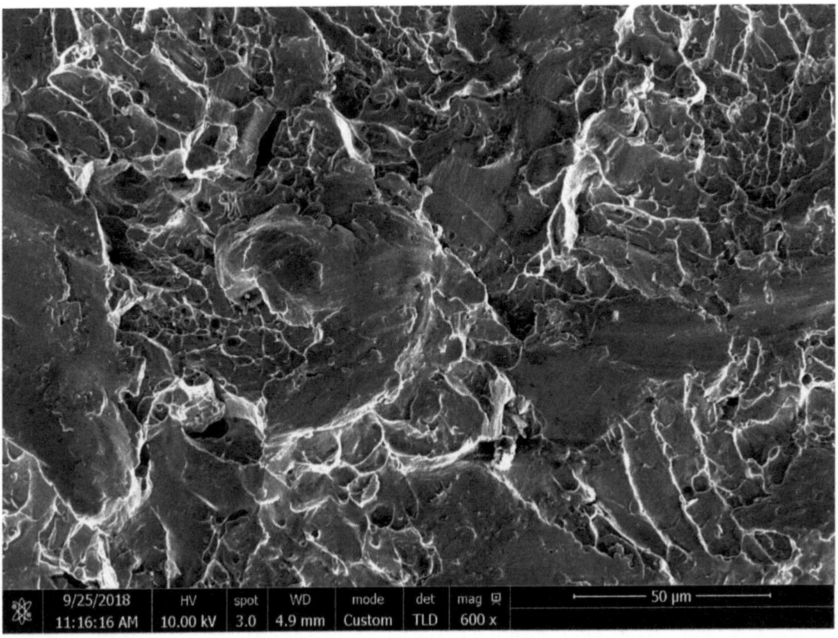

Fig. 1.7 SEM fractograph of a typical superelastic SMA bar

counterparts (fracture strain is typically less than 15% for the former). This is due to an insufficient microvoid growth and coalescence process, prohibiting the inelastic flow of the material around initiated microvoids (Kanvinde 2016; Anderson 2017). Figure 1.7 shows the scanning election microscope (SEM) fractograph of a fracture surface of an SMA bar, which fractures at a strain of around 10% with negligible necking. Both dimpled fracture pattern, indicating ductile behaviour, and river patterns with shiny facets, indicating brittle cleavage behaviour, are observed from the SEM fractograph. This suggests that the brittle failure of the SMA bar is preceded by certain plasticity and ductile crack growth. As a result of the subsequent transition to unstable, brittle propagation, the overall ductility of the SMA bar is limited. Moreover, large-scale NiTi SMA elements are not easily cold worked, and abrasive techniques are preferred when machining the material.

1.3 Shape-Memory Alloys for Civil Engineering

With an increasing need for more hazard resilient structural systems, the knowledge of SMA has disseminated dramatically to the civil engineering community over the last two decades. The potential of utilising SMAs for civil engineering applications was not recognised, until the initiation of two pioneering research projects,

Fig. 1.8 S. Giorgio Church, Italy (Indirli et al. 2001a)

i.e. 'Manside' and 'Istech' projects, in the 1990s (Chang and Araki 2016). The two research programmes led to successful implementations of NiTi SMA devices in the retrofitting of historic masonry structures in Italy. Compared with SME, SE seems to receive more attention for seismic applications due to spontaneous self-centring, stable hysteretic damping and free of external power source/electricity supply (Ozbulut et al. 2011). One of the most famous projects that incorporated the SMA technology was the retrofitting of the S. Giorgio Church (Fig. 1.8). The structure was an ancient chapel, which was built in 1302. It consists of a central 18.5 m-high bell tower and surrounding structures with heights up to approximately one-half of the height of the tower. The upper part of the bell tower was seriously damaged by a Richter Magnitude 4.8 earthquake, which occurred on 15 October 1996. After that earthquake, severe transverse, vertical, diagonal and even 'cutting-through' cracks were observed in the corner columns, masonry walls and mullioned windows of the tower, and as a result, the structure was tagged as unsafe and entrance forbade (Indirli et al. 2001a).

In order to restore the tower of the S. Giorgio Church, a series of SMA devices in series to prestressed steel tie bars were installed vertically along the height of the four corner column of the main tower (Fig. 1.9). Each SMA device includes 60 superelastic NiTi SMA wires of 1 mm diameter and could offer a load-carrying capacity (i.e. 'yield' resistance) of around 20 kN. The device enters into the forward transformation plateau when the applied load exceeds 20 kN. This is, in fact, an important benefit of the SMA solution where the maximum load applied to the masonry members can be well controlled. Several in situ dynamic tests revealed that the retrofitted tower exhibited higher stiffness and different modal shapes compared with the unrepaired damaged tower. Nevertheless, the efficacy of these SMA devices was not demonstrated until a new Richter Magnitude 4.5 seismic event occurring on 18 June 2000 with the same epicentre. With a careful inspection after the main shock, the tower showed no damage of any type. Adopting the same approach, the bell tower of Badia Fiorentina was strengthened by 18 SMA devices in 2006 (Chang and Araki 2016).

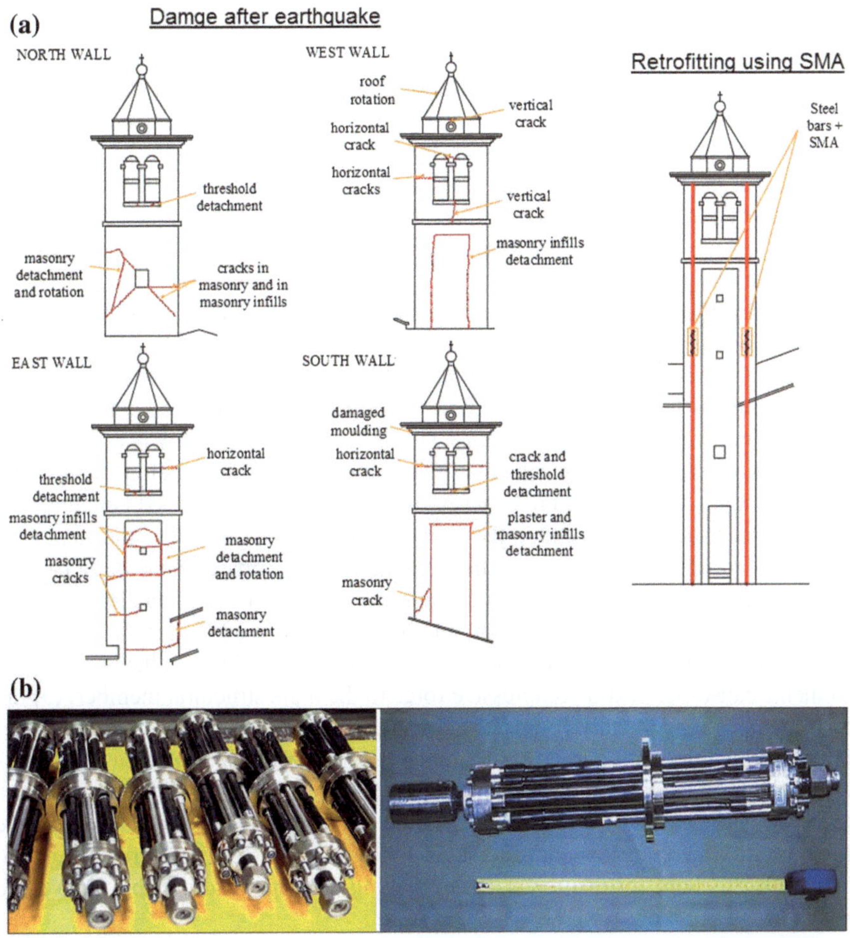

Fig. 1.9 Retrofitting of S. Giorgio Church with SMA devices: **a** damage pattern, **b** SMA devices (Indirli et al. 2001a)

Two other practical examples of using SMA devices are the restoring intervention programmes for the transept tympana of the St. Francis Basilica in Assisi and for the San Feliciano Cathedral facade at Foligno, both heavily damaged by the 1997 Marche and Umbria earthquake (Indirli et al. 2001b; Indirli and Castellano 2008). Being different from the installation method considered for the bell tower of the S. Giorgio Church, the SMA devices were placed perpendicularly to the masonry walls to prevent out-of-plane collapse caused by the inertia forces orthogonal to the façades. A total of 47 SMA devices were used to repair the façade of the transept tympana of the St. Francis Basilica (Croci 2001; Martelli 2008). Numerical investigations showed that the SMA devices could effectively improve the energy-dissipation performance and overall earthquake resistance of the structures. By 2008, at least 19 buildings in

Italy were retrofitted by SMA or other energy-dissipation devices (Benavent-Climent 2008).

The basic design intention for the abovementioned projects was to utilise the inherent energy-dissipation capacity of superelastic SMA when subjected to cyclic loading. The material has a clear yield (forward transformation) plateau which can help control and limit the maximum stress induced in the masonry walls and columns. Another key motivation for using superelastic SMA lies in its unique self-centring capability, such that the post-earthquake residual deformation and the associated structural damage could be reduced.

There are alternative solutions that provide self-centring capability for structures. Post-tensioning (PT) technique is one of them (Ricles et al. 2001, 2002; Garlock et al. 2008; Wolski et al. 2009; Lin et al. 2013a, b). The PT solution was typically implemented by the application of PT steel tendons/cables, anchored at the external columns, in parallel to the steel beams. The self-centring capability is provided via a connection gap opening mechanism that induces elastic elongation of the steel tendons/cables, where additional energy-dissipation devices are used concurrently. Due to the elastic stress stored in the elongated steel tendons/cables, this gap opening is closed upon the removal of the seismic load. Compared with the PT-based self-centring strategy, the SMA solution may have the following benefits: (1) SMA has a much larger recoverable strain, and hence can be used for resisting stronger earthquakes without compromising the self-centring capability; (2) the use of SMA elements causes less extra compressive force to the main structural member; (3) the SMA devices are generally compact in size and are easy to install; (4) SMA has negligible creep deformation in its elastic range (Raj and Noebe 2013), which indicates that the loss of preload is minimal during the service life, as long as the prestress level does not exceed σ_{Ms}; (5) SMA has good fatigue resistance and can experience long duration earthquakes and aftershocks with no need for replacement; and (6) SMA has excellent corrosion resistance, making it economically viable from a life cycle design point of view.

In parallel to the utilisation of superelastic SMAs for structural retrofitting, SME has also been considered in civil engineering practice. The basic principle lies in the fact that SME enables elongated SMA elements (e.g. bars) to recover to their original shape through heating, and this behaviour is of particular interest for the purpose of crack closure. The first field implementation using SME was on a highway bridge on Sherman Road in Michigan, USA (Soroushian et al. 2001). The concrete bridge girders had suffered cracks in the T-beams due to insufficient shear resistance. To strengthen the bridge girder, post-tensioning action was realised by applying a device incorporating a series of Fe–Mn–Si–Cr SMA rods with a diameter of 10.4 mm. The device was mounted crossing the cracks onto the surface of the girder (Fig. 1.10). Holes were first drilled through the web of the T-beam, and the holding angles were then anchored onto the web surface. By heating the SMA rods to 300 °C with electrical power, the recovery action of the SMA rods resulted in a reduction of the crack width by 40%. After air cooling, the restrained shape recovery force induced in each SMA rod reached 10 kN.

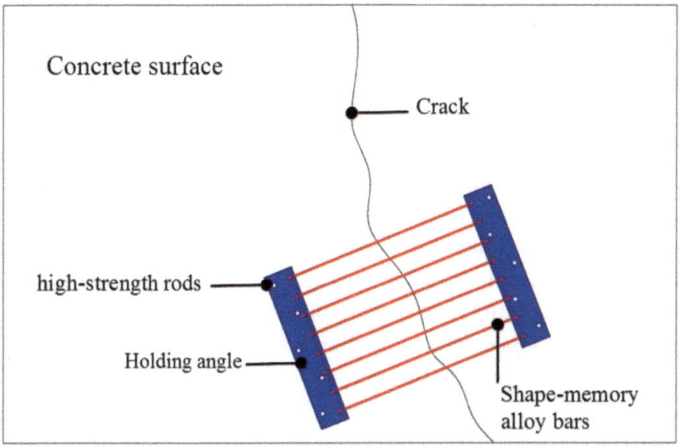

Fig. 1.10 Implementation of SME for crack closure for a highway bridge in Michigan, USA

After the initial successes of the Manside, Istech and other projects, significant follow-up research efforts have been made, and a large number of relevant research projects have been granted by major funders, especially those in seismic-active countries. Since 1990, the US National Science Foundation (NSF) has launched more than 15 research projects related to the application of SMA in civil engineering or relevant areas. The more recent ones launched after 2010 include:

- Multi-Hazard Response Mitigation Systems Using High Strength and Damping Capacity Shape-Memory Alloys (Award No.: 1538770);
- Self-Stressing Shape-Memory Alloys/Fibre-Reinforced Polymer (SMA/FRP) Composite Patches for Rehabilitation of Cracked Steel Structures (Award No.: 1100954);
- Innovative Seismic Retrofits for Resilient-Reinforced Concrete Buildings (Award No.: 1041607);
- Innovative Confinement Technology for Strong Main Shock–Aftershock Damage Mitigation (Award No.: 1055640);
- Increasing Resiliency and Sustainability of Reinforced Concrete against Ageing and Seismic Hazards through Novel Materials (Award No.: 1642488);
- Interaction of Smart Materials for Transparent, Self-regulating Building Skins (Award No.: 1548243).

There are also many other NSF projects with subtasks related to the application of SMA in civil engineering.

The National Natural Science Foundation of China (NSFC) has approved at least 10 relevant research grants after 2010, covering the use of SMAs for steel frames (Grant Nos. 51408437, 51208447), concrete structures (Grant Nos. 51478438, 51108426, 51008314, 51578289, 51678110), base isolators (Grant No. 51308195), pipelines (Grant No. 51308357), cable-stayed structures (Grant No. 51108269) and historic towers (Grant No. 51178388). SMA-related research is also encouraged in seismically non-reactive countries. In 2016, the Engineering and Physical Sciences Research Council (EPSRC) in the UK initiated a project that investigates the application of SMA in retrofitting concrete structures (EP/N029127/1).

The massive research-oriented investments made over the last two decades have dramatically accumulated the necessary knowledge that may prepare engineers for more widespread use of SMAs in civil engineering practice. Although there are still knowledge barriers across materials science and structural engineering fronts, a large number of novel SMA devices and members, including dampers, braces, beam-to-column connections, column feet, base isolators and smart concrete components, have already been developed. More detailed descriptions of some of these innovative studies are reserved for Chaps. 2 through 4.

1.4 Manufacturing and Annealing

1.4.1 Typical Manufacturing Procedures

The manufacturing process for SMA involves a number of necessary working steps, which all affect the final thermal–mechanical properties of the SMA products. The required procedures vary with different SMA classes and products. In this section, only the NiTi SMA is discussed, where the core manufacturing and heat treatment steps are introduced. The flowchart diagram of the typical working processes for commercial NiTi SMA products is summarised in Fig. 1.11 (Lecce and Concilio 2015). The melting process is the first step that ensures desired properties of SMA by strictly controlling the chemical composition and impurity content. The typical melting procedure includes vacuum induction melting (VIM), vacuum arc remelting (VAR), plasma arc remelting (PAR), etc. These methods are used either individually or in combination. The melting process is often followed by a hot-working process at 800–950 °C to reduce the size (e.g. diameter) of the starting cast ingot and to set up homogenisation microstructures. Subsequently, the SMA ingot could be further cold worked or hot worked, until the wanted dimensions and mechanical characteristics are achieved.

Many SMA classes, including NiTi, can be cold worked (i.e. reduction in cross-sectional area of the specimen due to drawing at room temperature) to final desired forms such as wires and bars. Although it is widely recognised that cold working could endow normal constructional metals, such as carbon steel, with higher strength and fatigue resistance due to increased dislocation densities, the conventional cold-working procedure may not be directly applicable to NiTi SMA elements which are supposed to have, apart from good strength and fatigue resistance, more unique monotonic and cyclic mechanical properties such as superelasticity (Speicher 2010). Alternatively, hot working, which is more economical than cold working (Tyber et al. 2007), is an attractive option, especially for large-size NiTi SMA elements such as

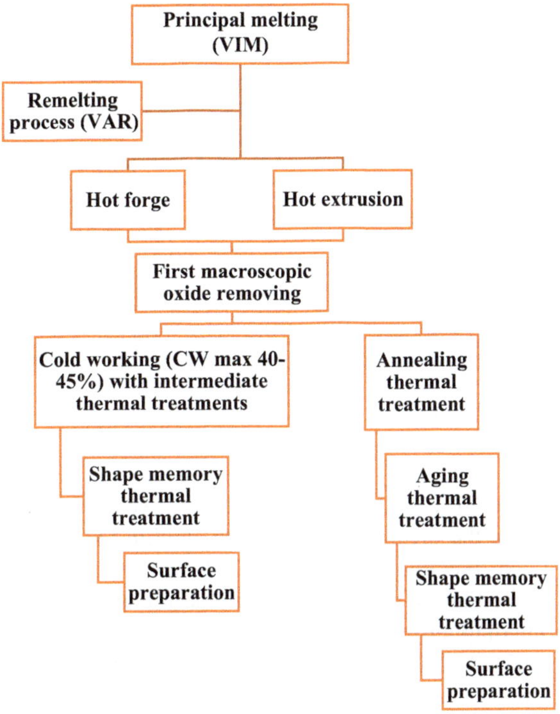

Fig. 1.11 Flowchart diagram of the typical working processes of commercial NiTi SMA products (Lecce and Concilio 2015)

bars. In this case, the ingots are rough hot rolled after melting and subsequently precision hot rolled (at 800–950 °C) down to a final wanted bar diameter, and then straightened at a lower temperature, e.g. 600 °C. The bars are often slightly oversized after precision hot rolling to allow for removal of surface oxidation (Tyber et al. 2007). The above hot-working process is also called high-temperature hot working.

1.4.2 Annealing

In some cases, the SMA products, after experiencing the abovementioned cold-working or high-temperature hot-working processes, are directly delivered to clients who are responsible for the subsequent low-temperature heat treatment (also known as annealing or ageing). Alternatively, the material suppliers can perform the low-temperature heat treatment based on their own experience which is often a trade secret. In either case, such a low-temperature heat treatment is extremely essential and can significantly affect the final thermal–mechanical behaviour of the SMA. As a result, optimum annealing protocol is of particular interest to engineers who look for reliable SE or SME functions offered by the SMA products.

Fig. 1.12 Annealing facilities for SMA products

For NiTi SMA bars and other large-scale elements, the importance of annealing lies in the precipitation of Ti_3Ni_4 within the microstructure. The existence of the Ti_3Ni_4 precipitates can greatly enhance the mechanical properties of SMA products by facilitating the martensitic transformation and hindering plastic flow in the surrounding austenite matrix (Frick et al. 2005; McCormick et al. 2007; Tyber et al. 2007). As a result of these microscopic mechanisms, the strain recovery capability is improved, and the unwanted permanent strain is effectively decreased. Ti_3Ni_4 is a metastable precipitate that forms in Ni-rich NiTi SMA (Ni-rich NiTi is used for most commercial NiTi products) from a solid solution of Ni at moderate annealing temperatures within a particular temperature range, i.e. 300–600 °C, over varying necessary durations from minutes to hours (Gall et al. 1998; 1999a, b, c). The SMA products are first put into a preheated environmental chamber (Fig. 1.12), which enables a constant elevated temperature to be maintained. After a predetermined annealing time, the SMA products are taken out and then immediately water quenched. If over aged, i.e. overly high annealing temperature or overly long annealing duration, $TiNi_3$, instead of Ti_3Ni_4, will precipitate through nonmartensitic processes from the austenite matrix (Frick et al. 2005). $TiNi_3$ precipitates have no beneficial effects on improving strain recovery property. On the other hand, if no annealing or insufficient annealing is performed, the inadequate precipitation of Ti_3Ni_4 leads to no or limited strain recovery capability. In this regard, the primary purpose of low-temperature heat treatment for NiTi SMA is the precipitation of a finely distributed Ti_3Ni_4 within the microstructure.

The effect of annealing on the microstructure of the SMA can be further revealed by transmission electron microscopy (TEM) images, as shown in Fig. 1.13. The obvious parallel dislocations found in the microstructures of the non-annealed SMA material are one of the main reasons behind the irreversible plastic deformation. On the other hand, the existence of numerous Ti_3Ni_4 precipitates presented in the matrix of the SMA after undergoing a 400 °C annealing temperature for 30 min makes the material easier to recover to the original shape after unloading.

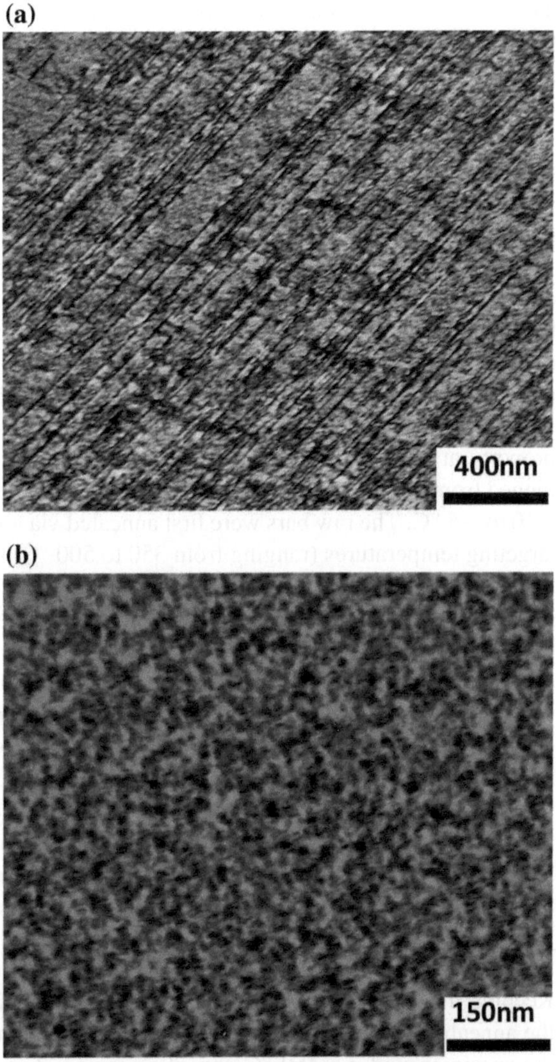

Fig. 1.13 Transmission electron microscopy (TEM) images of microstructure for SMA bars at room temperature: **a** as-received, **b** annealed at 400 °C for 30 min

Annealing can also affect the transformation temperatures of SMA (Khalil-Allafi et al. 2002; Michutta et al. 2004). This is another reason why appropriate annealing protocols are of great importance to engineers who need particular functions of SMA (either SE or SME) over a certain temperature range. From the seismic application standpoint, civil engineers often want SE at room temperature. This requires the austenite finish temperature (A_f) to be lower than the room temperature. Similarly, if one allows permanent deformation of SMA at room temperature but wants SME to be triggered when the material is heated, the austenite start temperature (A_s) should be sufficiently higher than the room temperature. One tricky issue is that the transformation temperatures of NiTi SMA are quite sensitive to the annealing strategy, and this phenomenon is linked to the Ti_3Ni_4 precipitate size which strongly affects the thermodynamics of the transformation (Tyber et al. 2007). Some studies indicated that the transformation temperatures in commercial Ni-rich NiTi SMA can be shifted by 50 °C due to the annealing process alone (Frick et al. 2005), and such change may just occur in the adjacent of the room temperature. This asks for a well-controlled annealing protocol that results in a necessary gap between the room temperature and the relevant transformation temperatures.

Regarding size effect, the required annealing temperature (or duration) for larger diameter SMA bars is often higher (or longer) than that required for small diameter bars. Figure 1.14 shows the incremental cyclic stress–strain responses of SMA dog-bone specimens with varying diameters and subjected to different annealing temperatures and durations (Wang et al. 2016). The specimens were made of commercial hot-rolled Ti–Ni 50.8 at.%, and were supposed to have superelastic property at room temperature. The original diameters of the as-received raw bars ranged from 8 mm to 30 mm, and the nominal A_s of the raw bars without annealing is −10 to −5 °C. The raw bars were first annealed via an electrical furnace to different targeting temperatures (ranging from 350 to 500 °C) for varying durations (ranging from 15 to 60 min), and then water quenched. After the annealing procedure, the raw bars were machined to dog-bone shape coupons to facilitate mechanical testing.

As can be seen from Fig. 1.14, the hysteretic behaviour of the SMA bars exhibits a strong dependence on the annealing procedure. Taking those with 20 mm diameter (measured from the raw bar rather than the shank diameter of the dog bone), for example, almost no self-centring capability is observed for the specimen without annealing. For the specimen annealed at 400 °C for 30 min, self-centring capability starts to show, although there are still considerable unwanted residual strains after 8% peak strain. A better hysteretic response is found for the specimen with the same annealing temperature but an increased annealing duration of 45 min.

An optimum stress–strain response seems to happen when the specimen is subjected to a 450 °C annealing temperature for 30 min, where apparent flag-shaped hysteresis is developed with limited residual strain being induced after the 8% peak strain. Keeping the same annealing temperature (450 °C), either decreasing or increasing the annealing duration could compromise the hysteretic response. In particular, for the specimen under a 450 °C annealing temperature for 15 min, the stress–strain response is similar to that of the as-received non-annealed specimen which shows no SE. If over aged, e.g. 450 °C annealing temperature for 60 min, the forward (A to M) and reverse (M to A) transformation plateaus tend to move downwards, leading to an increase in residual strain. In addition, the prolonged annealing duration leads to stress-induced R-phase transformation, which is signified by a small transformation strain shown in the linear elastic stage of the austenite (Frick et al. 2005). When the annealing temperature increases to 500 °C for 45 min duration, the specimens completely lose SE.

It is also confirmed from the results that the optimum annealing strategies for the SMA specimens are influenced by the bar diameter. When annealed at 450 °C for 30 min (a case leading to optimum hysteretic behaviour for the 20 mm-diameter

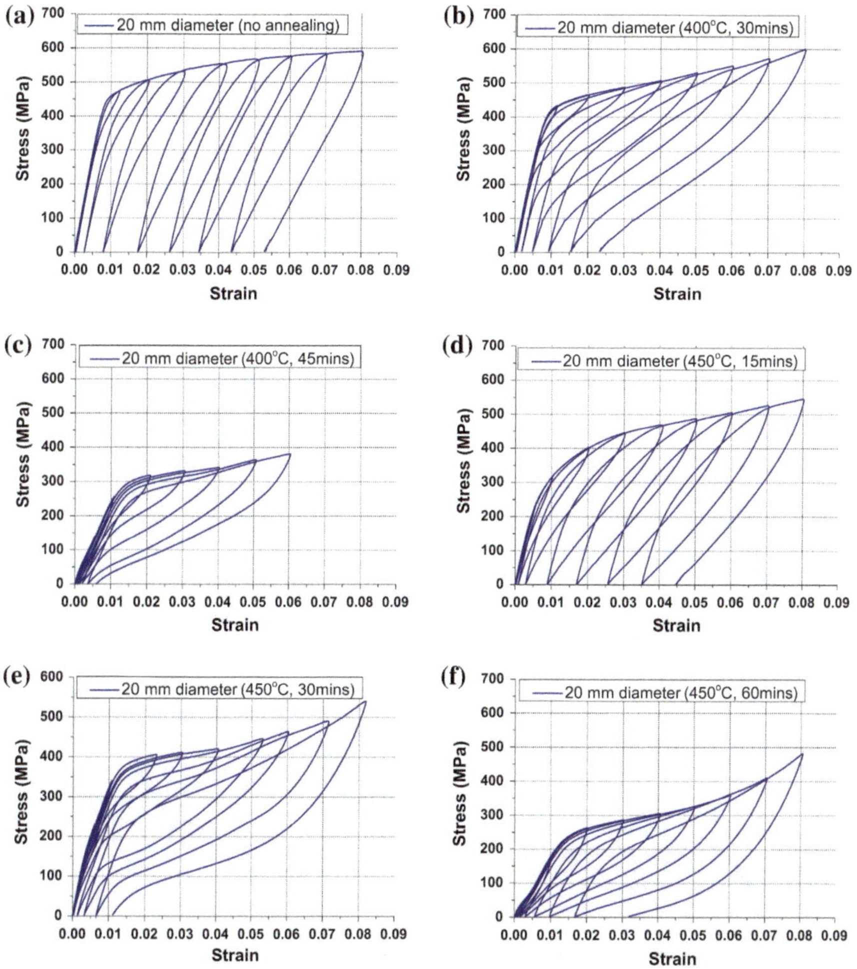

Fig. 1.14 Influence of different annealing schemes on mechanical properties of superelastic SMA bars: **a** 20 mm diameter—no annealing, **b** 20 mm diameter—400 °C and 30 min, **c** 20 mm diameter—400 °C and 45 min, **d** 20 mm diameter—450 °C and 15 min, **e** 20 mm diameter—450 °C and 30 min, **f** 20 mm diameter—450 °C and 60 min, **g** 20 mm diameter—550 °C and 45 min, **h** 8 mm diameter—400 °C and 30 min, **i** 16 mm diameter—400 °C and 30 min, **j** 16 mm diameter—450 °C and 30 min, **k** 30 mm diameter—450 °C and 40 min, **l** 30 mm diameter—450 °C and 60 min

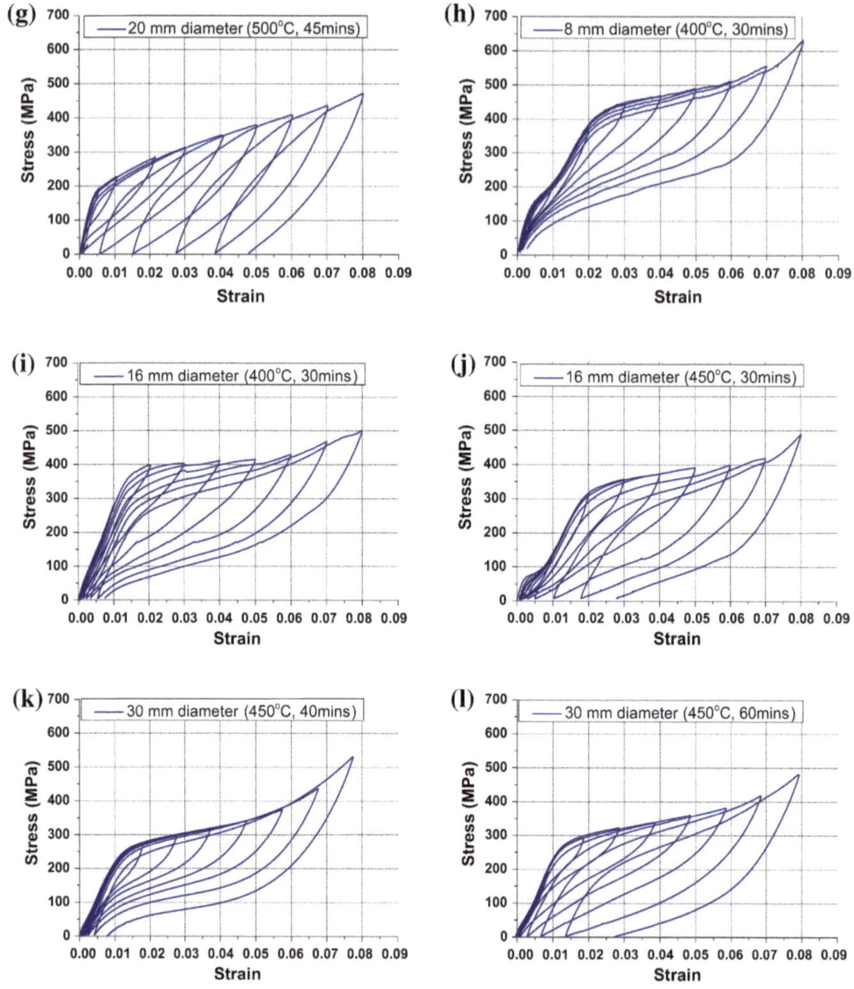

Fig. 1.14 (continued)

bars), the 16 mm-diameter bar exhibits less satisfactory self-centring response with increased residual strains. If the annealing temperature is reduced to 400 °C with the same duration, the hysteretic response is improved. This annealing scheme also results in satisfactory hysteretic behaviour for the 8 mm-diameter bars, although a more obvious R-phase transformation type stress–strain relationship is developed in the initial loading stage. It is also observed that a 450 °C annealing temperature with a duration of 40 min leads to satisfactory hysteretic behaviour of the 30 mm-diameter bar. However, when the duration is further increased to 60 min, less satisfactory behaviour is shown, implying over ageing.

The above tendencies suggest that in order to acquire an optimum mechanical property, the annealing temperature and duration of the SMA elements need to be carefully determined. It is intuitively deduced that higher annealing temperatures or longer durations are needed for larger size SMA elements where heat is more slowly transferred and uniform temperature distribution within the body of the specimen is more difficult to achieve.

1.5 Experimental Characterisation

From a practical design point of view, the basic thermal–mechanical parameters that are important to civil engineers include: (1) the characteristic temperatures that are associated with the transformation processes, i.e. martensitic start temperature M_s, martensitic finish temperature M_f, austenite start temperature A_s, and austenite finish temperature A_f; (2) the characteristic mechanical parameters during a complete SE stress–strain cycle. These include: forward transformation start stress σ_{Ms}, forward transformation finish stress σ_{Mf}, reverse transformation start stress σ_{As}, reverse transformation finish stress σ_{Af}, Young's Moduli (E_A and E_M), and maximum transformation strain ε_L; and (3) the characteristic stresses during a complete SME stress–strain cycle, including twinned martensite to detwinned martensite transformation start stress $\sigma_s{}^{cr}$ and finish stress $\sigma_f{}^{cr}$, transformation-induced residual strain ε_r, and strain recovery rate after heating the material to a temperature above A_f. Obtaining these parameters is adequate for civil engineers to understand and characterise the basic thermal–mechanical properties of SMA, and accordingly, to design SMA-based seismic resistant/mitigation devices. The relevant parameters can also be used in commercial finite element software for capturing the desirable behaviour of SMA elements. As civil engineers are often more interested in seismic passive control, the experimental characterisation methods discussed in this section mainly focus on SE-related parameters.

1.5.1 Transformation Temperatures

The transformation temperatures of SMA are important parameters as their location relative to the testing/working environment (e.g. room temperature) determines if the material shows SE or SME (or neither). For example, when one orders SMA products which are expected to exhibit SE, the material will be likely provided with A_f in a location reasonably below the room temperature.

To guarantee desirable mechanical properties of SMA at a certain temperature, it is essential to first know the transformation temperatures of the material. Differential scanning calorimetryer (DSC) testing (Fig. 1.15) is the most common approach used by researchers, engineers, and material suppliers to obtain the transformation temperatures of SMA. DSC captures heat flow with continuous changes of temperature

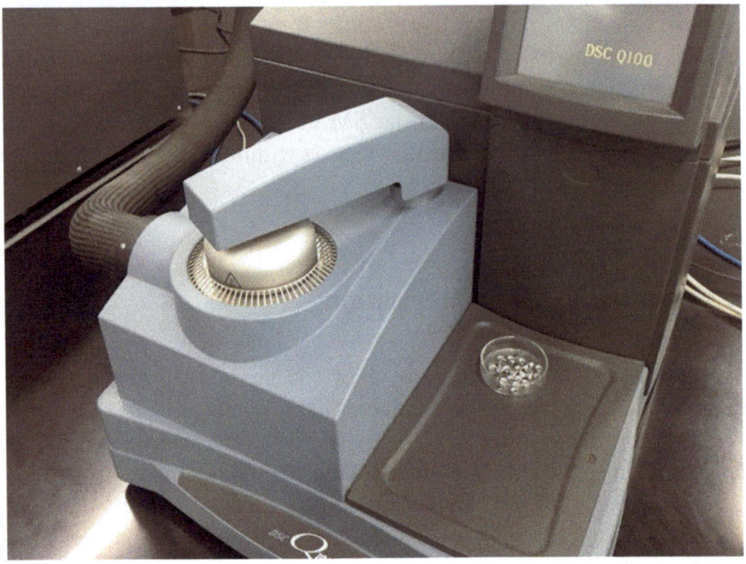

Fig. 1.15 Differential scanning calorimeter (DSC) testing instrument

of the material and provides an indication of the associated transformation temperatures as the material goes through endothermic and exothermic solid-state phase transformations. The sample material needs to be sealed and loaded into the DSC machine and is then subjected to a heating–cooling cycle (or cooling–heating cycle) with a typical rate of 10 °C/min.

Figure 1.16 illustrates representative DSC curves with distinct phase transformation peaks. The phase transformation-start and -finish temperatures can be determined based on the onset points of the slope changes in the DSC curves. The locations of these onset points are adequately identified by finding intersections of lines tangent to the baseline heat flows, as illustrated in Fig. 1.16 (Huang and Liu 2001; Lagoudas 2008). The pattern of the DSC curve depends on various factors, including the alloy's chemical composition, annealing scheme and the history of its mechanical deformation (Sadiq et al. 2010). A single peak in the heating or cooling DSC curve indicates that the alloy experiences phase transformation without passing through the R-phase, in which case the toes of the peaks correspond to the representative A to M and M to A transformation temperatures, i.e. M_s, M_f, A_s and A_f. In some cases, two successive peaks are seen in the cooling or heating cycle (or in both procedures), and such pattern indicates the existence of R-phase between the martensite and austenite phases. Sometimes not all the peaks in a DSC curve are easily identified, probably because of an insufficient amount of material loaded into the DSC machine or impeded mobility of the crystals due to internal stress fields and dislocations. As a result only approximated transformation temperatures can be obtained, which should be determined based on professional judgement.

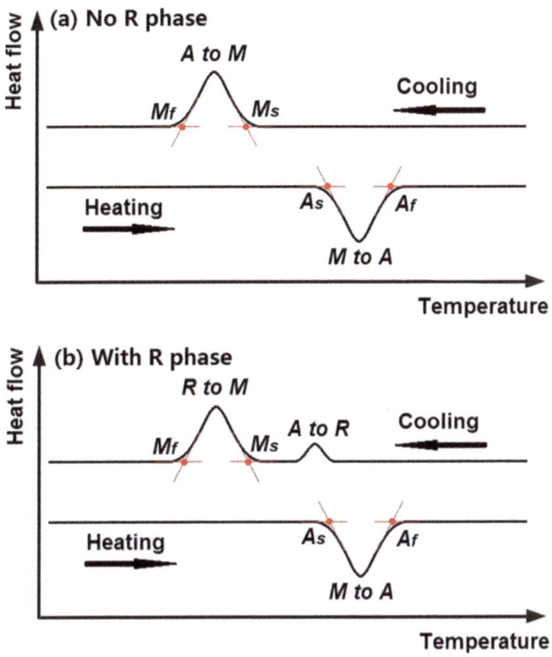

Fig. 1.16 Illustrative DSC curves: **a** no R-phase, **b** with R-phase

As mentioned in Sect. 1.4, the transformation temperatures of SMA are sensitive to the annealing scheme. Figure 1.17 shows the DSC results of a typical 0.8 mm-diameter Ti–Ni 50.8 at.% wire experiencing different annealing histories. The SMA wire that receives no annealing (e.g. as-received sample) experiences two-stage forward (cooling cycle) and reverse (heating cycle) transformations. Using the tangent method, the transformation temperatures for this wire sample during the cooling cycle are approximately 15.1, −9.2, −92.4 and −115.1 °C for R_s (rhombohedral phase start temperature), R_f (rhombohedral phase finish temperature), M_s and M_f, respectively; and the transformation temperatures during the heating cycle are approximately 3.5 and 19.7 °C for A_s and A_f, respectively.

The transformation temperatures are raised when the SMA wire is subjected to 400 °C annealing temperature for 15 min. Taking the austenite finish temperature for example, after being annealed at 400 °C, the A_f of the wire is increased to around 24.5 °C, compare with $A_f = 19.7$ °C for the as-received wire sample. With a further increase in the annealing temperature, however, the transformation temperatures tend to decrease again, where the measured A_f values of the wires subjected 450 and 500 °C annealing temperatures are approximately 11.4 and −15.3 °C, respectively. In other words, with the four considered annealing schemes, the A_f value can be varied by nearly 40 °C, i.e. from −15.3 to 24.5 °C. This warns that although an appropriate annealing procedure may help enhance the strain recoverability of SMA, the associated change in the transformation temperatures must be considered in the

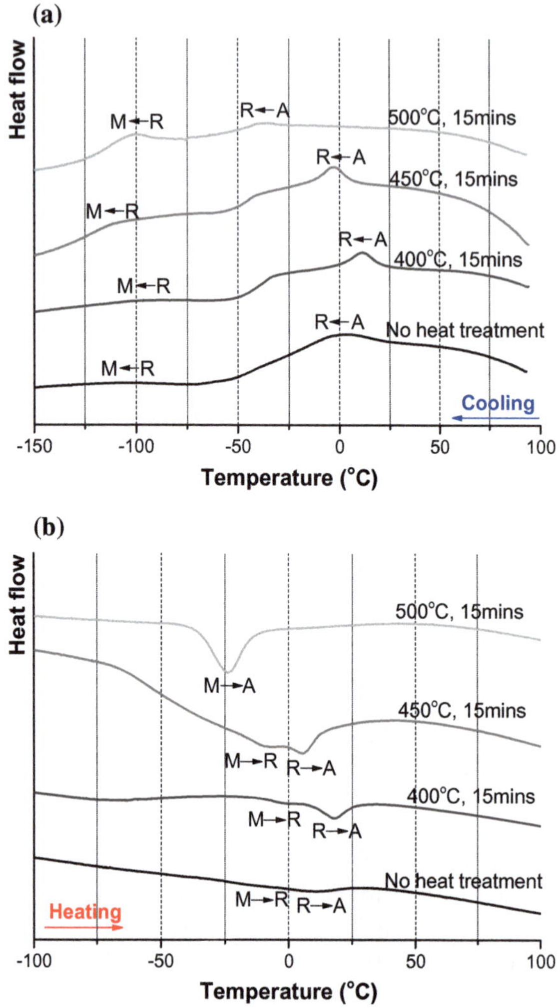

Fig. 1.17 Transformation temperatures of SMA subjected to varying annealing schemes: **a** cooling cycle, **b** heating cycle

design, especially when one wants to strictly control the transformation temperature threshold during the service life of SMA.

Another interesting phenomenon shown in Fig. 1.17 is that the two-stage transformation peaks gradually merge into a single peak during the heating cycle when the annealing temperature increases. This indicates that the sample is transformed directly from martensite to austenite with no presence of the R-phase. In addition, the transformation peaks of the DSC curves for the case of higher annealing temperatures are more pronounced than those observed for the samples receiving no annealing or lower annealing temperatures. As aforementioned, a less remarkable peak can be caused by internal stress fields and dislocations which are generated in SMA wires following a cold-working procedure during the fabrication process.

These defects impede the mobility of the crystals during the heating and cooling cycles (Sadiq et al. 2010). An increase in annealing temperature tends to release the internal stress fields and hence could make the peaks more remarkable.

1.5.2 Mechanical Properties—Uniaxial Tests

In order to acquire the mechanical parameters of superelastic SMA, the most common approach is uniaxial cyclic test, which is often performed isothermally on SMA elements, such as wires, at room temperature. The cross section of SMA wires is uniformly loaded, and as a result, the engineering stress–strain relationship can be accurately obtained, converted from the readings from the load cells and displacement transducers/extensometers. A Universal Test Machine (UTM) suffices for performing such tests, provided that the ends of the SMA wire sample are reliably gripped. Different loading protocols can be employed to enable a comprehensive understanding of the SMA wires subjected to various loading scenarios. Of course, the most 'accurate' loading protocol is that reflects the actual stress–strain histories of the SMA wires which are incorporated in SMA-based devices such as dampers. However, for most cases, the actual stress–strain history is not easy to obtain due to the uncertainty of seismic excitations (record-to-record variability). Therefore, standard loading protocols, including monotonic, cyclic loading with incremental strain, cyclic loading with constant strain, and pre-strained cyclic loading, can be employed, as typically shown in Fig. 1.18. These loading protocols can adequately capture the fundamental hysteretic responses, including the degradation phenomenon (as explained later), of the SMA samples.

Incidentally, it is worth mentioning that the hysteretic loops of SMA elements tend to be stabilised under repeated loading cycles. In other words, an increase in residual strain due to the degradation effect only occurs when the peak strain of a newly arrived cycle exceeds the peak strain that occurs in the previous loading history. If the historical peak strain is not exceeded, the hysteresis would be stable. Therefore, the stepwise loading protocols are in fact effective for characterisation purposes, even if the actual earthquake loading is random.

The typical stress–strain curve of an SMA wire subjected to pseudo-static cyclic loading (with a constant strain) is shown in Fig. 1.19. It is noticed that the forward transformation plateau gradually moves downwards with an increase in the number of cycles, especially during the first few cycles. This is accompanied by an accumulation of residual strain. The hysteretic loop becomes stable after a certain number of cycles. The above degradation phenomenon is called 'functional fatigue' or 'transformation induced fatigue' (Eggeler et al. 2004), which is different from the 'conventional' high cycle fatigue that also occurs in SMA. For the latter case where the strain remains within the elastic regime, the SMA can have a fatigue life of as high as ~10^7 cycles, and finally fails just like other typical engineering materials.

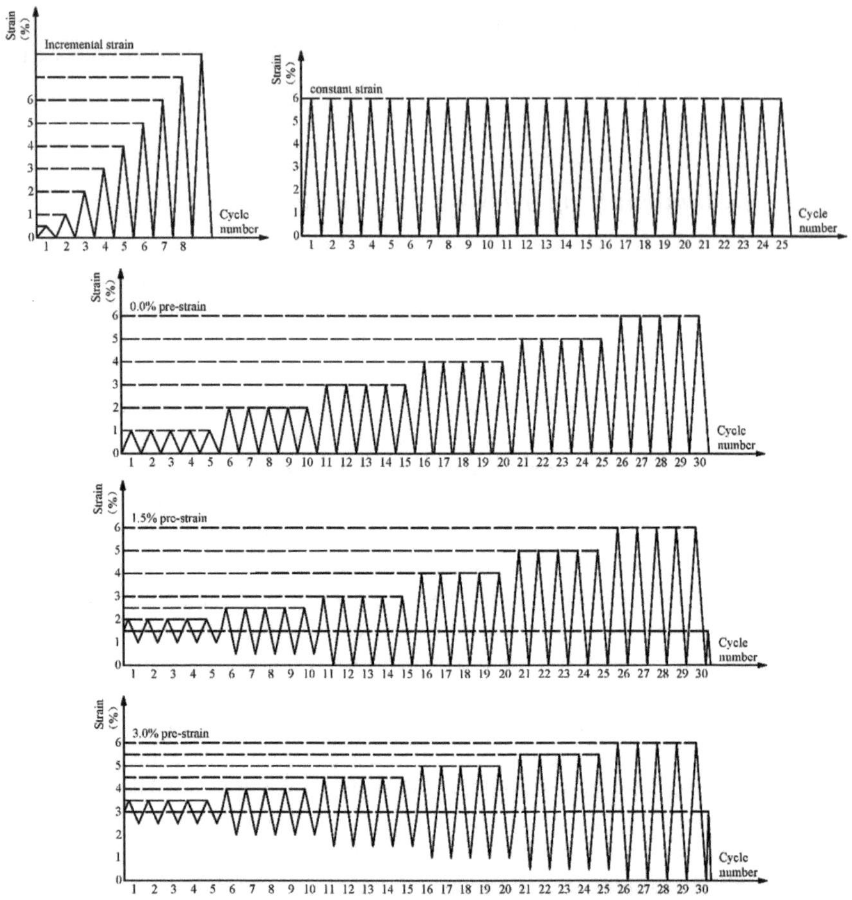

Fig. 1.18 Typical standard loading protocols for SMA wires

When the peak strain advances beyond the elastic regime, leading to partial or complete forward transformations, microstructural changes gradually occur, which is responsible for the degradation of the hysteretic response. The structural fatigue life also significantly decreases to the order of thousands of cycles (Lagoudas 2008). For seismic applications, a low-cycle structural fatigue life of thousands of cycles is adequate, so the main issue left for SMA is the loss of its function, i.e. development of residual strain and decrease of the forward transformation plateau. The significance of functional fatigue depends on a number of factors including chemical composition, fabrication process, annealing and environmental temperature.

In order to minimise the degree of degradation during the service life of SMA, a training process is often necessary to help stabilise the hysteretic response. The key mechanical parameters can, therefore, be extracted based on a stabilised stress–strain curve with the peak strain exceeding the transformation complete strain. To quantify this curve, a series of straight lines tangent to the loading and unloading paths may be

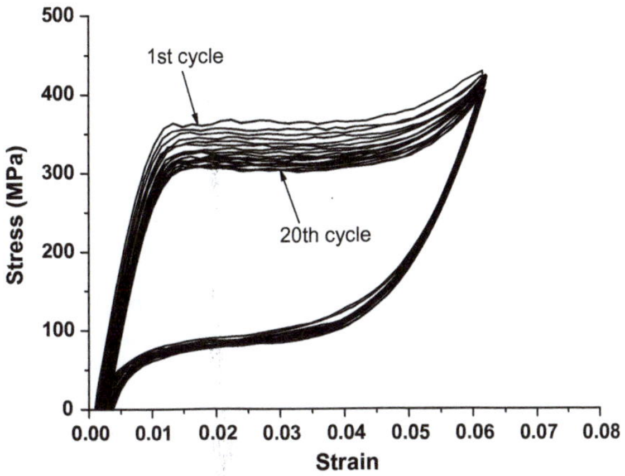

Fig. 1.19 Stress–strain relationship and degradation of SMA under constant cyclic loading

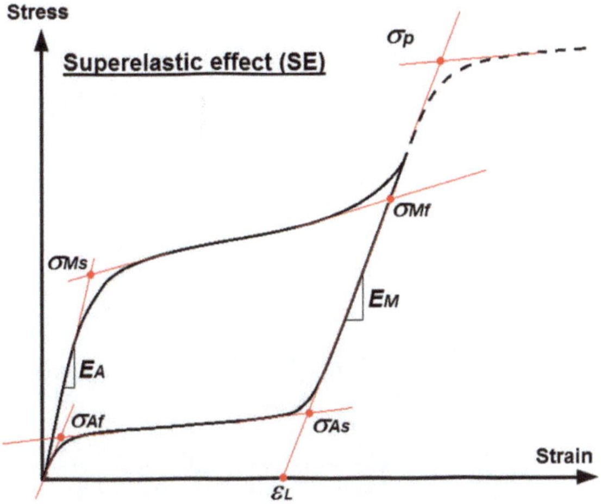

Fig. 1.20 Measurement of characteristic parameters of a SE stress-strain cycle

constructed (Fig. 1.20). The distinctive intersections of these straight lines provide a possible measure for the characteristic parameters, including forward transformation start stress σ_{Ms}, forward transformation finish stress σ_{Mf}, reverse transformation start stress σ_{As}, reverse transformation finish stress σ_{Af} and Young's Moduli (E_A and E_M), during a complete SE stress–strain cycle. The maximum transformation strain ε_L can be obtained by constructing a reverse extension line along the unloading path prior to the reverse transformation. If one wants to know the stress at the onset of plastic deformation (σ_p), a further increase in load can be applied beyond the forward transformation stage until a second change of the stress–strain slope occurs.

A similar experimental characterisation method can be applied to SMA bars. The raw bars are often machined to a dog-bone configuration to facilitate testing (Fig. 1.21). As SMAs do not have good machinability and are quite sensitive to temperature variations, the cutting process should be carefully conducted, and coolant

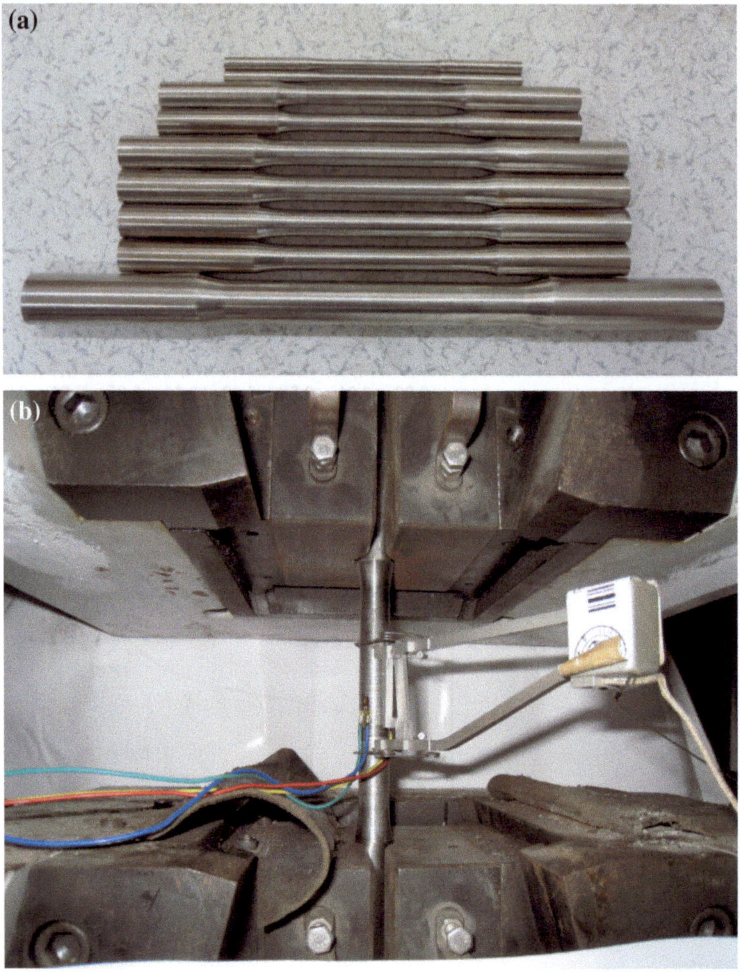

Fig. 1.21 Test setup for uniaxial tests of SMA bar specimens: **a** dog-bone specimens, **b** test arrangement

must be applied. Electrical discharge machining (EDM) can also be used, which is particularly useful for raw materials with arbitrary shapes. The dimension and configuration of the SMA tension coupons can be determined from material testing standards (e.g. ISO 2009). Again, a UTM fitted with a wedge grip that is compatible with the diameter of the gripped part of the dog bone can be used. An example of such tests is shown in Fig. 1.21 where the strain is measured by an extensometer and the stress is converted from the load recorded by the built-in load cell.

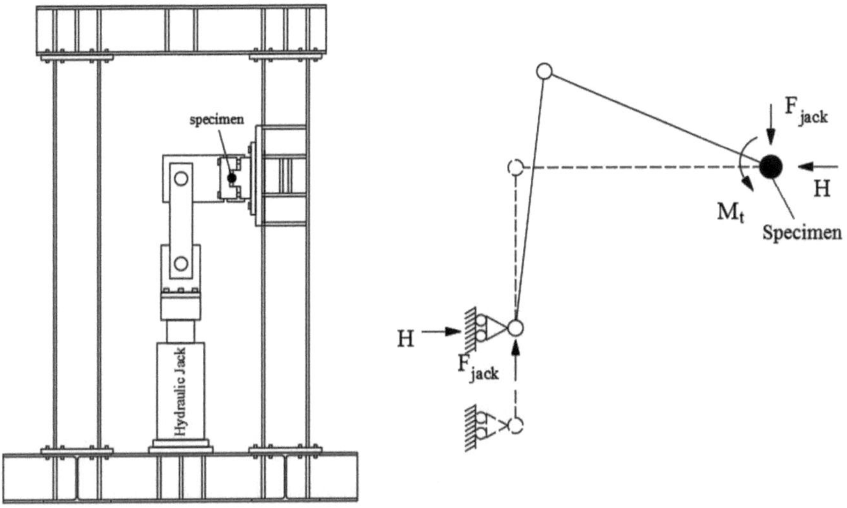

Fig. 1.22 Test setup for torsion tests on SMA bar specimens (Dolce and Cardone 2001)

1.5.3 Mechanical Properties—Shear Tests

Shear tests are required for a more in-depth understanding of the material behaviour of SMA under multiaxial loading conditions. SMA specimens in the form of tubes are usually used for this purpose. A sufficiently thin wall can ensure a near-uniform shear stress developed in the tube when subjected to torsion. A torsion test on a solid bar can also be performed to obtain its torsion–rotation response if the entire bar is intended to be used as a torsion member in a damping device (Dolce and Cardone 2001). The torsion action can be applied through a specially designed linking device such as that shown in Fig. 1.22.

Direct shearing is another type of test that examines the material behaviour under pure shear action (Fang et al. 2015). Direct shear (or that coupled with bending/tension) is a potential load-bearing scenario for some SMA elements, e.g. SMA bolts, used in beam-to-column connections and other connectors. A special testing device proposed by Wallaert and Fisher (1965) can be employed, which is typically used for testing the shear resistance of conventional high-strength bolts. As shown in Fig. 1.23, the SMA specimen is subjected to a double shear action via two cover plates and one mid-plate. Normal steel can be used for these plates, but much harder steel filler should be employed locally around the specimen to minimise the bearing deformation of the plate material and the potential bending effect of the SMA specimen during direct shear. The tensile loading exerted by the UTM is transferred through the gripped areas at the end of the mid-plates to the two cover plates. A cyclic load can be applied to examine the cyclic direct shear response of the specimen.

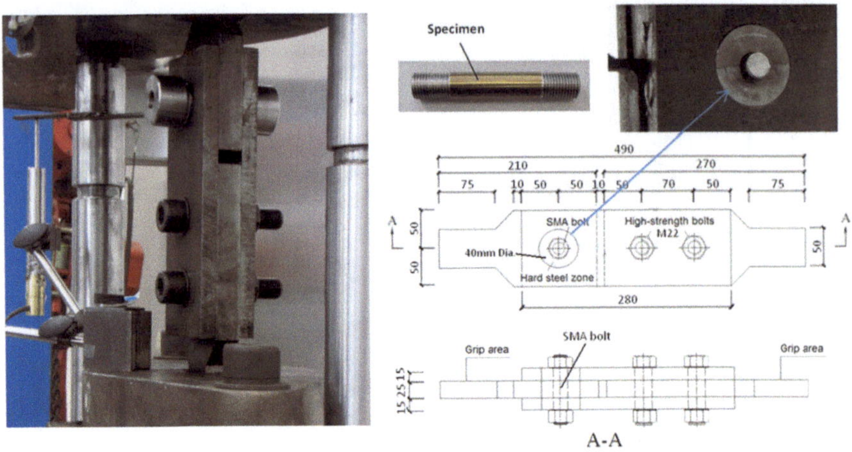

Fig. 1.23 Test setup for direct shear tests on SMA bar specimens (Fang et al. 2015)

1.6 Modelling of Shape-Memory Alloys

Over the past decades, a significant research process has been made on the development of constitutive models capable of simulating the SE and SME behaviour of SMA. The necessary knowledge gained from extensive experimental investigations has motivated fast development of numerical models that facilitate the design of SMA devices for civil engineering. Depending on the modelling strategy, the available constitutive models for SMA can be generally categorised into three types, namely microscopic thermodynamic models, micro–macro models and macroscopic models (Cisse et al. 2016).

Microscopic thermodynamic models involve the consideration of phase nucleation, interface motion, and martensite twin growth, and describe microstructural features at the grain or lattice levels. However, microscopic thermodynamic models are very complicated and computationally expensive, and are therefore impractical for use by civil engineers. The micro–macro models utilise micromechanics to describe the material behaviour at the micro or mesoscales, and this is followed by a scale transition for macroscopic constitutive equations (Cisse et al. 2016). In micro–macro models, a suite of observable variables, including temperature and stress (or strain), and internal variables, consisting of martensitic volume fraction and mean transformation strain, are required. Probably, the most suitable class of constitutive models for civil engineering application is the third type, i.e. the macroscopic model, which describes the behaviour of SMAs based on phenomenological considerations. These models can be derived from either the theory of plasticity or thermodynamic potentials.

Tanaka model (Tanaka and Nagaki 1982) is the first practical macroscopic phenomenological model that employed the energy balance principle and internal variables to describe the phase transformation phenomena associated with SE and SME in SMAs. Based on the basic laws of thermodynamics and boundary conditions, a general stress–strain (σ-ε) constitutive relationship is expressed by

$$\dot{\sigma} = E\dot{\varepsilon} + \Omega\dot{\xi}_{mv} + \theta_t\dot{T} \tag{1.1}$$

where E is the Young's modulus, Ω is the phase transformation strain tensor, ξ_{mv} is the martensite volume fraction, θ_t is the strain tensor related to the thermal coefficient of expansion and T is the temperature. Based on the strain recovery characteristics of SMA, this work was extended by Liang and Rogers (1990) who employed cosine functions to depict the relationship between the martensite volume fraction and the temperature and stress. The variations of the martensite volume fraction during the forward and reverse transformations are given by

Conversion from austenite to martensite

$$\xi_{mv} = \frac{1-\xi_0}{2}\cos\left[a_M\left(T - M_f - \frac{\sigma}{C_M}\right)\right] + \frac{1+\xi_0}{2}$$
$$\text{for} \quad C_M(T - M_s) < \sigma < C_M\left(T - M_f\right) \tag{1.2}$$

Conversion from martensite to austenite

$$\xi_{mv} = \frac{\xi_0}{2}\left\{\cos\left[a_A\left(T - A_s - \frac{\sigma}{C_A}\right)\right] + 1\right\}$$
$$\text{for} \quad C_A\left(T - A_f\right) < \sigma < C_A(T - A_s) \tag{1.3}$$

where ξ_0 is the martensite volume fraction prior to the current transformation, σ is the applied stress, T is the temperature, the constants C_M and C_A are material properties that describe the relationship between temperature and the critical stress to induce transformation, as previously shown in Fig. 1.6, and a_M and a_A are defined by

$$a_M = \frac{\pi}{M_s - M_f},\ a_A = \frac{\pi}{A_f - A_s} \tag{1.4}$$

Brinson (1993) extended this model by splitting the martensite volume fraction into a temperature-induced fraction ξ_T and a stress-induced fraction ξ_s, i.e. $\xi_{mv} = \xi_T + \xi_s$. Considering the subscript '0' as an initial state, the modified stress–strain constitutive relationship can be expressed by

$$\sigma - \sigma_0 = E(\xi_{mv})\varepsilon - E(\xi_0)\varepsilon_0 + \Omega(\xi)\xi_s - \Omega(\xi_0)\xi_{s0} + \theta(T - T_0) \tag{1.5}$$

$$\Omega(\xi_{mv}) = -\varepsilon_L E(\xi_{mv}) \tag{1.6}$$

where ε_L is the maximum recovered strain. Combining Eq. (1.5) and Eq. (1.6), and defining the austenite and martensite Young's moduli as E_A and E_M, respectively, the above equations can be expressed by

$$\sigma = E(\xi_{mv})[\varepsilon - \varepsilon_L \xi_s] + K_0 \tag{1.7}$$

in which

$$E(\xi_{mv}) = [\xi_{mv} E_M + (1 - \xi_{mv}) E_A] \tag{1.8}$$

$$K_0 = \sigma_0 - E(\xi_0)\varepsilon_0 - \Omega(\xi_0)\xi_{s0} + \theta(T - T_0) \tag{1.9}$$

Equations (1.7) through (1.9) are also called Tanaka–Liang–Brinson model. It is noted that the evolution equations [Eqs. (1.2 and 1.3)] should be modified to accommodate the definition of temperature-induced and stress-induced martensite volume fractions, and to allow for the shape-memory effect at temperatures below M_s. The evolution equations modified by Brinson (1993) are represented as

Conversion to detwinned martensite

for $T > M_s$ and $\sigma_s^{cr} + C_M(T - M_s) < \sigma < \sigma_f^{cr} + C_M(T - M_s)$:

$$\xi_s = \frac{1 - \xi_{s0}}{2} \cos\left\{ \frac{\pi}{\sigma_s^{cr} - \sigma_f^{cr}} \times \left[\sigma - \sigma_f^{cr} - C_M(T - M_s)\right] \right\} + \frac{1 + \xi_{s0}}{2} \tag{1.10}$$

$$\xi_T = \xi_{T0} - \frac{\xi_{T0}}{1 - \xi_{s0}} (\xi_s - \xi_{s0}) \tag{1.11}$$

for $T < M_s$ and $\sigma_s^{cr} < \sigma < \sigma_f^{cr}$:

$$\xi_s = \frac{1 - \xi_{s0}}{2} \cos\left[\frac{\pi}{\sigma_s^{cr} - \sigma_f^{cr}} \times \left(\sigma - \sigma_f^{cr}\right) \right] + \frac{1 + \xi_{s0}}{2} \tag{1.12}$$

$$\xi_T = \xi_{T0} - \frac{\xi_{T0}}{1 - \xi_{s0}} (\xi_s - \xi_{s0}) + \Delta_{T\xi} \tag{1.13}$$

where, if $M_f < T < M_s$ and $T < T_0$,

$$\Delta_{T\xi} = \frac{1 - \xi_{T0}}{2} \left\{ \cos\left[a_M\left(T - M_f\right)\right] + 1 \right\} \tag{1.14}$$

else, $\Delta_{T\xi} = 0$

Conversion to Austenite

for $T > A_s$ and $C_A\left(T - A_f\right) < \sigma < C_A(T - A_s)$:

$$\xi_{mv} = \frac{\xi_0}{2} \left\{ \cos\left[a_A \left(T - A_s - \frac{\sigma}{C_A} \right) \right] + 1 \right\} \tag{1.15}$$

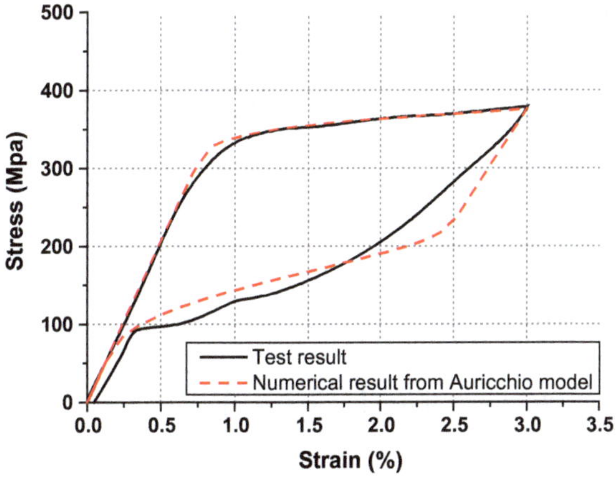

Fig. 1.24 Test result and numerical result predicted by Auricchio model

$$\xi_s = \xi_{s0} - \frac{\xi_{s0}}{\xi_0}(\xi_0 - \xi) \tag{1.16}$$

$$\xi_T = \xi_{T0} - \frac{\xi_{T0}}{\xi_0}(\xi_0 - \xi) \tag{1.17}$$

In the above equations, ${\sigma_s}^{cr}$ and ${\sigma_f}^{cr}$ are the critical stresses at the start and finish of the conversion of the martensitic variants. Some of the parameters related to the above evolution equations are based on experimental results from a particular SMA material, and therefore, some modifications to the equations may be made in order to better reflect the actual physical processes.

Apart from the Tanaka series model, the Auricchio model is also widely used in engineering practice. The early version of the Auricchio model (Auricchio et al. 1997) is quite simple yet effective in simulating the SE response of SMA (Fig. 1.24). A Drucker–Prager-type loading function was considered, which is a robust solution applicable to general finite element tools. The model assumes the following constitutive relationships:

$$\Delta\sigma = E(\Delta\varepsilon - \Delta\varepsilon^{tr}) = E\left(\Delta\varepsilon - \Delta\xi_s\varepsilon_L\frac{\partial F}{\partial\sigma}\right) \tag{1.18}$$

where $\Delta\sigma$ and $\Delta\varepsilon$ are incremental Kirchhoff stress and strain, E is the Young's modulus of SMA, F is the Drucker–Prager-type loading function (Auricchio 2001), ξ_s is the single variant martensite volume fraction, as expressed by

Forward transformation

$$\dot{\xi}_s = H^{AS}(1 - \xi_s)\frac{\dot{F}}{F - \sigma_{Mf}\left(\sqrt{\frac{2}{3}} + \alpha_m\right)} \tag{1.19}$$

Reverse transformation

$$\dot{\xi}_s = H^{SA}\xi_s \frac{\dot{F}}{F - \sigma_{Af}\left(\sqrt{\frac{2}{3}} + \alpha_m\right)} \tag{1.20}$$

in which α_m is a material parameter, H^{AS} and H^{SA} are scalar quantities as defined by

$$H^{AS} = 1 \text{ if } \sigma_{Ms}\left(\sqrt{\frac{2}{3}} + \alpha_m\right) < F < \sigma_{Mf}\left(\sqrt{\frac{2}{3}} + \alpha_m\right) \text{ and } \dot{F} > 0 \tag{1.21}$$

$$H^{SA} = 1 \text{ if } \sigma_{Af}\left(\sqrt{\frac{2}{3}} + \alpha_m\right) < F < \sigma_{As}\left(\sqrt{\frac{2}{3}} + \alpha_m\right) \text{ and } \dot{F} < 0 \tag{1.22}$$

For cases other than the conditions of Eqs. (1.21) and (1.22), H^{AS} and H^{AS} are taken as zero. The Auricchio model also allows the consideration of permanent plastic deformations when the stress level exceeds the plastic stress σ_p (Fig. 1.25), and this behaviour is called superelastic–plastic (SE-P) behaviour. Due to its high computational efficiency, the basic Auricchio model has been adopted as a built-in user-defined material model by many commercial finite element software packages such as ABAQUS (2010), ANSYS (2013), and Marc (2014). Taking ABAQUS, for example the number of variables that users can define is DEPVAR = 31 for SE-P models. The variables cover the considerations of transformation stresses, Young's moduli, Poisson's ratios, temperature-dependent parameters, annealing steps and plastic response, which are, according to the input sequence, summarised in Table 1.2. Some of the variables can be further understood via Fig. 1.26. In order to call the subroutine, the name parameter on *MATERIAL should start with a unique term, i.e. ABQ_SUPER_ELASTIC.

An upgraded version of the model was developed later by Auricchio and co-workers to enable a more comprehensive consideration of the SE and SME characteristics along with more complicated thermal–mechanical loading paths and training processes (Auricchio et al. 2003). It was assumed that the martensite volume fraction ξ_s consists of a residual part ξ_R and a reversible part ξ_V. The stress–strain relationship in a small deformation regime can be expressed by

$$\sigma = E\varepsilon^e = E(\varepsilon - \xi_s\beta_m + \xi_R(\beta_m - \kappa) - \psi(T - T_0)) \tag{1.23}$$

where ε^e is the elastic strain, $\xi_s\beta_m$ is the inelastic strain caused by phase transformation, β_m is an internal variable that describes martensite reorientation, κ is a training parameter, ψ is the coefficient of thermal expansion, T_0 is the reference temperature, E is the elastic modulus, as given by

$$E(\xi_s) = \frac{E_A E_s}{E_s + \xi_s(E_A - E_s)} \tag{1.24}$$

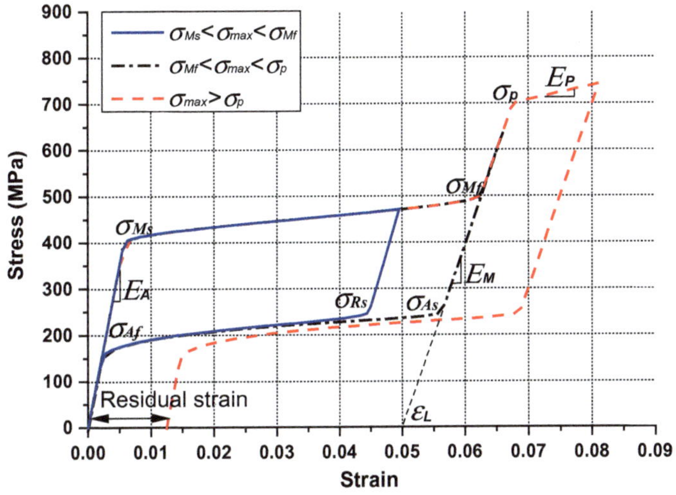

Fig. 1.25 Illustration of superelastic–plastic behaviour of SMA defined in Auricchio model

Table 1.2 Input to SE-P material model in ABAQUS

Symbol of variables	Descriptions
E_A	Young's modulus of austenite
v_A	Poisson's ratio of austenite
E_M	Young's modulus of martensite
v_M	Poisson's ratio of martensite
ε_L	Maximum transformation strain
C_M	Change of critical stresses that induce forward transformation with changing temperatures
σ_{Ms}	Forward transformation start stress
σ_{Mf}	Forward transformation finish stress
T_0	Reference temperature, which is usually set as '0'
C_A	Change of critical stresses that induce reverse transformation with changing temperatures
σ_{As}	Reverse transformation start stress
σ_{Af}	Reverse transformation finish stress
${\sigma_{CL}}^S$	Start of transformation stress during loading in compression, as a positive value (Fig. 1.26)
${\varepsilon_V}^L$	Volumetric transformation strain. If ${\varepsilon_V}^L = \varepsilon_L$, an associated flow algorithm is used, with ${\varepsilon_V}^L$ computed based on σ_{Ms} and ${\sigma_{CL}}^S$. For all other cases, a nonassociated flow algorithm is used, which produces an unsymmetric Jacobian matrix. The USYMM parameter is thus required on the *USER MATERIAL keyword

(continued)

Table 1.2 (continued)

Symbol of variables	Descriptions
N_A	Number of annealing to be performed during the analysis
N_{S1}-N_{SNA}	Step numbers at which all state dependent variables are set to zero
NP	Number of stress–strain pairs to define yield curve
${\sigma_1}^P, \varepsilon_1 \ldots {\sigma_{NP}}^P, \varepsilon_{NP}$	Stress–strain points in the yield curve

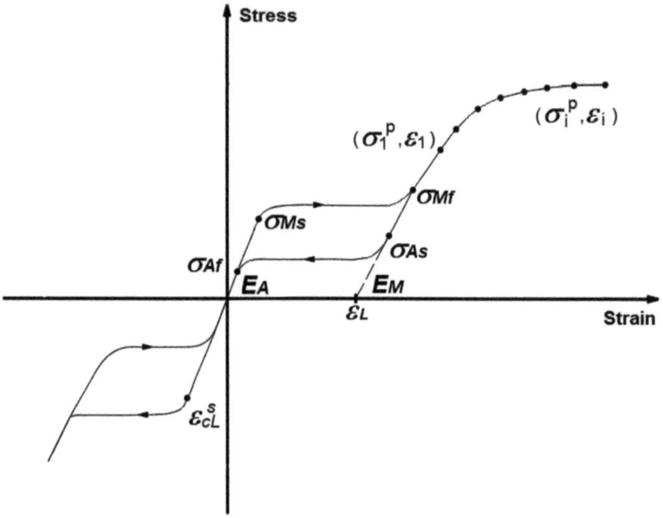

Fig. 1.26 Illustration of input to SMA material model in ABAQUS

where E_A and E_s are the Young's moduli of the austenite and single variant martensite, respectively. The kinetic rules are set as

$$\dot{\xi}_s = \dot{\xi}_s^{AS} + \dot{\xi}_s^{SA} \tag{1.25}$$

where $\dot{\xi}_s^{AS}$ and $\dot{\xi}_s^{SA}$ represent the single-variant martensite fraction volume rates occurring during the austenite to single variant martensite and single variant martensite to austenite phase transformations, respectively. The following evolution equations are proposed:

Conversion from austenite to martensite

$$\dot{\xi}_s^{AS} = \pi^{AS}(1 - \xi_s)\frac{\dot{G}^{AS}}{\left(S_f^{AS} - G^{AS}\right)^{\alpha^{AS}}}\rho^{AS} \tag{1.26}$$

$$G^{AS} = \eta - \frac{C^{AS}}{E}T \tag{1.27}$$

$$S_f^{AS} = \frac{\sigma_f^{AS} - C^{AS} T_S^{AS}}{E_S} + R_f^{AS} \tag{1.28}$$

$$S_S^{AS} = \frac{\sigma_S^{AS} - C^{AS} T_S^{AS}}{E} + R_S^{AS} \tag{1.29}$$

with

$$\text{if } \sigma \geq 0 \begin{cases} \eta = \varepsilon \\ R_S^{AS} = \xi_s \beta - \xi_R (\beta - \kappa) \\ R_f^{AS} = \beta - \xi_R (\beta - \kappa) \end{cases} \tag{1.30}$$

$$\text{if } \sigma < 0 \begin{cases} \eta = -\varepsilon \\ R_S^{AS} = -\xi_s \beta + \xi_R (\beta - \kappa) \\ R_f^{AS} = -\beta + \xi_R (\beta - \kappa) \end{cases} \tag{1.31}$$

Conversion from martensite to austenite

$$\dot{\xi}_s^{SA} = -\pi^{SA} (\xi_s - \xi_R) \frac{\dot{G}^{SA}}{\left(S_f^{SA} - G^{SA}\right)^{\alpha^{SA}}} \rho^{SA} \tag{1.32}$$

$$G^{SA} = \eta - \frac{C^{SA}}{E} T, \tag{1.33}$$

$$S_f^{SA} = \frac{-C^{SA} T_f^{SA}}{E} + R_f^{SA} \tag{1.34}$$

$$S_S^{SA} = \frac{-C^{SA} T_S^{SA}}{E} + R_S^{SA} \tag{1.35}$$

with

$$\text{if } \sigma \geq 0 \begin{cases} \eta = \varepsilon \\ R_S^{SA} = \xi_s \beta - \xi_R (\beta - \kappa) \\ R_f^{SA} = \xi_R \kappa \end{cases} \tag{1.36}$$

$$\text{if } \sigma < 0 \begin{cases} \eta = -\varepsilon \\ R_S^{SA} = -\xi_s \beta + \xi_R (\beta - \kappa) \\ R_f^{SA} = -\xi_R \kappa \end{cases} \tag{1.37}$$

In the above equations, π^{AS}, π^{SA}, α^{AS}, α^{SA} are material constants, C^{AS} and C^{SA} are the Clausius–Clapeyron constants for the phase transformations, ρ^{AS} and ρ^{SA} are activation factors.

$$\rho^{AS} = \begin{cases} 1 \text{ when } \begin{cases} \dot{G}^{AS} > 0 \\ S_S^{AS} \leq G^{AS} \leq S_f^{AS} \end{cases} \\ 0 \qquad \text{otherwise} \end{cases} \tag{1.38}$$

$$\rho^{SA} = \begin{cases} 1 \text{ when } \begin{cases} \dot{G}^{SA} < 0 \\ S_f^{SA} \leq G^{SA} \leq S_S^{SA} \end{cases} \\ 0 \qquad \text{otherwise} \end{cases} \tag{1.39}$$

The model also enables the consideration of cyclic training effect and permanent inelasticity, with more details given in Auricchio et al. (2003, 2007). Other typical macroscopic phenomenological models include those developed by Boyd and Lagoudas (1996), Zaki and Moumni (2007), Leclercq and Lexcellent (1996), etc. These models were developed based on either the theory of plasticity or thermodynamic potentials. However, some of them are quite computational expensive and therefore are not widely adopted by civil engineers and finite element program developers. An up-to-date review of representative constitutive models for SMAs has been made by Cisse et al. (2016).

References

ABAQUS (2010) 6.10 analysis user's manual. Dassault Systemes Simulia Corp, Providence, RI

Anderson TL (2017) Fracture mechanics: fundamentals and applications, 4th edn. CRC Press, Boca Raton

ANSYS (2013) 14.5 analysis user's manual. ANSYS Inc., Canonsburg, PA

Auricchio F (2001) A robust integration-algorithm for a finite-strain shape-memory-alloy superelastic model. Int J Plasticity 17(7):971–990

Auricchio F, Marfia S, Sacco E (2003) Modeling of SMA materials: training and two way memory effects. Comput Struct 81(24):2301–2317

Auricchio F, Reali A, Stefanelli U (2007) A three-dimensional model describing stress-induced solid phase transformation with permanent inelasticity. Int J Plasticity 23(2):207–226

Auricchio F, Taylor RL, Lubliner J (1997) Shape-memory alloys: macromodelling and numerical simulations of the superelastic behavior. Comput Method Appl M 146(3):281–312

Bellouard Y (2008) Shape memory alloys for microsystems: a review from a material research perspective. Mat Sci Eng A Struct 481–482:582–589

Benavent-Climent A (2008) Development and application of passive structural control systems in the moderate-seismicity Mediterranean area: the case of Spain. In: Proceedings of the 14th world conference on earthquake engineering, Beijing, China, 2008

Boyd JG, Lagoudas DC (1996) A thermodynamical constitutive model for shape memory materials. Part I. The monolithic shape memory alloy. Int J Plasticity 12(6):805–842

Brinson LC (1993) One-dimensional constitutive behavior of shape memory alloys: thermomechanical derivation with non-constant material functions and redefined martensite internal variable. J Intel Mat Syst Struct 4(2):229–242

Buehler WJ, Gilfrich JV, Wiley RC (1963) Effect of low-temperature phase changes on the mechanical properties of alloys near composition TiNi. J Appl Phys 34(5):1475–1477

Chang WS, Araki Y (2016) Use of shape-memory alloys in construction: a critical review. Proc Inst Civil Eng-Civ Eng 169(2):87–95

Cisse C, Zaki W, Zineb TB (2016) A review of constitutive models and modeling techniques for shape memory alloys. Int J Plasticity 76:244–284

Cladera A, Weber B, Leinenbach C, Czaderski C, Shahverdi M, Motavalli M (2014) Iron-based shape memory alloys for civil engineering structures: an overview. Constr Build Mater 63:281–293
Croci G (2001) Strengthening the Basilica of St Francis of Assisi after the September 1997 earthquake. Struct Eng Int 11(3):207–210
Dolce M, Cardone D (2001) Mechanical behaviour of SMA elements for seismic applications—part 1 martensile and austenite NiTi bars subjected to torsion. Int J Mech Sci 43(11):2631–2656
Eggeler G, Hornbogen E, Yawny A, Heckmann A, Wagner MFX (2004) Structural and functional fatigue of NiTi shape memory alloys. Mat Sci Eng A Struct 378(1–2):24–33
Fang C, Yam MCH, Ma HW, Chung KF (2015) Tests on superelastic Ni–Ti SMA bars under cyclic tension and direct-shear: towards practical recentring connections. Mater Struct 48(4):1013–1030
Frick CP, Ortega AM, Tyber J, Maksound AElM, Maier HJ, Liu YN, Gall K (2005) Thermal processing of polycrystalline NiTi shape memory alloys. Mat Sci Eng A Struct 405(1–2):34–49
Fugazza D (2005) Experimental investigation on the cyclic properties of superelastic NiTi shape-memory alloy wires and bars. Rose School, European school for advanced studies in reduction of seismic risk, Pavia
Gall K, Sehitoglu H, Chumlyakov YI, Kireeva IV, Maier HJ (1999a) The influence of aging on critical transformation stress levels and martensite start temperatures in NiTi: part I—aged microstructure and micro-mechanical modeling. J Eng Mater-T ASME 121(1):19–27
Gall K, Sehitoglu H, Chumlyakov YI, Kireeva IV, Maier HJ (1999b) The influence of aging on critical transformation stress levels and martensite start temperatures in NiTi: part II—discussion of experimental results. J Eng Mater-T ASME 121(1):28–37
Gall K, Sehitoglu H, Chumlyakov YI, Kireeva IV (1999c) Tension-compression asymmetry of the stress-strain response in aged single crystal and polycrystalline NiTi. Acta Mater 47(4):1203–1217
Gall K, Sehitoglu H, Chumlyakov YI, Zuev YL, Karaman I (1998) The role of coherent precipitates in martensitic transformations in single crystal and polycrystalline Ti-50.8at%Ni. Scripta Mater 39(6):699–705
Garlock MEM, Ricles JM, Sause R (2008) Influence of design parameters on seismic response of post-tensioned steel MRF systems. Eng Struct 30(4):1037–1047
Hodgson DE, Wu MH, Biermann RJ (1990) Shape memory alloys. ASM handbook committee properties and selection: nonferrous alloys and special-purpose materials, 1st edn. ASM International, Ohio, pp 897–902
Huang W (2002) On the selection of shape memory alloys for actuators. Mater Design 23(1):11–19
Huang W, Toh W (2000) Training two-way shape memory alloy by reheat treatment. J Mater Sci Lett 19(17):1549–1550
Huang X, Liu Y (2001) Effect of annealing on the transformation behavior and superelasticity of NiTi shape memory alloy. Scripta Mater 45(2):153–160
Indirli M, Castellano MG (2008) Shape memory alloy devices for the structural improvement of masonry heritage structures. Int J Archit Herit 2(2):93–119
Indirli M, Castellano MG, Clemente P, Martelli A (2001a) Demo-application of shape-memory alloy devices: the rehabilitation of the S. Giorgio Church Bell Tower. In: Liu SC (ed) Proceedings of SPIE's 8th annual international symposium on smart structures and materials, Newport Beach, CA, USA, 2011. Smart structures and materials 2001: smart systems for bridges, structures, and highways, vol 4330, p 262–272
Indirli M, Forni M, Martelli A, Spadoni B, Venturi G, Alessandri C, Bertocchi A, Cami R, Capelli C, Baratta A, Procaccio A, Clemente P, Canio GD, Carpani B, Bonacina G, Franchioni G, Viani S, Cesari F, Mucciarella M, Meucci C (2001b) Further new projects in Italy for the development of innovative techniques for the seismic protection of cultural heritage. In: Proceedings of the 7th international seminar on seismic isolation, passive energy dissipation and active control of vibrations of structures, Assisi, Italy, 2001
ISO 6892-1:2009 Metallic materials—tensile testing, part 1: method of test at ambient temperature

Jani JM, Leary M, Subic A, Gibson MA (2014) A review of shape memory alloy research, applications and opportunities. Mater Design 56:1078–1113
Janke L, Czaderski C, Motavalli M, Ruth J (2005) Applications of shape memory alloys in civil engineering structures—overview, limits and new ideas. Mater Struct 38(279):578–592
Kanvinde A (2016) Predicting fracture in civil engineering steel structures: state of the art. J Struct Eng-ASCE 143(3):03116001
Kauffman GB, Mayo I (1997) The story of nitinol: the serendipitous discovery of the memory metal and its applications. Chem Educator 2(2):1–21
Khalil-Allafi J, Ren XB, Eggeler G (2002) The mechanism of multistage martensitic transformations in aged Ni-rich NiTi shape memory alloys. Acta Mater 50(4):793–803
Lagoudas DC (2008) Shape memory alloys: modeling and engineering applications. Springer, TX, USA
Lecce L, Concilio A (eds) (2015) Shape memory alloy engineering: for aerospace, structural and biomedical applications. Elsevier
Leclercq S, Lexcellent C (1996) A general macroscopic description of the thermomechanical behavior of shape memory alloys. J Mech Phys Solids 44(6):953–980
Liang C, Rogers CA (1990) One-dimensional thermomechanical constitutive relations for shape memory materials. J Intel Mat Syst Struct 1(2):207–234
Lin YC, Sause R, Ricles JM (2013a) Seismic performance of a steel self-centering, moment-resisting frame: hybrid simulations under design basis earthquake. J Struct Eng-ASCE 139(11):1823–1832
Lin YC, Sause R, Ricles JM (2013b) Seismic performance of a large-scale steel self-centering moment-resisting frame: MCE hybrid simulations and quasi-static pushover tests. J Struct Eng-ASCE 139(7):1227–1236
Ma J, Karaman I, Noebe RD (2010) High temperature shape memory alloys. Int Mater Rev 55(5):257–315
Marc MSC (2014) Vol. A: theory and user information. MSC Software Corporation
Martelli A (2008) Recent progress of application of modern anti-seismic systems in Europe—part 2: energy dissipation systems, shape-memory alloy devices and shock transmitters. In: Proceedings of the 14th world conference on earthquake engineering, Beijing, China, 2008
McCormick J, Tyber J, DesRoches R, Gall K, Maier HJ (2007) Structural engineering with NiTi. II: mechanical behavior and scaling. J Eng Mech-ASCE 133(9):1019–1029
Michutta J, Carroll MC, Yawny A, Somsen C, Neuking K, Eggeler G (2004) Martensitic phase transformation in Ni-rich NiTi single crystals with one family of Ni_4Ti_3 precipitates. Mat Sci Eng A-Struct 378(1–2):152–156
Ölander A (1932) An electrochemical investigation of solid cadmium-gold alloys. J Am Chem Soc 54(10):3819–3833
Otsuka K, Wayman CM (1998) Shape memory materials. Cambridge University Press, New York
Ozbulut OE, Hurlebaus S, DesRoches R (2011) Seismic response control using shape memory alloys: a review. J Intel Mat Syst Struct 22(14):1531–1549
Penar BW (2005) Recentering beam-column connections using shape memory alloys. Master thesis, Georgia Institute of Technology
Perkins J, Hodgson D (1990) The two-way shape memory effect. In: Duerig TW, Melton KN, Stöckel D, Wayman CM (eds) Engineering aspects of shape memory alloys, p 195–206
Qiu CX, Zhu SY (2014) Characterization of cyclic properties of superelastic monocrystalline Cu-Al-Be SMA wires for seismic applications. Constr Build Mater 72:219–230
Raj SV, Noebe RD (2013) Low Temperature creep of hot-extruded near-stoichiometric NiTi shape memory alloy part I: isothermal creep. Mat Sci Eng A-Struct 581(5):145–153
Ricles JM, Sause R, Garlock MEM, Zhao C (2001) Post-tensioned seismic-resistant connections for steel frames. J Struct Eng-ASCE 127(2):113–121
Ricles JM, Sause R, Peng SW, Lu LW (2002) Experimental evaluation of earthquake resistant posttensioned steel connections. J Struct Eng-ASCE 128(7):850–859
Sadiq H, Wong MB, Al-Mahaidi R, Zhao XL (2010) The effects of heat treatment on the recovery stresses of shape memory alloys. Smart Mater Struct 19(3):035021

Song GC, Ma N, Li HN (2006) Applications of shape memory alloys in civil structures. Eng Struct 28(9):1266–1274

Soroushian P, Ostowari K, Nossoni A, Chowdhury H (2001) Repair and strengthening of concrete structures through application of corrective posttensioning forces with shape memory alloys. Transp Res Rec J Trans Res Board 1770(1):20–26

Speicher MS (2010) Cyclic testing and assessment of shape memory alloy recentering systems. PhD thesis, Georgia Institute of Technology

Tanaka K, Nagaki S (1982) A thermomechanical description of materials with internal variables in the process of phase transitions. Ing Archiv 51(5):287–299

Tyber J, McCormick J, Gall K, DesRoches R, Maier HJ, Maksoud AEA (2007) Structural engineering with NiTi. I: basic materials characterization. J Eng Mech-ASCE 133(9):1009–1018

Vernon LB, Vernon HM (1941) Process of manufacturing articles of thermoplastic synthetic resins. US Patent 2,234,993, 1941

Wallaert JJ, Fisher JW (1965) Shear strength of high-strength bolts. J Struct Div 91(3):99–125

Wang W, Fang C, Liu J (2016) Large size superelastic SMA bars: heat treatment strategy, mechanical property and seismic application. Smart Mater Struct 25(7):075001

Wolski M, Ricles JM, Sause R (2009) Experimental study of a self-centering beam-column connection with bottom flange friction device. J Struct Eng-ASCE 135(5):479–488

Zaki W, Moumni Z (2007) A 3D model of the cyclic thermomechanical behavior of shape memory alloys. J Mech Phys Solids 55(11):2427–2454

Chapter 2
Shape-Memory Alloy Elements

Abstract In this chapter, a series of SMA elements, including SMA wires, cables, bars, bolts, helical springs, washer springs and ring springs, are described in detail. These elements are important additions to the 'arsenal' against the seismic hazard. For each type of element, the manufacturing process, testing scheme, key mechanical behaviour and analytical expressions are presented. The various SMA elements are shown to have their own unique load–deformation, self-centring and energy dissipation characteristics, catering to different design purposes. The main advantages and possible limitations of these elements are also discussed, with a particular focus on their failure mode and material utilisation efficiency.

2.1 Introduction to Basic SMA Elements

'SMA elements' are fundamental forms of SMA products that constitute more diverse self-centring devices and members used in structural systems. As mentioned in Chap. 1, civil engineers prefer to utilise the SE property of SMA to achieve a spontaneous self-centring driving mechanism. Another important motivation for using superelastic SMAs is their inherent energy dissipation capability attributing to reversible martensite interfacial motion, a damping mechanism which differs from the dislocation based plasticity exhibited by other metals such as steel.

An SMA element with an anticipated function is the key to success for the development of more comprehensive SMA-based self-centring devices/members. The basic SMA elements include SMA monofilament wires, cables, bars, as well as different types of springs (Fig. 2.1). These elements are conventionally made of high-strength steel or other metals, and as a result, their recoverability and energy absorbing capacity are often limited by the small elastic strain range of the material. However, when these elements are endowed with the SE property, the scope of their application can be extensively broadened.

C. Fang and W. Wang, *Shape Memory Alloys for Seismic Resilience*, https://doi.org/10.1007/978-981-13-7040-3_2

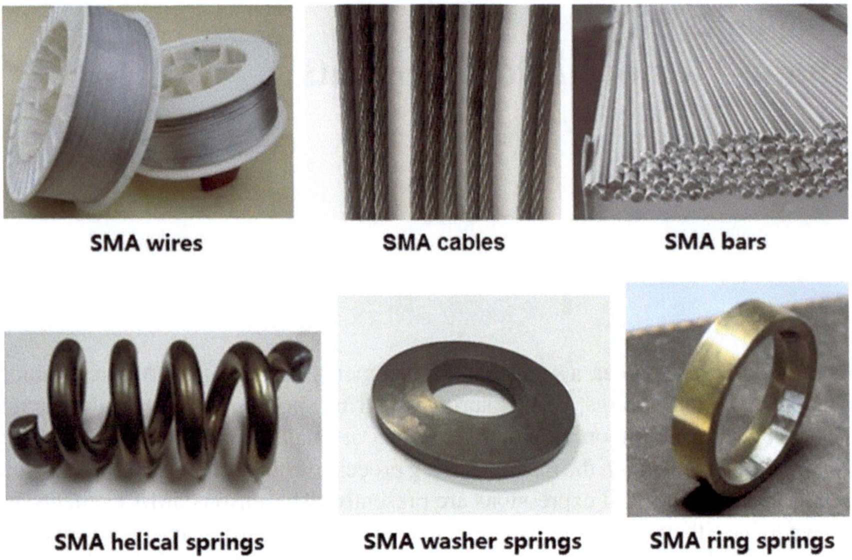

Fig. 2.1 Typical SMA elements

Depending on the geometrical configurations, different forms of SMA elements vary in mechanical behaviour. Each type of element has its own advantages and possible limitations and is therefore suited to different fields of application. Generally speaking, any SMA element selected for civil engineering use should preferably exhibit the following properties: (1) high material utilisation efficiency from an economic point of view, (2) low risk of damage and (3) ease of implementation and installation. This chapter introduces typical SMA elements that exhibit desirable load–deformation responses catering to civil engineering practice, especially for the field of seismic engineering. To be more specific, the following forms of SMA element are discussed in detail in this chapter: SMA wires and cables, SMA bars and bolts, SMA helical springs, SMA Belleville washer springs, and SMA ring springs. Their load–deformation response, ductility, self-centring capability, energy dissipation, and possible failure modes, are presented in detail. Unless specified otherwise, the SMAs discussed in the remaining of this chapter refer to NiTi SMAs, and only the SE of the material is utilised.

2.2 SMA Wires and Cables

2.2.1 Monofilament SMA Wires

The monofilament wire form of SMA is the most widely considered candidate of SMA products for practical use due to its high material utilisation efficiency and desirable deformation mode (i.e. uniaxial tension). Monofilament SMA wires are tension-only elements, and their tensile load–elongation responses are identical to the engineering stress–strain relationships of the material. The mechanical behaviour of individual SMA wires has been the subject of numerous studies in the past several decades and is now well understood. A common finding from the previous studies is that the performance of superelastic SMA wires is sensitive to the maximum strain, the number of repeated loading cycles, temperature and loading rate. For example, through a series of cyclic uniaxial tests performed on 2.54 mm-diameter wires, Lim and McDowell (1995) found that the forward transformation plateau gradually decreases, the residual strain accumulates, and the hysteretic energy dissipation decreases, with an increase in the number of loading cycles. Similar responses were reported by Wolons et al. (1998) and Tamai and Kitagawa (2002). With the consideration of temperature variation, Piedboeuf et al. (1998) and Dolce and Cardone (2001a) confirmed that an increase in temperature leads to upward shifts of the hysteretic loops. It was also suggested that pre-straining should be applied to SMA wires to ensure stabilisation of the hysteretic response. Tobushi et al. (1998) found that an overly high temperature or a very large strain rate can cause increased residual strains of SMA wires. Fugazza (2005) observed that an increase in strain rate leads to a decrease in equivalent viscous damping. Most of the above findings are related to the inherent properties of the SMA material, which has been discussed in Chap. 1.

Industrial experience suggests that with an appropriate manufacturing and cold working process, small diameter SMA wires can exhibit outstanding recoverability, ductility, hysteretic stability and fatigue resistance. Given these advances, SMA wires have been widely considered as kernel elements for SMA-based devices such as dampers. However, there are still practical issues related to their real-life implementation. One problem is the difficulty of gripping or end fixing, especially for relatively large-diameter wires, e.g. diameter larger than 1 mm. The conventional grippers such as U-connectors are found to be ineffective for smooth and large-diameter SMA wires. In addition, providing a sufficient level of load carrying capacity means that a large number of wires should be included in a device, which further complicates the arrangement of the wires and the gripping strategy.

2.2.2 SMA Cables

Apart from monofilament SMA wires, SMA cables can be a good alternative that combines the advantage of conventional steel cables and the unique mechanical

Fig. 2.2 Typical manufacturing process of SMA cables

properties thanks to the SE (or SME). A structural cable typically consists of a group of helically wrapped strands, and each strand is a bundled assembly of helically wrapped monofilament wires. Taking a typical 7 × 7 cable construction, for example, each cable consists of seven helically wrapped strands, and each strand consists of seven helically wrapped monofilament SMA wires. A more complex pattern is required for SMA cables involving more SMA wires. The typical manufacturing process, i.e. the formations of SMA strands and cables for a 12 × 12 SMA cable construction, is shown in Fig. 2.2.

SMA cables are able to reasonably 'scale up' the satisfactory properties of monofilament SMA wires with stiffness and strength better meeting the civil engineering requirements. In contrast to SMA bars, as discussed later, SMA cables are more cost-effective and easy to handle and spool, and are adequately stiff in axial tension whereas having no buckling issue in compression (Fang et al. 2019). The construction of SMA cables is very flexible with diverse helix angles and layups, catering to different design purposes.

However, SMA cables are newer and less explored compared to SMA wires, and the research started relatively late. Reedlunn et al. (2013) carried out an experimental study on superelastic SMA cables with different layups and confirmed that the SMA cable with a 7 × 7 construction could have mechanical responses comparable to those

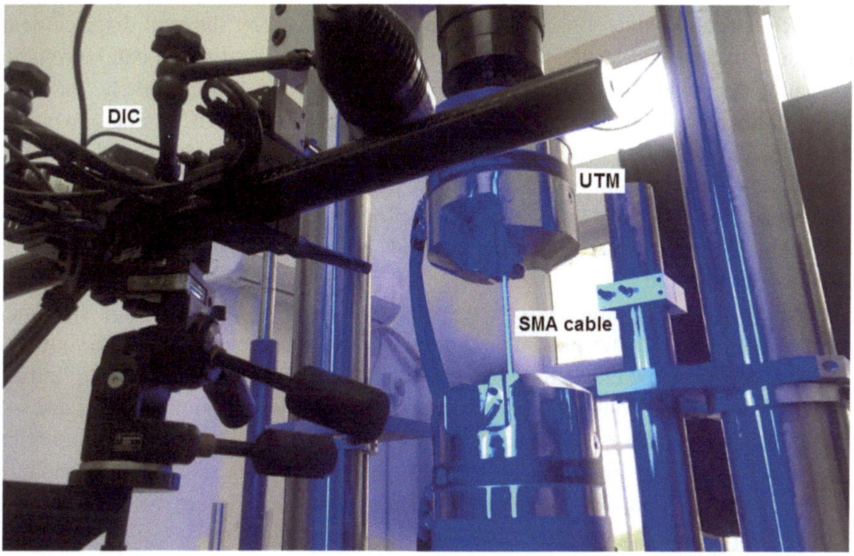

Fig. 2.3 Typical test setup for 7 × 7 SMA cables

of monofilament wires. Similar studies were later carried out by Ozbulut et al. (2015) and Sherif and Ozbulut (2018), where the focus was on the fatigue characteristics. Carboni et al. (2014) established a phenomenological model for SMA cables, where the prediction agreed well with the available test results. Biggs and Shaw (2016) tried to explore the potential of SME of SMA cables for actuator application. They experimentally characterised the SMA cables under repeated thermal cycles, and it was again confirmed that a 7 × 7 construction worked well. Mas et al. (2017) presented the application of SMA cables for reinforced concrete structures. Pull-out tests were conducted to examine the bonding strength between the cables and the surrounding concrete, and small-scale concrete beams with SMA cable reinforcement were physically tested. In parallel to those experimental works, numerical studies were performed to further explore the application fields such as high-rise towers (Sun and Lv 2016) and bridge bearings (Zheng et al. 2018).

The mechanical behaviour of SMA cables can be conveniently measured via a UTM. Figure 2.3 shows the typical test setup for examining the uniaxial stress–strain behaviour a 7 × 7 × 0.8 mm SMA cable (0.8 mm is the diameter of the individual SMA wire). Special attention should be paid to end fixing. If no special end gripping device is fabricated, the ends of the SMA cable specimen may be directly clamped by the top and bottom hydraulic wedge grips that match the external diameter of the cable. In this case, it is suggested that the ends of the cable were coated with a thin layer of epoxy to avoid stress concentration (Fang et al. 2019).

The stress of the cable could be calculated by dividing the measured load by the sum of the cross-sectional area of the, say, 49 monofilament wires for the 7 × 7 cable construction. The global strain can be determined by the grip displacement, provided that the end fixing slippage is negligible. The local strain can be reliably measured by non-contact laser displacement transducers. Alternatively, the local strain field can be monitored by a Digital Image Correlation (DIC) camera, which is an optical, non-contact technique that measures surface displacements of an object by tracking specular patterns on the surface of a specimen (Sutton et al. 2009). If no extra annealing is applied and as a result, the form setting ability of the SMA cable is not enabled (as elaborated later), the two ends of the cable specimen must be firmly constrained to avoid unravelling.

The typical global stress–strain responses of the 7 × 7 SMA cable specimen (shown in Fig. 2.3) under incremental or constant strain cyclic loading are shown in Fig. 2.4. A flag-shaped hysteretic behaviour with satisfactory strain recoverability is clearly observed, although the 'yield' strength and transformation plateaus are less recognisable than those observed in the typical stress–strain curve of a monofilament SMA wire (e.g. Figure 1.19). The relatively smooth stress–strain behaviour of the SMA cables is attributed to their bundled construction, where different wires in the cable are loaded in different and complex deformation modes, including combined tension, torsion, and bending, with diverse stress distributions. As a result, the axial load is not uniformly shared among the different wires, and some wires could undergo forward transformation earlier than others. The non-synchronisation of stress development leads to a more gradual evolution of the global stress–strain responses of the entire cable.

It is also observed from Fig. 2.4 that certain degradations of the yield strength, accompanied by an accumulation of residual strain, are induced in the SMA cables. The residual strain is mainly from two sources, i.e. transformation-induced fatigue of the material itself and structural relaxation. The latter is a common mechanical phenomenon observed in almost all kinds of cables. While it is often difficult to tell the exact proportions of the contributing factors, proper training can mitigate both 'relaxation' effects. This can be particularly confirmed by observing the stress–strain curves of the specimen subjected to a constant strain amplitude, where the first loading cycle leads to the most pronounced residual strain, whereas the hysteretic loops become stabilised in the subsequent loading cycles.

The SMA cables may be additionally annealed at different targeting temperatures for varying durations, and in this way, some thermal–mechanical properties such as transformation temperature, stiffness and form setting (shape setting) ability, can be tuned. In particular, 'form setting' refers to the process used to form an SMA product into a specific and often complex geometric shape. This process is often realised by temporarily constraining the material into a new desired shape and then performing a form setting annealing process followed by water quenching. In fact, annealing and form setting are often intentionally done simultaneously. For practical use of the SMA cables, it may be necessary to maintain the twisted form and to avoid unravelling at room temperature, unless reliable end constraints are permanently applied to the cable segments.

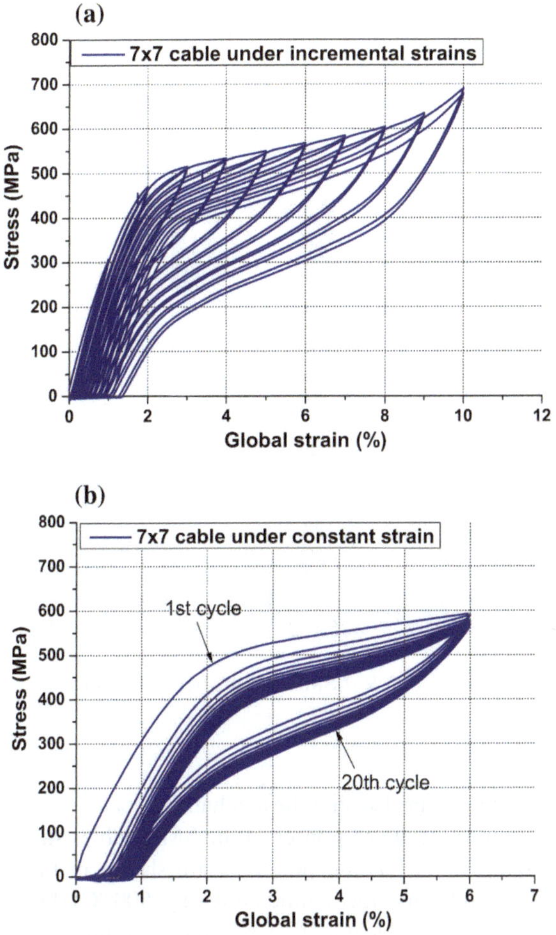

Fig. 2.4 Typical global stress–strain responses of 7 × 7 SMA cables: **a** incremental amplitudes, **b** constant amplitude

Figure 2.5 shows the different cable specimens subjected to varying annealing temperatures for 15 min. After being cut from the mid-length, no form setting ability is observed for the specimen without annealing, and as a result, the cable completely unravelled after being cut. A moderate form setting ability is gained for the specimen subjected to an annealing temperature of 350 °C, where partial unravelling near the cutting location is observed. Enhanced form setting ability is enabled when the annealing temperature increases to 400–500 °C, a range which is generally adequate to avoid unravelling.

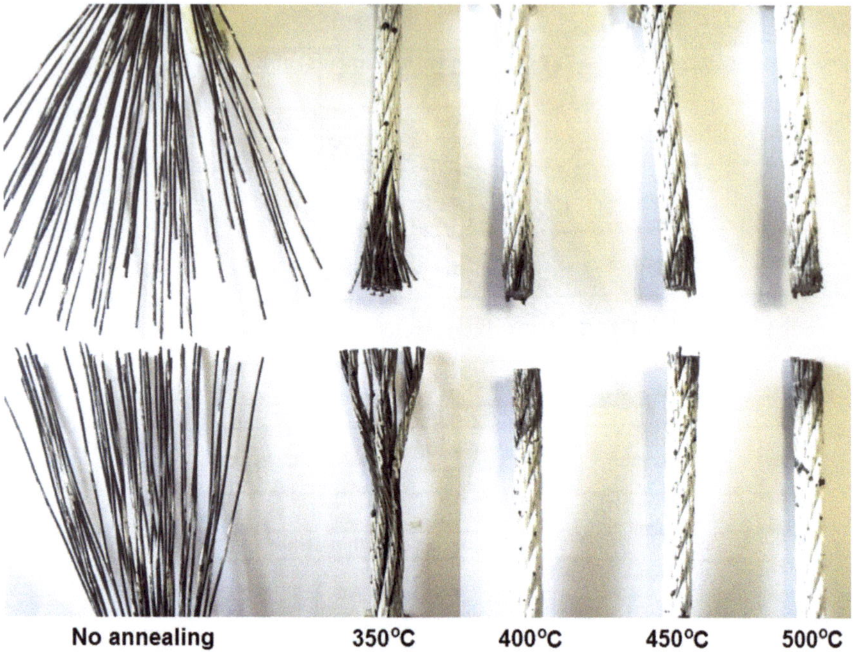

Fig. 2.5 Form setting property of 7 × 7 SMA cables after being subjected to various annealing temperatures for 15 min

The annealing process can also alter the mechanical behaviour of the SMA cables. The representative stress–strain responses of the SMA cables subjected to different annealing temperatures and loading protocols are shown in Fig. 2.6. The cables receiving a relatively high annealing temperature (e.g. 500 °C) show an 'over ageing' behaviour with more evident degradation and more pronounced accumulation of residual strain.

The initial stiffness changes with varying annealing schemes. Figure 2.7 shows the elastic moduli of different SMA cables as a function of cycle numbers. The strain levels for incremental cyclic loading are also indicated in the figure. It is seen that the elastic moduli of the cables range from 30 to 45 GPa, where a higher annealing temperature tends to cause an increase in the elastic modulus. This tendency is related to the form setting properties of the cables. The twisted wires after experiencing a form setting process can more closely interact with each other, and in addition, the change of the elastic moduli of the material itself due to the different annealing schemes could also attribute to the difference in cable stiffness. It is noted that in practice, SMA cables are often used in a prestrained state such that a near infinite initial stiffness can be achieved prior to decompression. In other words, the relatively low initial stiffness of the SMA cables is not a practical problem if a sufficient prestrain is applied.

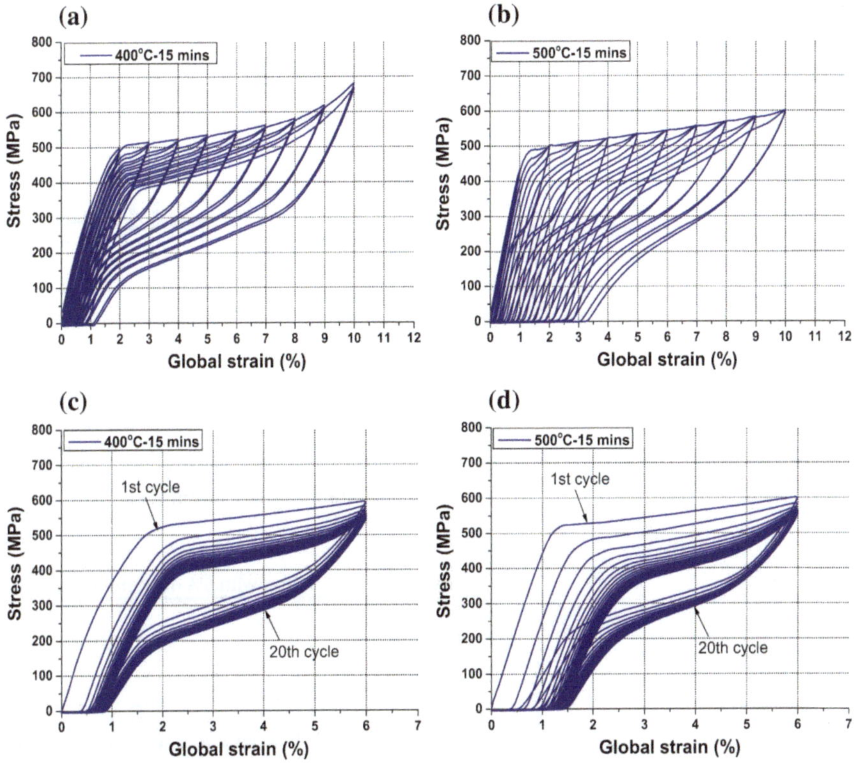

Fig. 2.6 Typical global stress–strain responses of 7 × 7 SMA cables after being subjected to various annealing temperatures for 15 min: **a** 400 °C—incremental amplitudes, **b** 500 °C—incremental amplitudes, **c** 400 °C—constant amplitude, **d** 500 °C—constant amplitude

Figure 2.8 shows the accumulated residual strain of the SMA cables as a function of cycle numbers. When subjected to incremental cyclic loading, the residual strain is gradually accumulated with an increase in the maximum strain. Except for the cable being subjected to an annealing temperature of 500 °C, the final accumulated residual strains are generally below 1.5%, corresponding to a recovery rate, i.e. the ratio of the recovered strain to the peak strain, of more than 85%. When constant cyclic loading is considered, the residual strain is most quickly accumulated during the first few loading cycles, and subsequently, the residual strain is developed much more slowly. This suggests that for practical use of the SMA cables, a 'pre-training' with several loading cycles at the target strain level can effectively stabilise the hysteretic behaviour of the cables and limit further development of the residual strain. Again, an overly high annealing temperature of 500 °C deteriorates the self-centring capability.

SMA cables are also effective energy dissipative elements. They can provide major or supplementary energy dissipation for a structural system. The energy dissipation

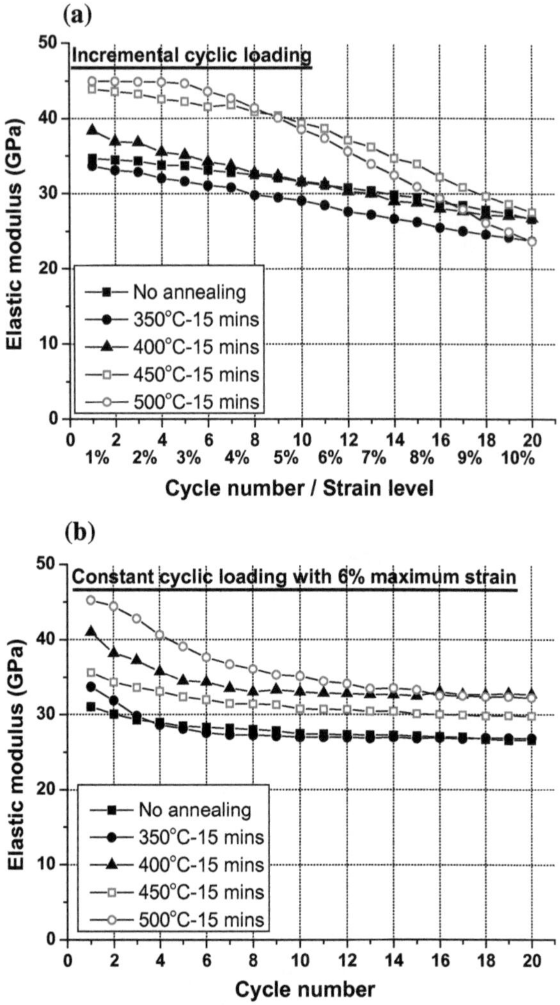

Fig. 2.7 Elastic modulus of 7 × 7 SMA cables after being subjected to various annealing temperatures for 15 min: **a** incremental amplitudes, **b** constant amplitude

capability of a member or device is normally presented by a dimensionless index, namely equivalent viscous damping ratio (EVD), as defined by

$$EVD = \frac{W_D}{4\pi W_E} \tag{2.1}$$

where W_D is the energy loss per cycle, which is essentially the area enclosed by a stress–strain hysteretic loop and W_E is the strain energy stored in a corresponding linear system. For the SMA cables discussed above, the EVD is found to be up to 6%, and it depends on several factors including the loading protocol, maximum strain, and annealing scheme, as shown in Fig. 2.9. At the same maximum strain level, the cables

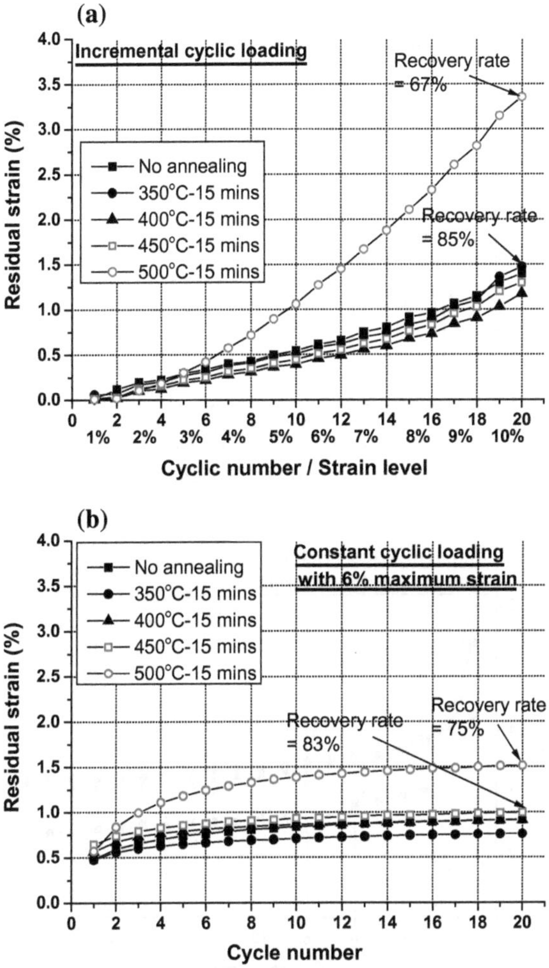

Fig. 2.8 Self-centring capability of 7 × 7 SMA cables after being subjected to various annealing temperatures for 15 min: **a** incremental amplitudes, **b** constant amplitude

receiving an annealing temperature of 350–450 °C generally show higher EVDs than the cables without annealing. This echoes the wider hysteretic shapes of the former. Again, the increased EVD is related to the form setting behaviour of the annealed cables. It is noted that due to the non-synchronisation of stress development within the cross section, the SMA cables tend to exhibit smaller EVDs than the monofilament SMA wires.

From a ductility point of view, SMA cables are able to provide large inelastic deformation before fracture. Figure 2.10 shows the typical stress–strain relationship of a well-trained 7 × 7 SMA cable under monotonic loading. The cable behaviour looks quite similar to the monotonic wire behaviour. After the completion of forward

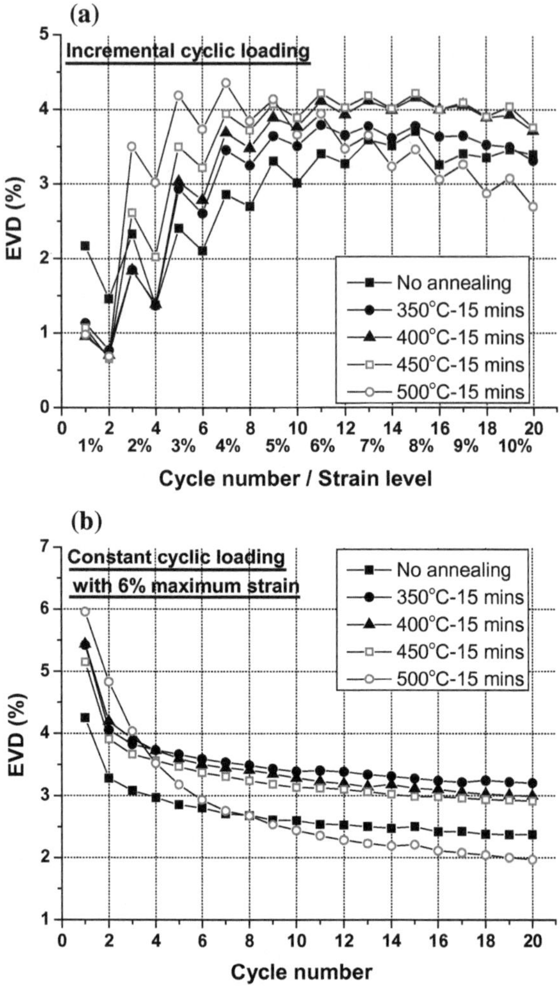

Fig. 2.9 Energy dissipation capability of 7 × 7 SMA cables after being subjected to various annealing temperatures for 15 min: **a** incremental amplitudes, **b** constant amplitude

transformation, the cable advances into the martensitic hardening stage, and then plastic deformation is induced. Once the cable enters the plastic deformation plateau, only part of the strain can be recovered. The final fracture strain of SMA cables is typically larger than 15%. It is also noticed that a certain residual load resistance is maintained after the occurrence of initial fracture, which is due to the less radical rupture behaviour of the wire groups. This successive-type failure mode is unique for SMA cables, and one may take advantage of such ductile and robust behaviour against accidental loadings.

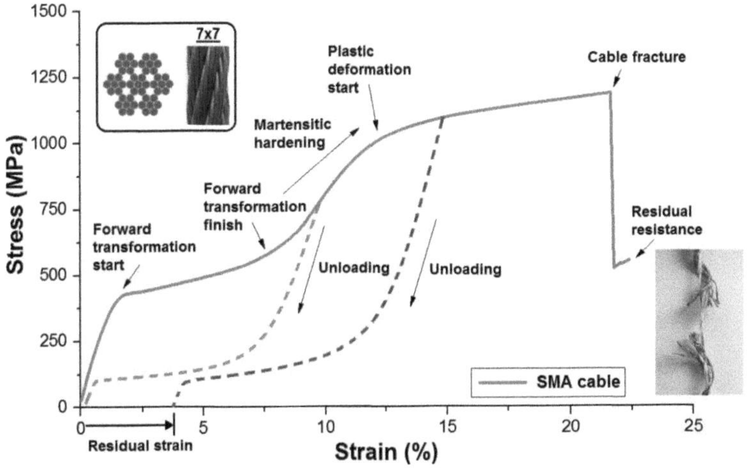

Fig. 2.10 Typical fracture behaviour of a 7 × 7 SMA cable

2.3 SMA Bars and Bolts

2.3.1 *Uniaxial Behaviour of SMA Bolts*

Large-scale SMA bars or tendons could provide an adequate level of load resistance for engineering applications, although early studies revealed that SMA bars have inferior SE properties in comparison with SMA wires (Dolce and Marnetto 1999). DesRoches et al. (2004) compared the hysteresis behaviour of SMA bar specimens with different diameters ranging from 1.8 to 25.4 mm. Fugazza (2005) examined the size effect of SMA bars with various diameters up to 8 mm. A common finding obtained from the existing studies is that both SMA wires and bars are able to exhibit sound SE behaviour if they are properly manufactured and annealed.

To facilitate installation and implementation, SMA bars are often machined down to the bolt form. Figure 2.11 shows the typical test setup used to examine the uniaxial behaviour of the SMA bolts. To realistically consider the behaviour of the threaded parts, the threaded length of the SMA bolts can be screwed to specially designed steel heads which are then fitted into the hydraulic wedges. The typical stress–strain relationships of SMA bolts are shown in Fig. 2.12. These specimens were subjected to quasi-static loading with a loading rate of 0.02 mm/s. The composition of all the bolts was nearly identical, with 55.9% nickel by weight and the balance weight contributed by titanium (Fang et al. 2015a). The geometrical dimensions of the bolts are given in Table 2.1.

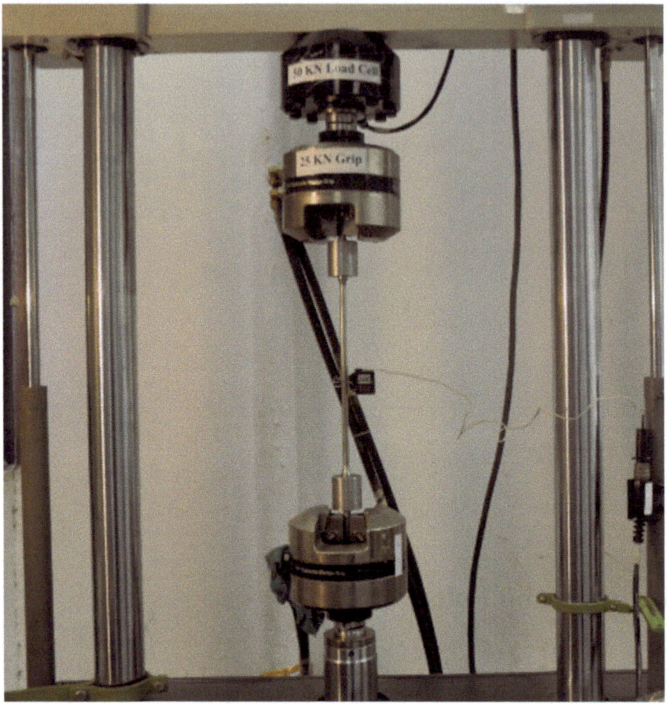

Fig. 2.11 Specially designed test setup for SMA bolts

For the three SMA bolts, the forward transformation stress is between 350 MPa and 400 MPa under the first few loading cycles, but it gradually drops with an increase in the number of loading cycles. Similar behaviour is observed for the reverse transformation plateau, leading to downward movements of the entire cyclic loop. The residual strain induced by each cycle, as well as the accumulated residual strain, are shown in Fig. 2.13. Limited residual strain is induced up to a peak strain of 4%, and the residual strain starts to increase more evidently afterwards.

The absolute energy dissipation during each loading cycle depends on the bolt diameter and maximum strain. As shown in Fig. 2.14a, the absolute energy loss per cycle during the first two cycles (1% strain level) is very limited, as the SMA bolts mainly underwent elastic deformation. Then the energy loss increases significantly, and the maximum energy loss is more than 150 J per cycle. As shown in Fig. 2.14b, the EVD increases abruptly from 1 to approximately 3.5–4% at the third cycle and tends to stabilise between 5 and 7% during the subsequent cycles. Size effect seems to be unclear from an EVD point of view, as the EVD values and trends for the SMA bolts with varying diameters are quite similar.

Caution should be exercised that the threaded parts of SMA bolts are susceptible to fracture, as shown in Fig. 2.15. Although the net diameters of the threaded parts are designed to be larger than the diameters of the shank parts (Table 2.1),

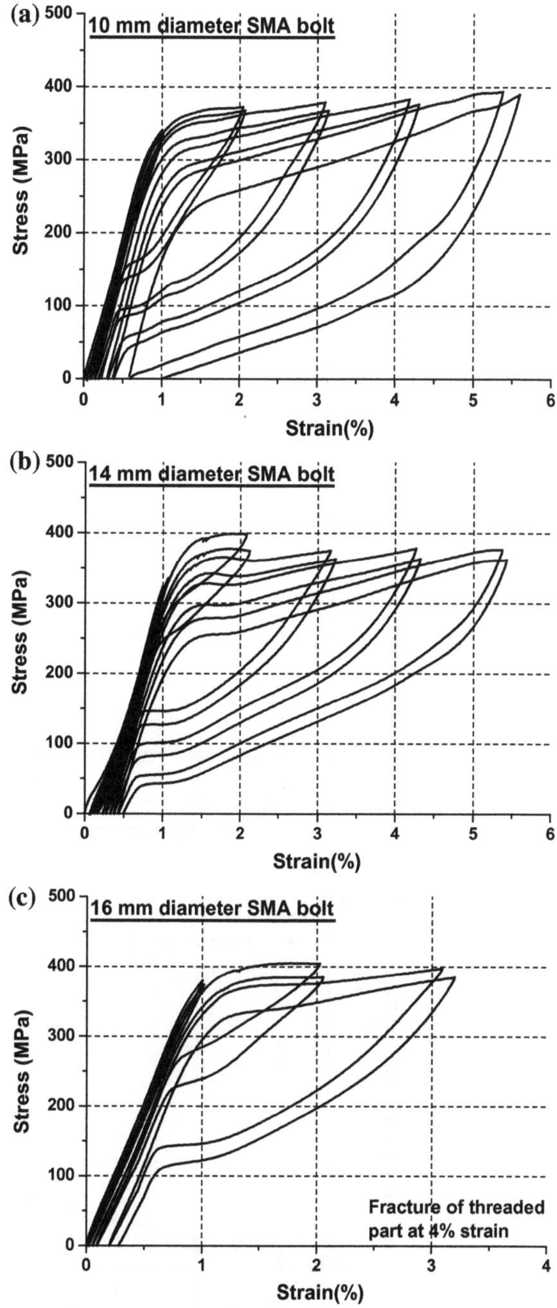

Fig. 2.12 Typical stress–strain responses of SMA bolts under cyclic tension: **a** 10 mm diameter, **b** 14 mm diameter, **c** 16 mm diameter (Fang et al. 2015a)

Table 2.1 Dimensions of SMA bolts under cyclic tension (Fang et al. 2015a)

Shank diameter (mm)	Net threaded diameter (mm)	Overall length (mm)	Threaded length (mm)
10.0	11.0	200	35
14.0	17.0	200	35
16.0	17.0	200	35

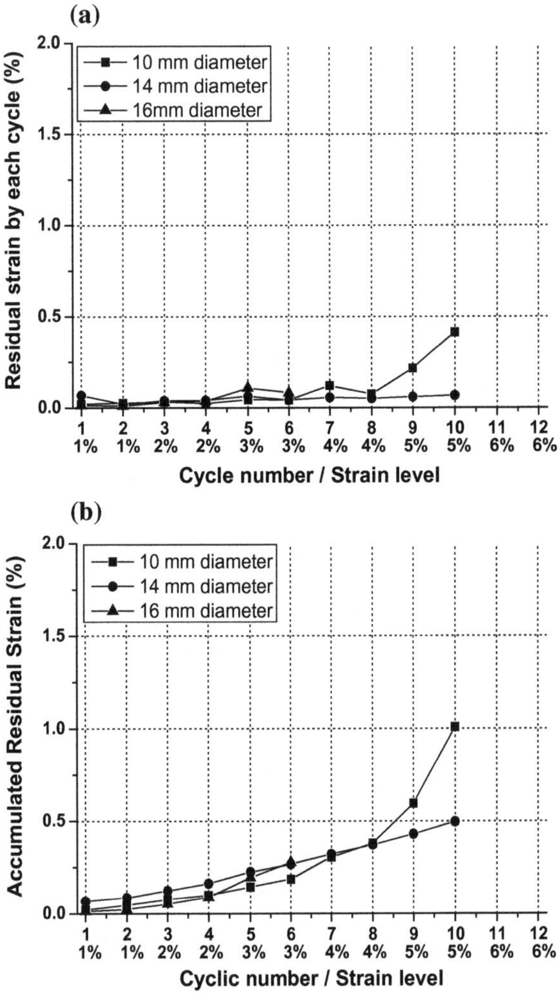

Fig. 2.13 Residual strain of SMA bolts under cyclic tension: **a** residual strain by each cycle, **b** accumulated residual strain

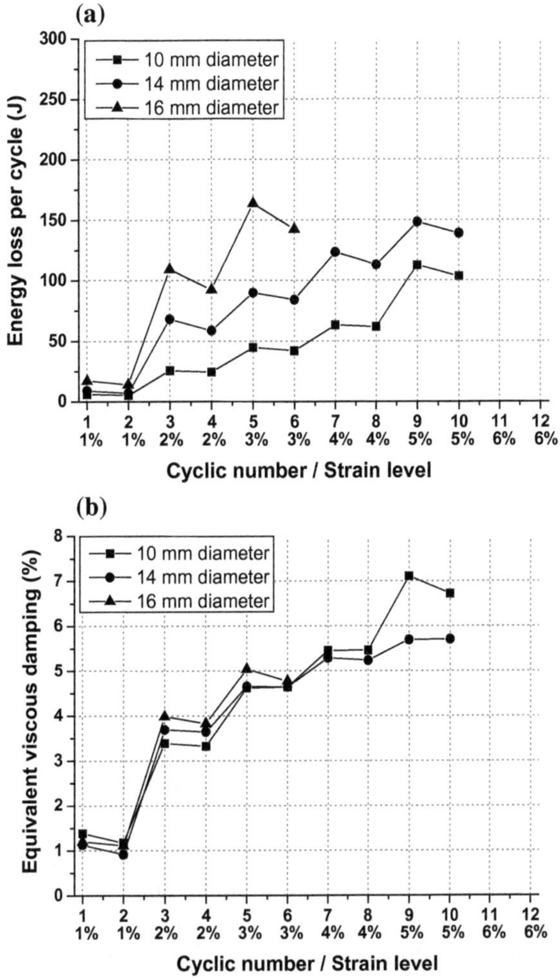

Fig. 2.14 Energy dissipation capacity of SMA bolts under cyclic tension: **a** energy loss per cycle, **b** equivalent viscous damping

fracture over the threaded area still occurs quite early. In particular, a relatively low net threaded-to-shank diameter ratio was considered for the 16 mm-diameter bolt, and as a result, the fracture strain was only 4%, which is much less than the available maximum strain reported in SMA wire tests. In view of this, special concern should be given to the threaded part which is potentially a vulnerable zone. In other words, an adequate net threaded-to-shank diameter ratio of the SMA bolts should be ensured to avoid premature fracture over the threaded area. According to Fang et al. (2014a), a net threaded-to-shank diameter ratio of approximately 1.0 leads to average fracture strains of around 3.6 and 5.2% for the 16 and 10 mm-diameter SMA bolts, respectively. With a similar net threaded-to-shank diameter ratio, Tamai et al. (2003) reported a 2.5% fracture strain for the SMA bolts used in column bases. Speicher

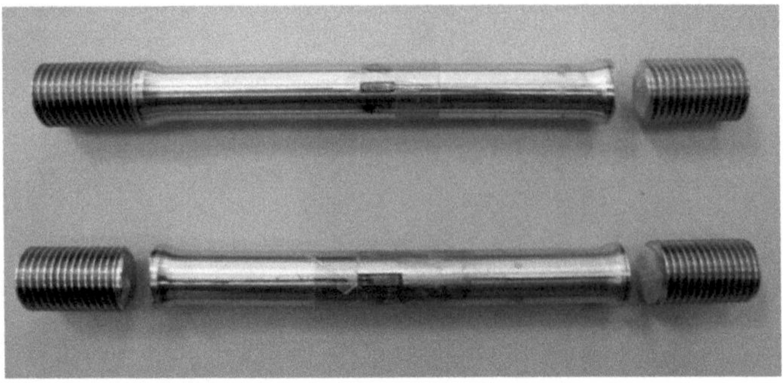

Fig. 2.15 Typical failure mode of SMA bolts under tension

et al. (2011) examined a series of SMA bolts with a net threaded-to-shank diameter ratio of 1.40, in which case no fracture was observed beyond 7% maximum strain. More recently, Fang et al. (2017) revealed that a net threaded-to-shank diameter ratio of 1.25 can ensure a fracture-free strain of at least 6%. Based on these limited test data, a net threaded-to-shank diameter ratio of 1.30 may be recommended to ensure sufficient ductility of the SMA bolts. A larger net threaded-to-shank diameter ratio has to be realised at the cost of more considerable material wastage during the machining process, which is not economical.

2.3.2 Shear Behaviour of SMA Bolts

For the shear behaviour of SMA elements, Dolce and Cardone (2001b) undertook torsion tests on superelastic SMA bars with round and hexagonal cross sections. A moderate EVD of 5–6% was observed, and a good self-centring capability was confirmed. Moreover, the torsional cyclic behaviour of the SMA bars was shown to be stable and repeatable.

SMA bolts may also be subjected to a direct shear action, e.g. when they are used in shear connectors. The typical direct shear force–displacement responses of SMA bolts are shown in Fig. 2.16 (Fang et al. 2015a). Three different diameters are considered with the detailed dimensions given in Table 2.2. The special test device discussed in Chap. 1 was used, and a 0.1 mm displacement increment in the shear direction was adopted. To enable a clearer demonstration, the hysteretic loops of the 5th, 10th, 15th, 20th, 25th, 30th, 35th, and 40th cycles are presented in the figure, with the accumulated residual displacement reflected in the abscissa.

Compared with the tension test results, the behaviour of the SMA bolts under direct shear is quite different. Considerable residual displacements are shown. No obvious transformation stress plateau is observed during the loading and unloading stage, while the peak force of each cycle tends to increase linearly with the incremental displacements.

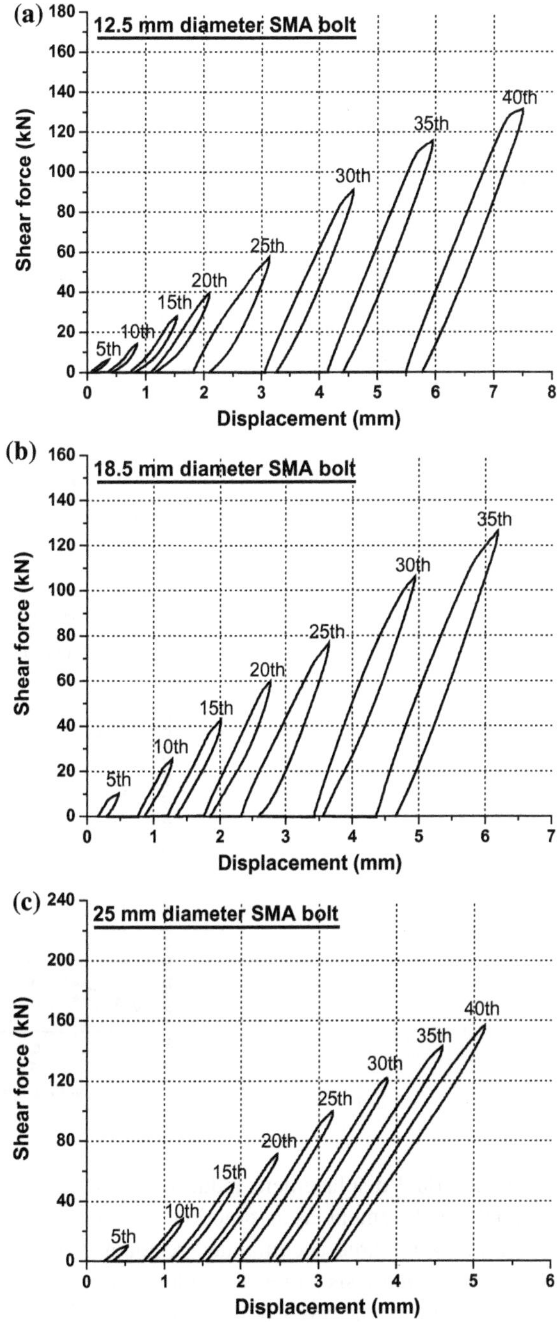

Fig. 2.16 Hysteretic stress–strain responses of SMA bolts under cyclic direction shear: **a** 12.5 mm diameter, **b** 18.5 mm diameter, **c** 25 mm diameter (Fang et al. 2015a)

Table 2.2 Dimensions of SMA bolts under cyclic direct shear (Fang et al. 2015a)

Shank diameter (mm)	Net threaded diameter (mm)	Overall length (mm)	Threaded length (mm)
12.5	10.5	120	25
18.5	16.5	120	27.5
25.0	23.0	120	32

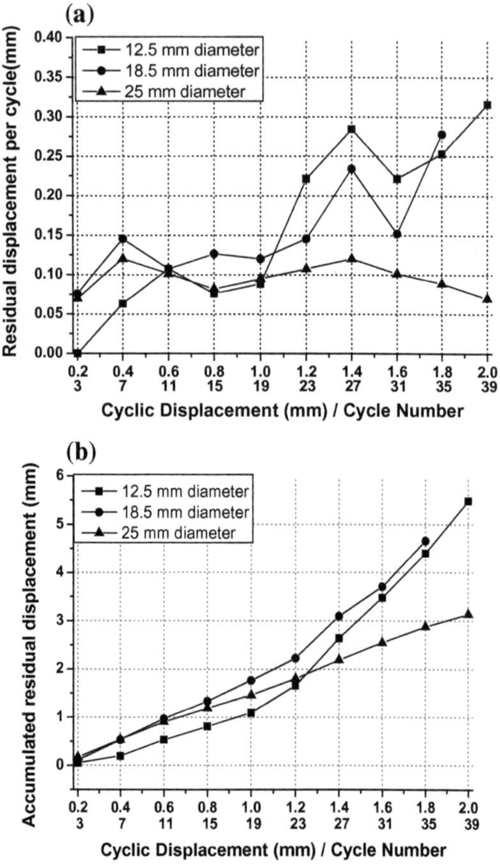

Fig. 2.17 Residual displacement of SMA bolts under cyclic direct shear: **a** residual displacement by each cycle, **b** accumulated residual displacement

The residual displacement induced by each loading cycle and the accumulated residual displacement as a function of cycle numbers are further illustrated in Fig. 2.17. Very large accumulated residual displacements are generally observed at the end of the tests (after 40 cycles). Figure 2.18 shows the energy dissipation characteristics of the SMA bolts under cyclic direct shear. The stabilised EVDs are between 1.5 and 3%, which are much lower than the values obtained from the tension tests.

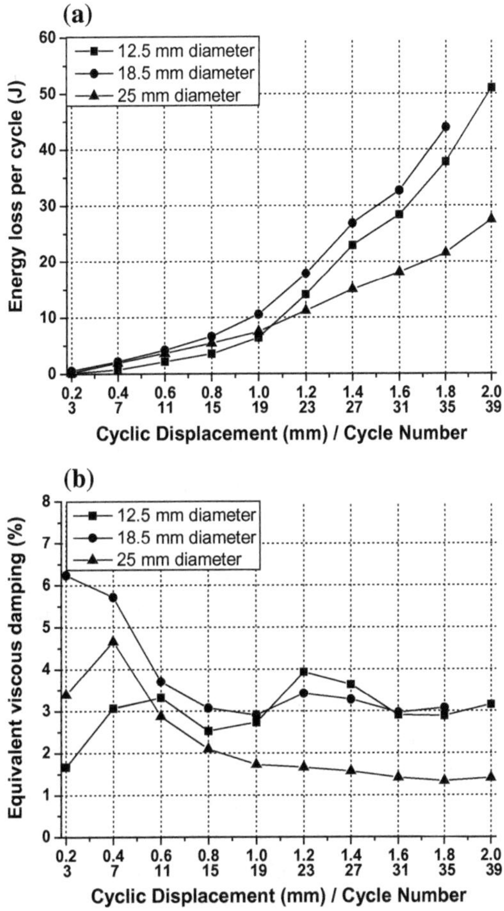

Fig. 2.18 Energy dissipation characteristics of SMA bolts under cyclic direct shear: **a** energy loss per cycle, **b** equivalent viscous damping

2.4 SMA Helical Springs

Commercial small-scale SMA helical springs are typically produced to exhibit SME response and are often used as sensor elements. Although the seismic-oriented application of SMA helical springs is still in its early stage, many researchers and damper developers start to recognise that the SE behaviour of SMA may provide ideal damping and deformation capabilities for helical springs and hence enables them to be used for passive control. Speicher et al. (2009) proposed a new prototype damper incorporating an SMA helical spring as the kernel element. A maximum load resistance of approximately 8kN was achieved for the spring with a 12.5 wire diameter. Mirzaeifar et al. (2011) developed analytical solutions for describing the mechanical behaviour of SMA helical springs, where a good agreement was observed between the available test results and the analytical predictions. Savi et al. (2015) provided

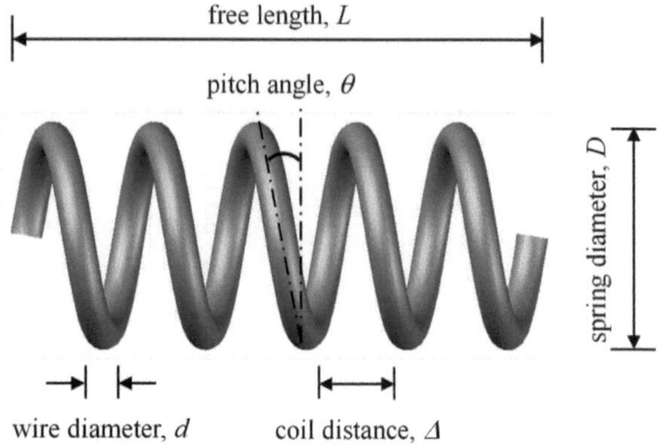

Fig. 2.19 Geometrical configuration of SMA helical springs (Qiu 2016)

insight into the influence of spring dimensions on their mechanical behaviour, and it was concluded that appropriately designed SMA helical springs can have improved energy dissipation capacity and much-increased deformability compared with normal steel helical springs.

Like normal steel helical springs, an SMA helical spring can be geometrically characterised by four governing parameters, namely, wire diameter d, spring diameter D, pitch angle θ and the number of active coils N indicating the ratio of free length to coil distance (i.e. L/Δ), as shown in Fig. 2.19. In addition, the spring index is defined as $C = D/d$. Helical springs are generally categorised by two types according to the details of their ends, namely plain-end and closed-end springs. A spring with plain ends has a non-interrupted helicoid, and its ends are the same as the case if a long spring has just been cut into smaller segments. Contrarily, a spring with closed ends is obtained by deforming the ends to a zero-degree helix angle. Due to loading eccentricity, the springs with plain ends are unstable in compression, and hence they are often referred to as tension springs. On the other hand, the springs with closed ends have stable tension and compression behaviour and are therefore named as tension–compression springs.

SMA helical springs can be fabricated from SMA wires or tubes with the following steps: (1) mechanical conformation of the springs, (2) appropriate annealing and form setting, (3) water quench or air cooled. Figure 2.20 clearly demonstrates these procedures. Annealing, which evidently affects the mechanical properties of SMA springs, is probably the most important step during the manufacturing process. Morgan and Broadley (2004) tried different annealing schemes and recommended that the proper annealing temperature should range from 450 to 550 °C. After a series of trials, Savi et al. (2015) claimed that an annealing temperature of 500 °C for 30 min can be the best means to produce an ideal superelastic SMA spring. Other independent researchers used annealing temperatures ranging from 500 to 550 °C for their spring products (Wu 2002; Jee et al. 2008). In view of the above, it can be seen that the appropriate annealing temperature for form setting of SMA helical springs generally falls within the range of 450–550 °C; however, for a particular

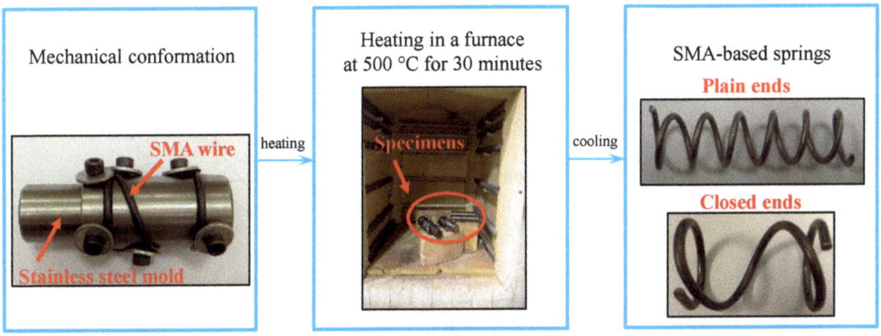

Fig. 2.20 Procedures of making SMA helical springs (Qiu 2016)

Table 2.3 Parameters of SMA helical spring specimens

End type	Specimen	Parameters					
		D (mm)	d (mm)	C	N	Δ (mm)	L (mm)
Plain-end	S1	16	2	8	6	8	62
	S2	16	2	8	2	18	42
	S3	32	2	16	6	8	62
	S4	40	4	10	2	20	52
	S5[a]	40	4	10	2	20	52
Closed-end	S6	40	4	10	1	40	52

[a]circular hollow section, thickness = 1 mm

batch of SMA spring products, the optimum annealing protocol may be obtained through different trials with different temperatures and durations.

For demonstration purpose, small-scale NiTi SMA helical springs were produced by Qiu (2016) using the above fabrication procedures with an annealing temperature of 500 °C for 30 min. These springs are divided into two groups according to their end types, and the parameter matrix is given in Table 2.3. Specimen S1 is a typical plain-end spring, and specimens S2 and S3 are compared with S1 to observe the effects of the number of active coils (N) and spring index (C), respectively. Specimens S4 and S5 use a larger wire diameter and are made of solid bars and hollow tubes, respectively. Specimen S6 is a closed-end spring. All the specimens were subjected to water quenching after annealing.

Figure 2.21 shows the typical flag-shaped force–displacement hysteresis response of specimen S1. 'Yielding' occurs at the point where the tangent stiffness starts to decrease, i.e. at a displacement of around 32 mm. The corresponding yield strength and initial spring stiffness are around 63.3 N and 1.97 N/mm, respectively. Contrary to conventional elastic springs, the SMA helical spring shows a highly nonlinear behaviour, where the deformation is essentially a summation of elastic deformation and martensitic transformation-induced deformation. In addition, stable and

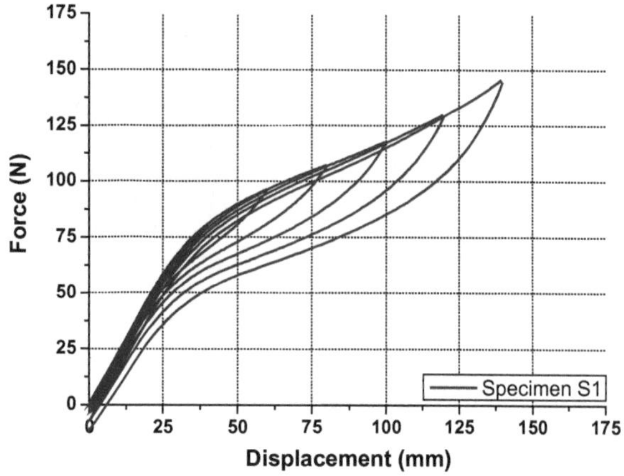

Fig. 2.21 Cyclic behaviour of plain-end spring (Qiu 2016)

repeatable hysteretic loops without obvious transformation-induced fatigue (TIF) degradation are observed.

Figure 2.22 further plots the EVD and residual displacement of specimen S1 as a function of loading amplitude. Energy dissipation is generally small with EVDs ranging from 1 to 3%. The EVD increases with loading amplitude because of the development of martensite fraction. The more pronounced forward transformation with an increase in loading amplitude causes the accumulation of residual displacement. Nevertheless, only 6 mm residual displacement is detected at the 140 mm loading cycle, which implies that the SMA helical spring has good deformability and satisfactory self-centring capability.

The number of active coils, N, is an important index affecting the mechanical behaviour of helical springs. A comparison is made between specimens S1 ($N = 6$) and S2 ($N = 2$), as shown in Fig. 2.23. It can be seen that an increase in N leads to a larger 'yielding' displacement, but does not affect the 'yielding' force. Another important parameter is the spring index, C, which refers to the ratio of spring diameter to wire diameter. Specimens S1 and S3 have different C values with the remaining parameters identical. Figure 2.24 plots the force–displacement responses of the two specimens. A larger spring diameter leads to a lower yield force, a larger yield displacement, and a smaller stiffness. The large yield displacement of S3 can be explained by the force–displacement relationship as expressed by (Liang and Rogers 1997)

$$y_e = \frac{8NC^3}{Gd} F_e \tag{2.2}$$

where y_e is the yield displacement, F_e is the yield force and G is the shear modulus of SMA wire. Equation (2.2) clearly indicates that increasing C produces a significantly larger yield displacement.

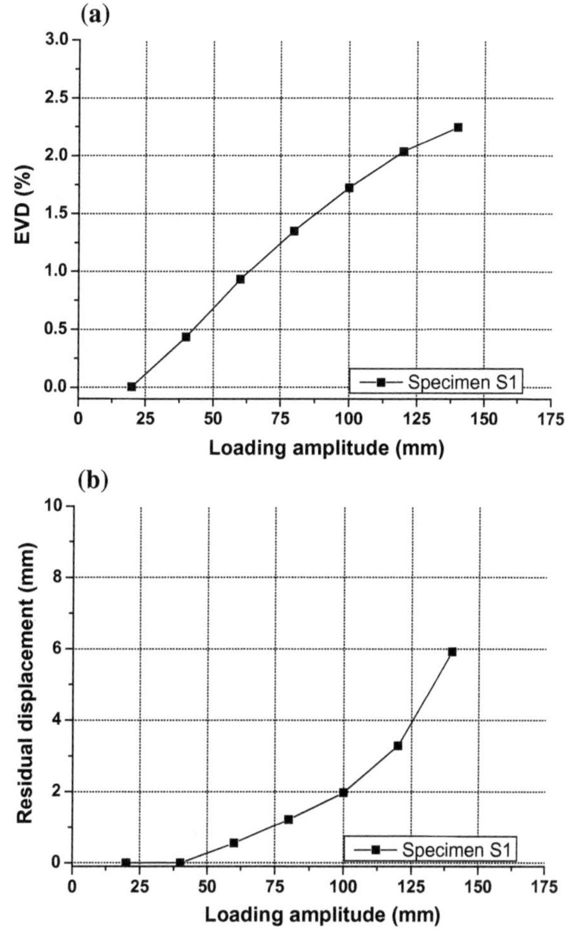

Fig. 2.22 Key performances of specimen S1: **a** equivalent viscous damping, **b** residual displacement (Qiu 2016)

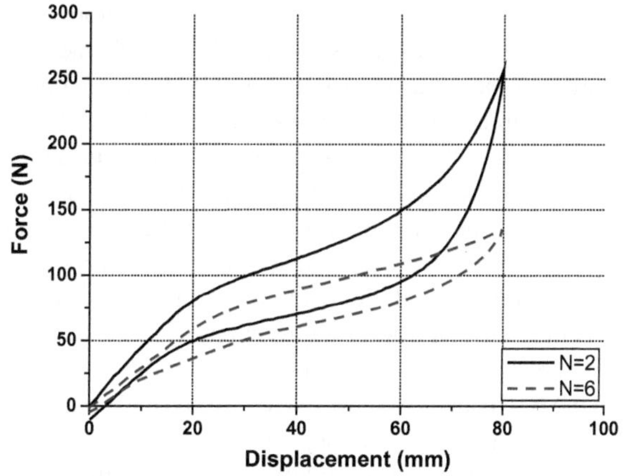

Fig. 2.23 Effect of number of active coils on cyclic behaviour, S1 ($N = 6$) versus S2 ($N = 2$) (Qiu 2016)

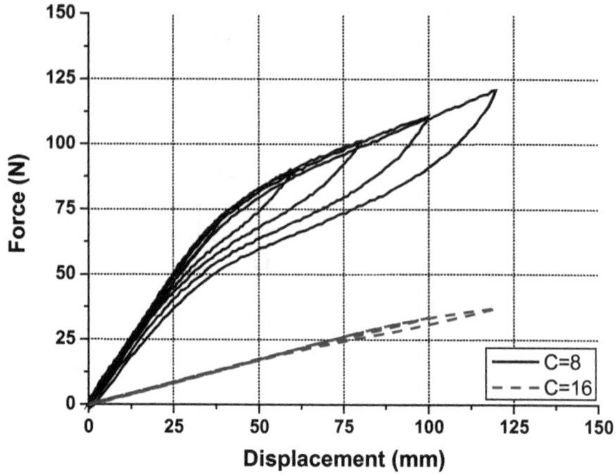

Fig. 2.24 Effect of spring index on cyclic behaviour, S1 ($C = 8$) versus S3 ($C = 16$) (Qiu 2016)

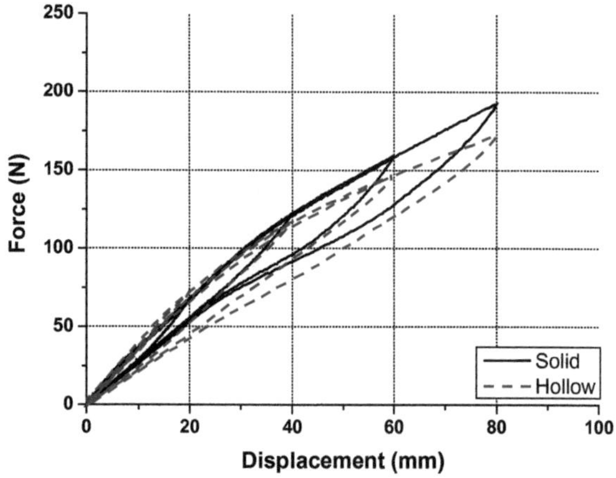

Fig. 2.25 Effect of wire types on cyclic behaviour, S4 (solid) versus S5 (hollow) (Qiu 2016)

The effect of wire types on the cyclic behaviour of SMA helical springs is evaluated by comparing specimens S4 and S5. The former is made of 4 mm-diameter solid wire and the latter is made of a hollow tube whose outer and inner diameters are 4 and 2 mm, respectively. As shown in Fig. 2.25, the two specimens exhibit very similar load–displacement responses in the elastic regime, although the spring with the hollow tube has slightly lower load resistance than its solid tube counterpart at larger displacements. For seismic applications, hollow springs may be preferred because of less material consumption. The cyclic behaviour of SMA helical springs can be economically adjusted by changing the wall thickness of the SMA tubes to meet the desired design requirement.

The closed-end spring (i.e. specimen S6) can sustain bidirectional forces. Specimen S6 is made of a 4 mm SMA wire and has a pitch angle of 26° which is significantly larger than that of ordinary springs (Fig. 2.26). The free height of the specimen is 52 mm, and the corresponding maximum compression stroke is 40 mm. Figure 2.27

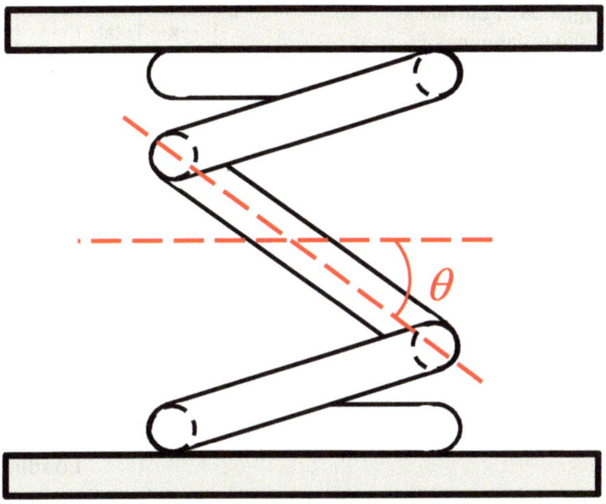

Fig. 2.26 Geometric configuration of the closed-end springs (S6)

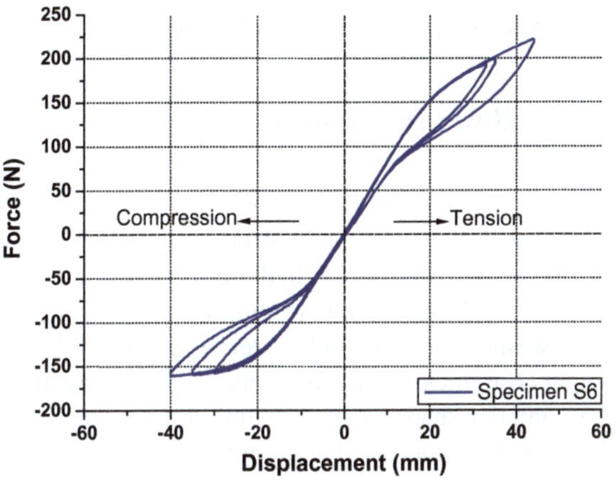

Fig. 2.27 Tension–compression behaviour of closed-end spring (S6) (Qiu 2016)

plots the cyclic behaviour of specimen S6 with varying tensile and compressive loading amplitudes. As observed, the specimen successfully carries the compression load without causing any instability problem. In addition, the spring shows stable flag-shaped hysteresis in tension and compression, and the residual displacement is negligible within the available deformation range. Nevertheless, the asymmetrical tension–compression behaviour is notable. This behaviour may be due to the variations of the pitch angle at different loading states, i.e. the pitch angle is decreased in compression but is increased in tension (Shigley et al. 1989).

Figure 2.28 shows the EVD of specimen S6 as a function of loading displacement amplitude. The EVD is calculated for both compression and tension behaviour. At the 40 mm loading cycle, the EVD value is approximately 1.2% and 2.3% when subjected to tension and compression, respectively, leading to a total EVD of 3.5% under tension–compression reversal loading.

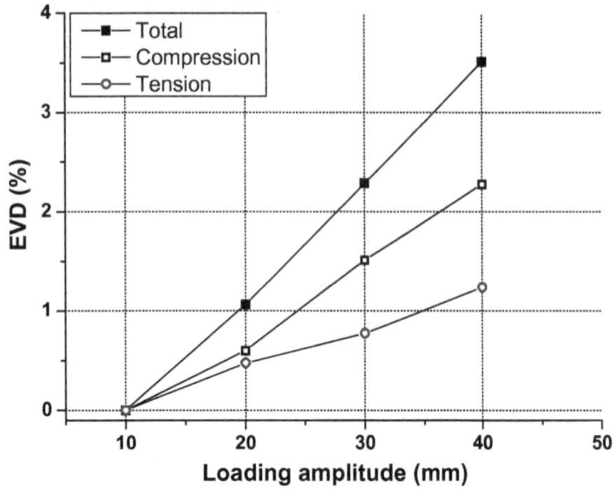

Fig. 2.28 Equivalent viscous damping of specimen S6 (Qiu 2016)

2.5 SMA Belleville Washer Springs

2.5.1 *Basic Information*

Another possible form of SMA elements is SMA Belleville washer springs. Conventional Belleville washer springs, also known as conical washer springs or just 'washers', are well-established mechanical components that are capable of sustaining large loads with a small installation space. The governing geometrical parameters for a washer spring include external diameter (D_e), internal diameter (D_i), overall height (H), thickness (t) and cone angle (θ), as illustrated in Fig. 2.29. Because of their annular shape, force transmission is relatively even and concentric, and thus, they can have more stable compression behaviour compared with helical springs. Importantly, the various combinations (e.g. in parallel and in series) of the washers can diversify their performances and applications. When the washers are endowed with the superelastic property, the field of application may be further expanded.

Early studies on SMA washers were mainly for electrical applications with the main purpose of electric circuit stabilisation (Labrecque et al. 1996). The potential of superelastic SMA washers for dampers was first investigated by Speicher et al. (2009). The washers showed satisfactory self-centring capability, but the load carrying capacity decreased significantly beyond the peak load due to an 'invert' tendency which compromised the overall damping performance of the washer groups. It was argued that the snap-through behaviour was due to the relatively high cone angle (θ) considered for the test washers. A later numerical investigation (Fang et al. 2014b), considering different geometric shapes of the SMA washers, showed improved self-centring and energy dissipation behaviour when the dimension of the washers was appropriately designed. Maletta et al. (2013) introduced a new thermo–mechanical process for producing SMA washers through disk cutting and a successive form set-

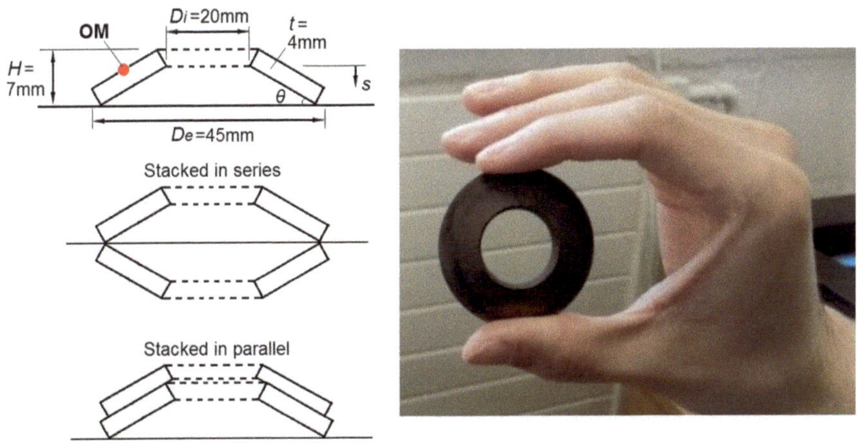

Fig. 2.29 SMA Belleville washers and governing geometrical dimensions

Table 2.4 Material properties of SMA washer model

Material properties	Value
Forward transformation (start) stress σ_{Ms}	430 MPa
Forward transformation (finish) stress σ_{Mf}	480 MPa
Reverse transformation (start) stress σ_{As}	250 MPa
Reverse transformation (finish) stress σ_{Af}	150 MPa
Young's modulus (austenite) E_A	60 GPa
Young's modulus (martensite) E_M	40 GPa
Maximum transformation strain ε_L	5%
Poisson's ratio (austenite) ν_A	0.33
Poisson's ratio (martensite) ν_M	0.33

ting via annealing. A semi-analytical model, which was found to correlate well with the FE predictions, was also developed by Sgambitterra et al. (2014).

2.5.2 *Numerical and Analytical Investigations*

The load resistance of a washer spring is offered via the development of hoop stress. Figure 2.30 shows the typical hoop stress states of a superelastic SMA washer at varying deformation states at room temperature. The result is obtained via a numerical model established in ABAQUS, where the key material properties are listed in Table 2.4. The dimension of the sample washer model is given in Fig. 2.29.

Non-uniform hoop stress distributions are generally observed, where the maximum tensile stress occurs near the lower part of the outer edge (location 3) whilst the maximum compressive stress is developed near the upper part of the inner edge (loca-

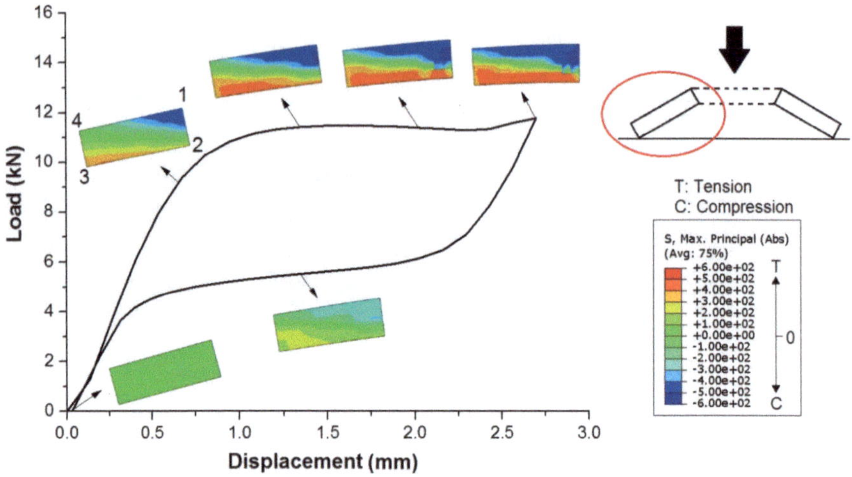

Fig. 2.30 Hoop stress distributions of SMA washer at different deformation stages (stress unit: MPa)

tion 1). The area crossing the 'diagonal zone' linking locations 2 and 4 remains at low-stress levels throughout the loading process. For the considered model, the maximum tensile and compressive stresses at the maximum (fully flattened) deformation amplitude are approximately 500 MPa and 530 MPa, respectively, which indicates that a full forward transformation is completed at these locations, recalling that σ_{Mf} is 480 MPa according to Table 2.4. During unloading, the stresses decrease rapidly, and ideally, no residual stress is developed upon completion of the loading cycle.

For the engineering design purpose, the yield resistance and initial stiffness of SMA washers can be predicted using the theoretical elastic solutions originally proposed for normal industrial steel washers, as detailed in the relevant handbook (Fromm and Kleiner 2003). It is recommended that the condition of a washer can be monitored by the stress level at a reference location (point OM as marked in Fig. 2.29), and it can be assumed that the yield resistance $F_{w,y}$ of an SMA washer is reached when the stress at this location (σ_{OM}) reaches σ_{Ms}. It should be noted that point OM is a typical high-stress zone rather than the peak stress location. Selecting the peak stress location for determining $F_{w,y}$ can lead to a significant underestimation of $F_{w,y}$. The development of the critical stress σ_{OM} as a function of washer deformation is expressed by

$$\sigma_{OM} = -\frac{12Ets}{\left(1 - 0.33^2\right)K_1 D_e^2 \pi} \tag{2.3}$$

$$K_1 = \frac{\left((\delta - 1)/\delta\right)^2}{\pi\left((\delta + 1)/(\delta - 1) - 2/\ln\delta\right)} \tag{2.4}$$

where s is the compressive deformation of the washer, E is Young's modulus of the SMA material, $\delta = D_e/D_i$, and the other symbols related to the geometric dimension are illustrated in Fig. 2.29. If the material properties such as σ_{Ms} are known, the corresponding compressive deformation s upon yielding of the washer (i.e. when $\sigma_{OM} = \sigma_{Ms}$) can be deduced. With the obtained yield deformation s, the yield resistance $F_{w,y}$ of the SMA washer can be obtained from the equation below

$$F_w(s) = \frac{4Et^3 s}{\left(1 - 0.33^2\right) K_1 D_e^2}\left[\left(\frac{H_0}{t} - \frac{s}{t}\right)\left(\frac{H_0}{t} - \frac{s}{2t}\right) + 1\right] \tag{2.5}$$

where H_0 is the maximum available deformation of each washer. The initial stiffness of the SMA washer can be calculated by the derivation of Eq. (2.5).

2.5.3 Experimental Investigation

A better understanding of the entire hysteretic responses of SMA Belleville washer springs is enabled via physical testing. A total of ten specimens with the same dimension as that shown in Fig. 2.29 were produced and tested by Fang et al. (2016). The considered geometric configuration leads to a cone angle (θ) of 13.5°. The testing parameters, as summarised in Table 2.5, include temperature, the arrangement of washers, and maximum compressive deformation (s_{max}). The austenite finish temperature A_f is approximately 5 °C such that the washers exhibit SE at room temperature. The composition of the SMA washers is nearly identical, with 55.87% nickel and 44.13% titanium by weight. An additional commercial stainless steel washer was tested for comparison purpose. The dimension of the stainless steel washer is: $D_e =$

Table 2.5 Test parameters for washer specimens (Fang et al. 2016)

Specimen code	s_{max} (mm)	Temperature (°C)	Stack arrangement	Material
SMA1-a	2.7	23	Single	SMA
SMA1-b	2.7	23	Single	SMA
SMA1-H	1.3	23	Single	SMA
SMA2-S	5.4	23	Two washers in series	SMA
SMA2-P	2.7	23	Two washers in parallel	SMA
SMA1-T(10)	2.7	10	Single	SMA
SMA1-T(40)	2.7	40	Single	SMA
SMA1-T(60)	2.7	60	Single	SMA
SMA1-T(0–23)	2.7	0–23	Single	SMA
SMA1-T(−20~23)	2.7	−20~23	Single	SMA
SS1	1.0	23	Single	Stainless steel

50 mm, $D_i = 22$ mm, $t = 5$ mm and the allowable deformation = 1 mm. These are slightly different from the SMA washer specimens.

The SMA washer specimens were tested isothermally under various temperatures ranging from minus (−) 20 to 60 °C. Single (individual) washers and those stacked in series or in parallel are investigated. For easy identification, each specimen is assigned with a specimen code according to its material, washer arrangement, temperature, and loading condition, as shown in Table 2.5. It should be noted that for some specimens, e.g. specimen SMA1-T(−20~23), two temperatures are included in the specimen code, in which case the test was first conducted under a temperature below A_f and then the temperature was increased beyond A_f (back to the room temperature) for further testing.

The specimens were subjected to cyclic load with a constant amplitude. The loading frequency was 0.02 Hz under displacement control. The maximum deformation s_{max} of each standard cycle was 2.7 mm, such that the washer could be deformed to an almost flattened state at s_{max}. For the specimens tested under the temperature above A_f, the loading–unloading process was repeated for 50 cycles. For those tested below A_f, where the superelastic behaviour could not be fully developed, 10 cycles were carried out first, and then the temperature was increased, and additional 50 loading cycles were performed at room temperature. These stepped-temperature tests were used to investigate the influence of low temperature (e.g. cold winter) weather conditions as well as the ability of the washers to regain superelastic behaviour upon reheating. The cyclic tests were performed using a UTM with an environmental chamber. A bespoke testing device, as shown in Fig. 2.31, was designed for holding the washers in position throughout the experiment. The arrangement of the stainless steel washer test was basically the same but with only 1 mm maximum deformation being applied.

Figure 2.32 shows the load–deformation relationships of all the SMA washer specimens. For those tested at temperatures above A_f, flag-shaped hysteretic curves with recognisable forward and reverse transformation plateaus are typically exhibited. The load at the maximum deformation tends to increase with increasing temperature. This is due to the C_M and C_A slopes of the material (Churchill et al. 2009), which has been discussed in Chap. 1. Degradations of the hysteretic responses, accompanied by residual deformations, are observed, especially during the first few cycles. The hysteretic loops become stable with subsequent loading cycles.

A minor 'snap-through' type load–deformation response, i.e. duck-head-shaped curve, is observed in some specimens in their initial cycle, but this behaviour quickly disappears in the subsequent cycles due to degradation. The SE performance of the SMA washers tested at room temperature (23 °C) or 10 °C is good. At lower temperatures (i.e. 0 and −20 °C), SE could not be fully triggered, and hence, significant residual deformation is induced immediately after the first cycle. This suggests that the SMA washers may be more suited to indoor usage where the expected lowest working temperature is above the austenite finish temperature. Nevertheless, the residual deformation is recovered after reheating, and stable hysteretic responses are shown during the subsequent 50 cycles. In general, the specimen being directly tested

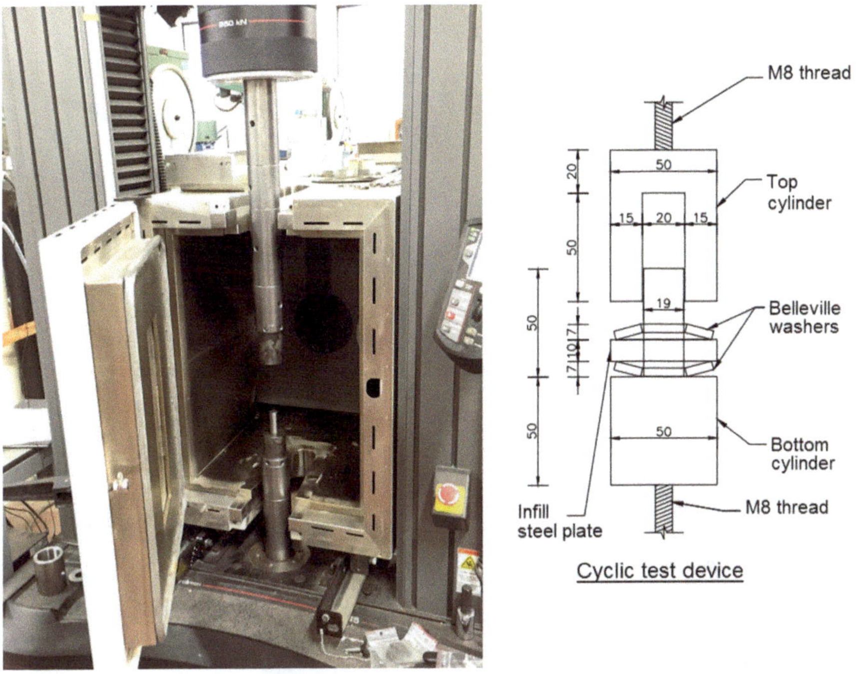

Fig. 2.31 Photo and details of test setup for SMA washers (unit in mm)

at ambient temperature and those subjected to the two-phase (low temperature and then ambient) tests exhibit comparable hysteretic behaviour.

On the other hand, when the temperature increases beyond the room temperature (i.e. to 40 and 60 °C), the load resistance increases, a phenomenon which is related to the C_M and C_A slopes. The increased residual deformation is caused by plastic (permanent) stresses of the material under elevated temperatures, where the required forward transformation stress may have exceeded the true yield (plastic) stress.

The peak load and secant stiffness responses of the SMA washer specimens with changing loading cycles are shown in Figs. 2.33 and 2.34, respectively. The peak load is almost constant throughout the 50 loading cycles. The secant stiffness for each cycle is the ratio of the peak load over the 'effective' deformation. The two washers stacked in parallel can lead to a secant stiffness four times of that of the washers stacked in series.

Figure 2.35 shows the accumulated residual displacements of the SMA washers with increasing cycle numbers. For the specimens at 10 °C or 23 °C, the maximum accumulated residual displacement after the completion of the loading process is 0.4 mm, corresponding to 14.8% of the maximum compressive displacement. In other words, more than 85% of the compressive displacement is recovered after a significant number of repeated loading. With the increase in temperature, the residual displacement is also increased. The accumulated residual displacement could be up

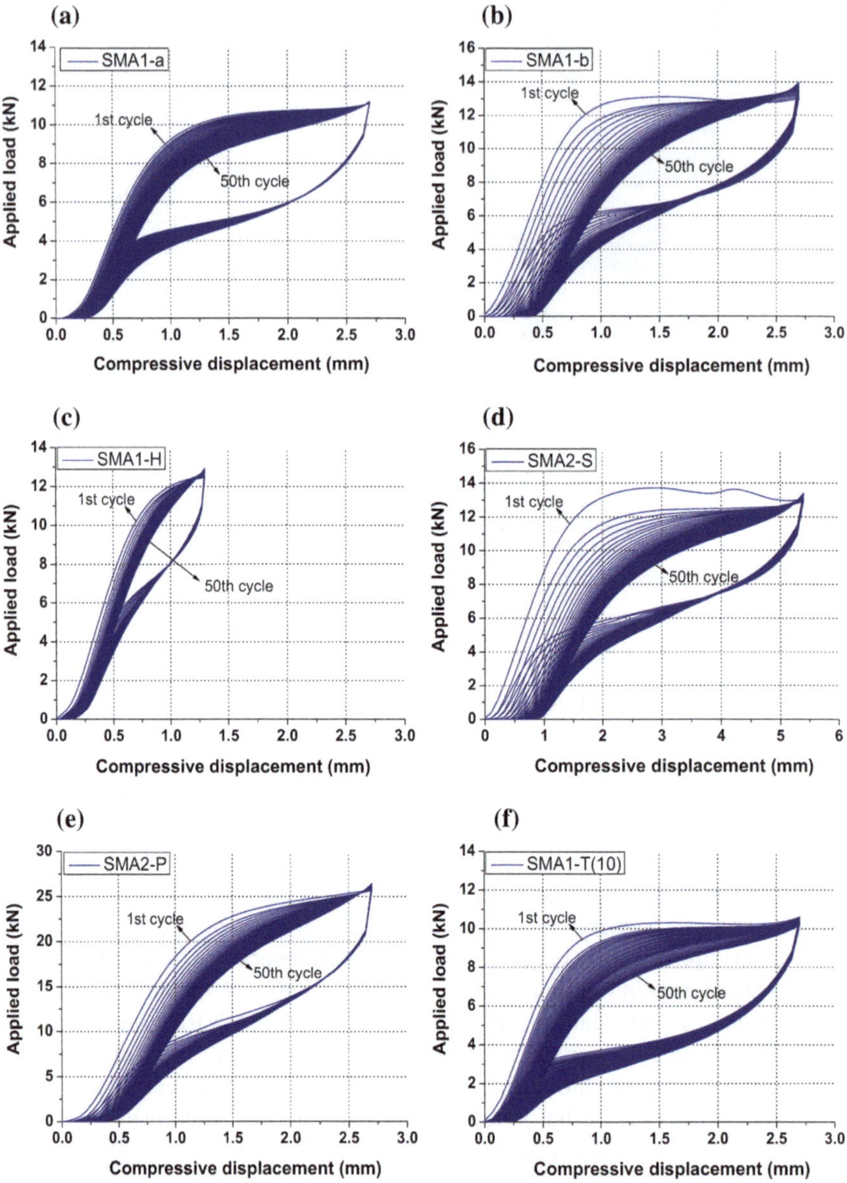

Fig. 2.32 Load–deformation response of SMA washer specimens: **a** SMA1-a, **b** SMA1-b, **c** SMA1-H, **d** SMA2-S, **e** SMA2-P, **f** SMA1-T(10), **g** SMA1-T(40), **h** SMA1-T(60), **i** SMA1-T(0–23), **j** SMA1-T(−20~23)

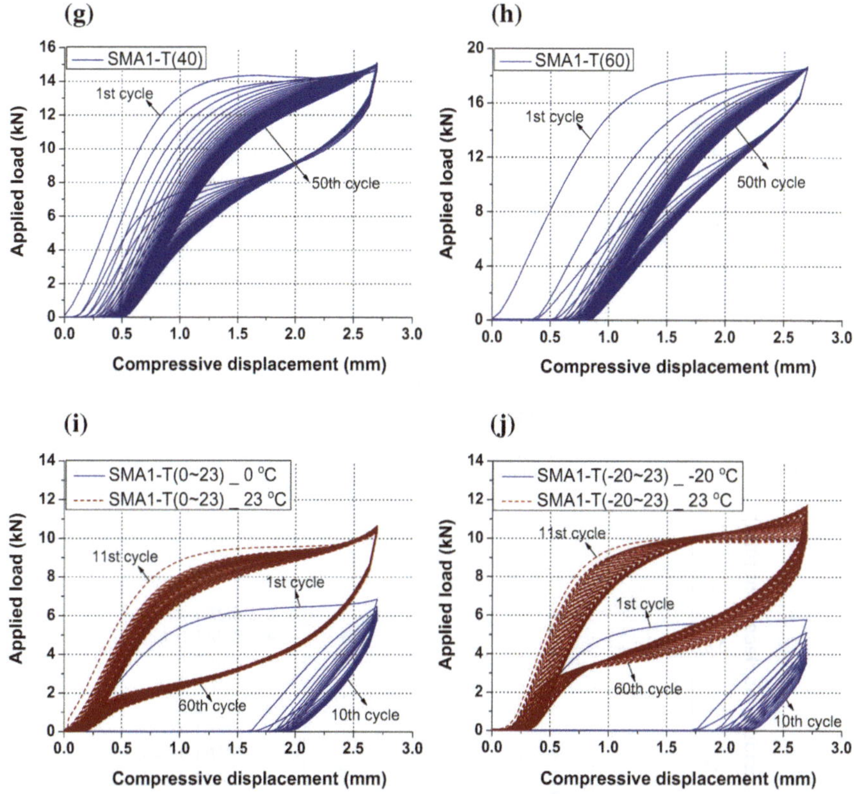

Fig. 2.32 (continued)

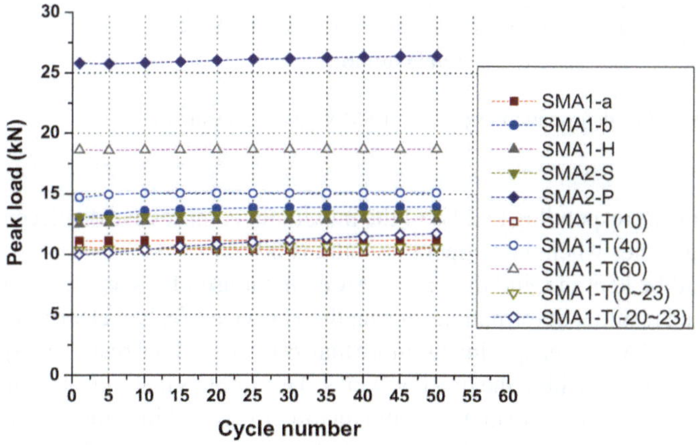

Fig. 2.33 Peak loads of SMA washer specimens

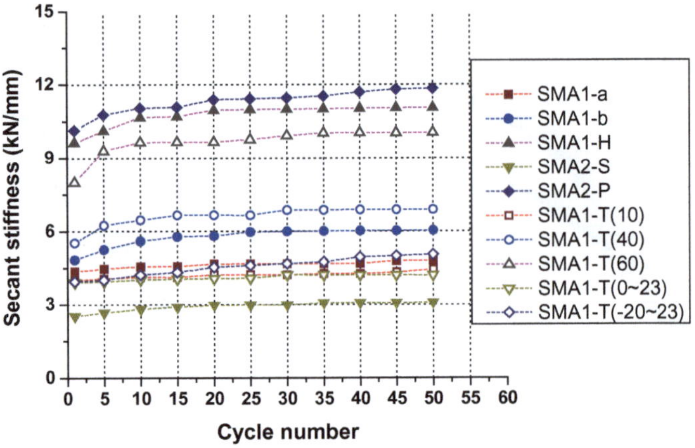

Fig. 2.34 Secant stiffness of SMA washer specimens

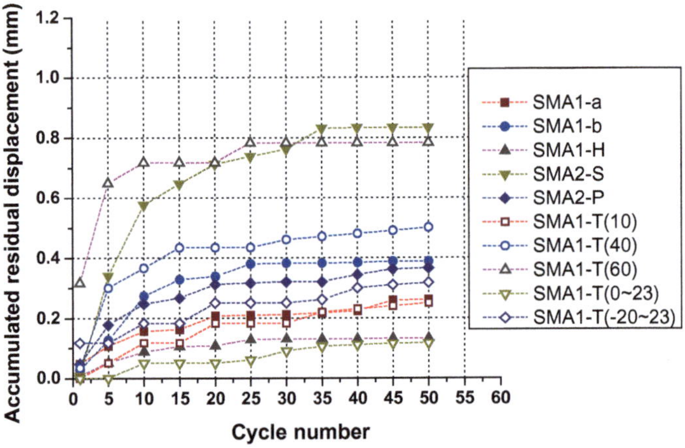

Fig. 2.35 Residual deformation responses of SMA washer specimens

to 0.78 mm, corresponding to 29% of the maximum compressive displacement, after 50 loading cycles when the temperature attains 60 °C.

It should be kept in mind that any SMA element used for seismic control should have limited residual deformations during their service life. It is clearly seen that for most of the SMA washer specimens, more than 60% of the final residual displacement has been developed within the first ten cycles. This implies that a pre-training process with a certain number of loading cycles can effectively stabilise the hysteresis of the washers. In fact, the SMA washers are not always fully flattened during the seismic action, so in practice only one or two fully compressed training cycles are sufficient.

The hysteretic behaviour of the SMA washers also clearly displays their energy dissipation capability. The energy loss per cycle is shown in Fig. 2.36a. Due to the TIF

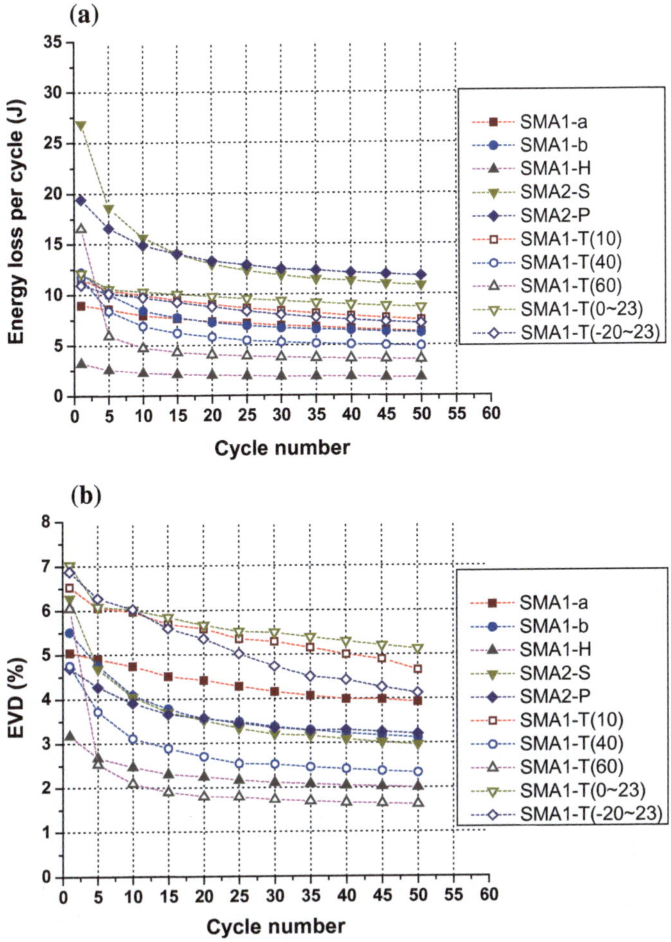

Fig. 2.36 Energy dissipation characteristics of SMA washer specimens: **a** energy loss per cycle, **b** equivalent viscous damping

degradation, the energy dissipation capacity decreases with loading cycles. When the working temperature is high, increased energy dissipation is exhibited only at the first cycle due to the increase of the forward transformation plateau. However, the energy loss per cycle decreases quickly in the subsequent loading cycles due to the 'narrowed' hysteretic shapes. For the SMA washers tested at ambient temperature, the energy loss per cycle is more stable. When half of the maximum deformation is applied, i.e. specimen SMA1-H, the energy loss per cycle is significantly decreased. This implies that a full consumption of the maximum available deformation of SMA washers/washer groups is desirable such that the energy dissipation potential can be more efficiently exploited. Therefore, it is recommended that the arrangement

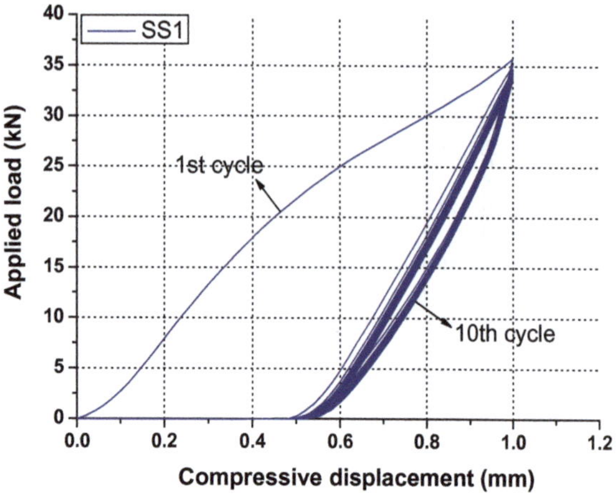

Fig. 2.37 Load–deformation response of stainless steel washer specimen

of SMA washers should be carefully determined to match the target deformation (e.g. inter-storey drift) of the structural system. The test results also confirm that the variable stack combinations of the SMA washers further diversify their energy dissipation properties. When stacked in parallel (e.g. specimen SMA2-P), the energy loss per cycle is almost doubled. When stacked in series (e.g. specimen SMA2-S), doubled energy loss per cycle is also attained but with a corresponding doubled maximum deformation.

As shown in Fig. 2.36b, an EVD of up to 7% is achieved. The EVD decreases with cycle numbers due to the decrease in the amount of energy loss per cycle. The relatively large EVD values exhibited by specimens SMA1-T(10), SMA1-T(0–23), and SMA1-T(−20~23) echo their relatively full hysteretic shapes. The typical EVD values of the SMA washers seem to be comparable to those of uniaxially loaded SMA bars. This indicates that the SMA washers can be considered as an alternative solution to SMA bars, especially when compressive loading resistance is desirable and when the installation space is small.

Finally, for the stainless steel washer (specimen SS1), significant residual displacement is induced immediately after the initial loading cycle, as shown in Fig. 2.37. No deformation recoverability, as expected, is exhibited during unloading, and then the stainless steel washer completely loses its energy dissipation capability. The distinct difference between the stainless steel and SMA washers clearly demonstrates the unique role played by the SMA material.

2.6 SMA Ring Springs

2.6.1 Basic Information

Although it is believed that properly designed SMA helical springs and Belleville washers are capable of providing satisfactory self-centring and energy dissipation capabilities, these elements are in fact less cost-effective than SMA wires or bars because part of the material within these springs is kept at low-stress levels throughout the loading process. In other words, their material efficiency is not fully motivated, and as a result, the overall load resistance may be inadequate for civil engineering practice. In order to overcome this issue and to increase the 'performance-to-cost' ratio, an alternative type of superelastic SMA spring, namely SMA ring spring system, is introduced herein.

As shown in Fig. 2.38, an SMA ring spring system includes a series of inner high-strength steel (HSS) rings and outer superelastic SMA rings stacked in alternation with mating taper faces. The key geometrical parameters for an SMA outer ring include external diameter (D_e), internal diameter (D_i), ring height (H), ring thickness (t) and taper angle (α), as illustrated in Fig. 2.39. When the system is in compression, the wedging action tends to expand the outer rings and concurrently squeeze the inner rings (Fig. 2.40). The forward phase transformation enables the SMA outer rings to be expanded significantly under the horizontal component of the contacting force over the taper face, and the vertical component of the contacting force contributes to the load resistance of the system. The frictional effect over the taper face can further increase the overall load resistance. When the external load is removed, the deformation of the whole system can be recovered because of the superelastic

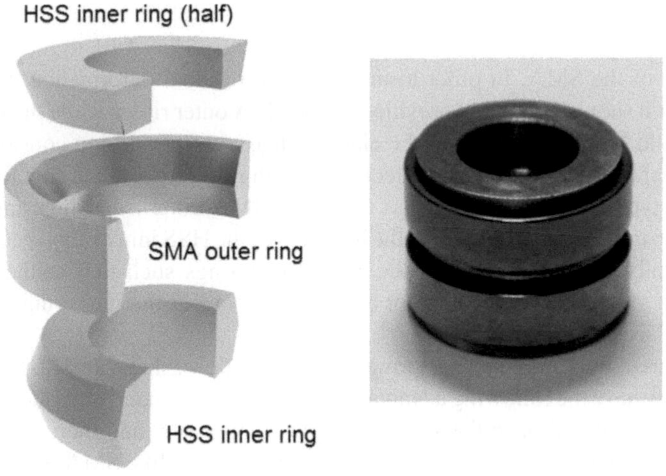

Fig. 2.38 Geometrical property of SMA ring spring system

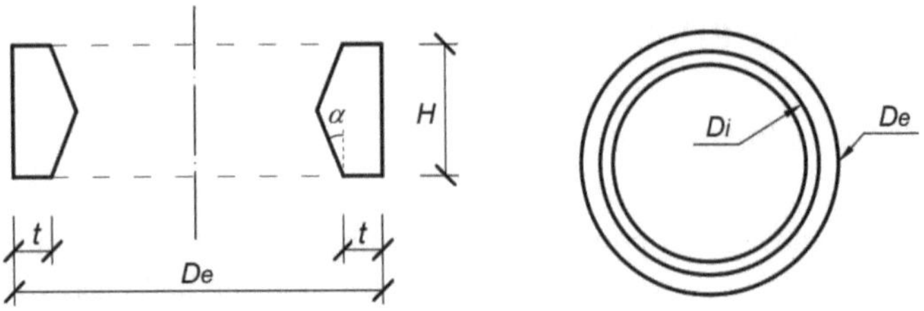

Fig. 2.39 Geometrical parameters of SMA outer rings

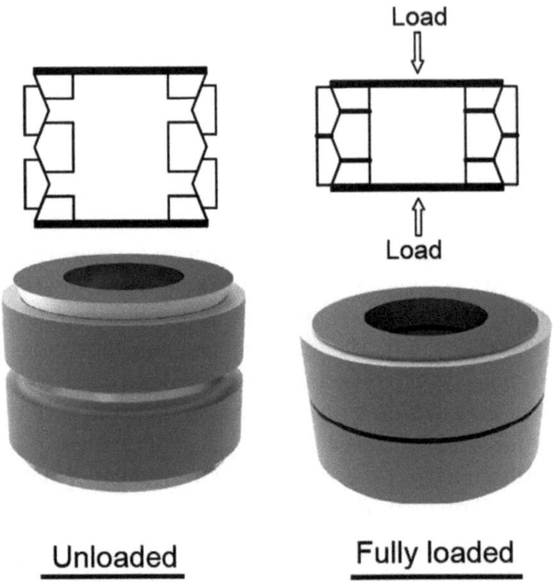

Fig. 2.40 Illustration of SMA ring spring system in compression

behaviour of the SMA. In order to minimise the residual deformation, the system deformation should be mainly provided by the SMA outer rings, while the HSS inner rings should be designed within the elastic range. Another design consideration is the material hardness which may affect the wearing condition of the contacting taper face. The typical Rockwell hardness (HRC) of SMA outer rings after annealing is HRC42, so an ideal situation is that the hardness of the HSS inner rings is comparable to or slightly lower than that of the SMA outer rings such that neither of them undergoes severe abrasion, or minor abrasion happens in the inner rings which are easily replaced (Wang et al. 2019).

Compared with SMA helical springs and Belleville washers, an SMA ring spring system may have the following advantages: (1) excessive expansion of the SMA outer ring is not permitted once the system is fully compressed, and therefore the SMA ring springs are protected from accidental fracture; (2) unlike SMA wires and bars, gripping is not required for ring spring systems; (3) when appropriately designed, the SMA outer rings can have a uniform hoop stress distribution over the ring thickness.

This enables the material to deform almost in a uniaxial manner, in which case one can make the best use of its self-centring and energy dissipation properties. For the same reason, a large load resistance can be achieved due to the effective utilisation of the material; (4) energy dissipation can be provided in a hybrid fashion, where the combined superelastic-friction action significantly increases the damping capacity of the system; and (5) the large recoverable strain of the SMA material enables considerable deformation offered by only a few SMA outer rings, which makes the entire system highly compact.

2.6.2 Numerical Investigation

The fundamental load–deformation behaviour and the typical stress distributions of SMA ring springs can be first understood via a numerical study. Figure 2.41 shows a representative FE model built in ABAQUS, together with the dimensions (Fang et al. 2015b). The considered material parameters for the SMA are given in Table 2.6. Each model includes two SMA outer rings and three HSS inner rings (or half inner rings). The parameter matrix includes two ring thicknesses (t), namely 3 and 5 mm; three taper angles (α), which are 11.3°, 21.8° and 31.0°, corresponding to the tangent values (tanα) of 0.2, 0.4 and 0.6, respectively; and four coefficients of friction, i.e. 0.0, 0.1, 0.2 and 0.3. The varying coefficients of friction are used to investigate the influence of different contacting conditions over the taper face. For each contacting pair between the inner and outer rings, the allowable axial deformation is 2.5 mm, and therefore, the overall allowable deformation for the system is 4×2.5 mm $= 10$ mm. For ease of identification, each model is assigned with a model code, which starts with the ring thickness, followed by the taper angle, and ends with the coefficient of friction. For example, T5A21.8F0.3 represents the model of 5 mm ring thickness, 21.8° taper angle and coefficient of friction $= 0.3$.

The hysteretic behaviour of the SMA ring spring system models is shown in Fig. 2.42. Flag-shaped hysteretic responses are generally exhibited, and the shape of the hysteretic loop is greatly influenced by the varying parameters. As anticipated, an increase in the SMA ring thickness can effectively increase the load resistance, but the shape of the hysteretic curve is not evidently changed. An increase in the taper angle also evidently increases the load resistance. Increasing the coefficient of friction tends to raise the forward transformation plateau and lower the reverse transformation plateau, leading to a widened hysteretic loop and therefore an increased energy dissipation. However, an overly large coefficient of friction compromises the self-centring capability of the system, especially for the models with a small taper angle, i.e. $\alpha = 11.3°$. For example, the load applied to the T3A11.3 and T5A11.3 series models can decrease to negative values during unloading when the coefficient of friction exceeds 0.2. This indicates that the restoring force is not sufficient to overcome the friction action, and as a result, the ring spring system would get stuck when the load is removed. The resulting residual displacement is not induced by the

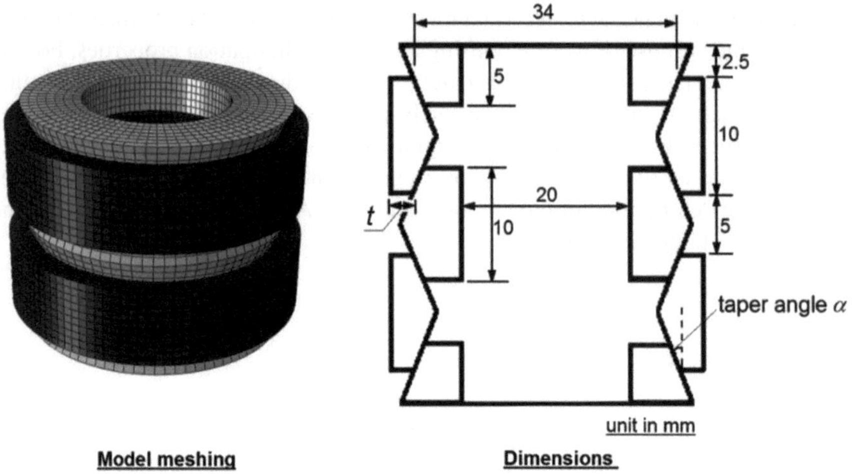

Fig. 2.41 Illustration of representative FE model and dimensions of SMA ring spring system

Table 2.6 Material properties used for SMA outer rings

Material properties	Value
Forward transformation stress σ_{Ms}	400 MPa
Forward transformation stress σ_{Mf}	500 MPa
Reverse transformation stress σ_{As}	250 MPa
Reverse transformation stress σ_{Af}	150 MPa
Plastic stress σ_p	700 MPa
Young's modulus (austenite) E_A	70 GPa
Young's modulus (martensite) E_M	40 GPa
Modulus of Plasticity E_p	3 GPa
Maximum transformation strain ε_L	5%
Poisson's ratio (austenite) v_A	0.33
Poisson's ratio (martensite) v_M	0.33

plastic strain of the material and hence is not permanent. A load with an opposite direction (i.e. tensile load) may pull the system back.

The restoring tendency is enhanced for the T3A21.8 and T5A21.8 series models upon unloading, due to the increased taper angle. The friction action can be more easily overcome, resulting in no residual displacement when the applied load is completely removed. When the taper angle is further increased to 31.0°, different hysteretic responses are observed. It is seen that the resisting load starts to increase more rapidly when the displacement exceeds 7 mm during the loading stage. This phenomenon is due to the 'martensitic hardening effect' when the stress exceeds σ_{Mf}. When the peak stress further increases beyond σ_p, plastic strain is induced, which causes permanent strain of the material. As can be seen in Fig. 2.42, more

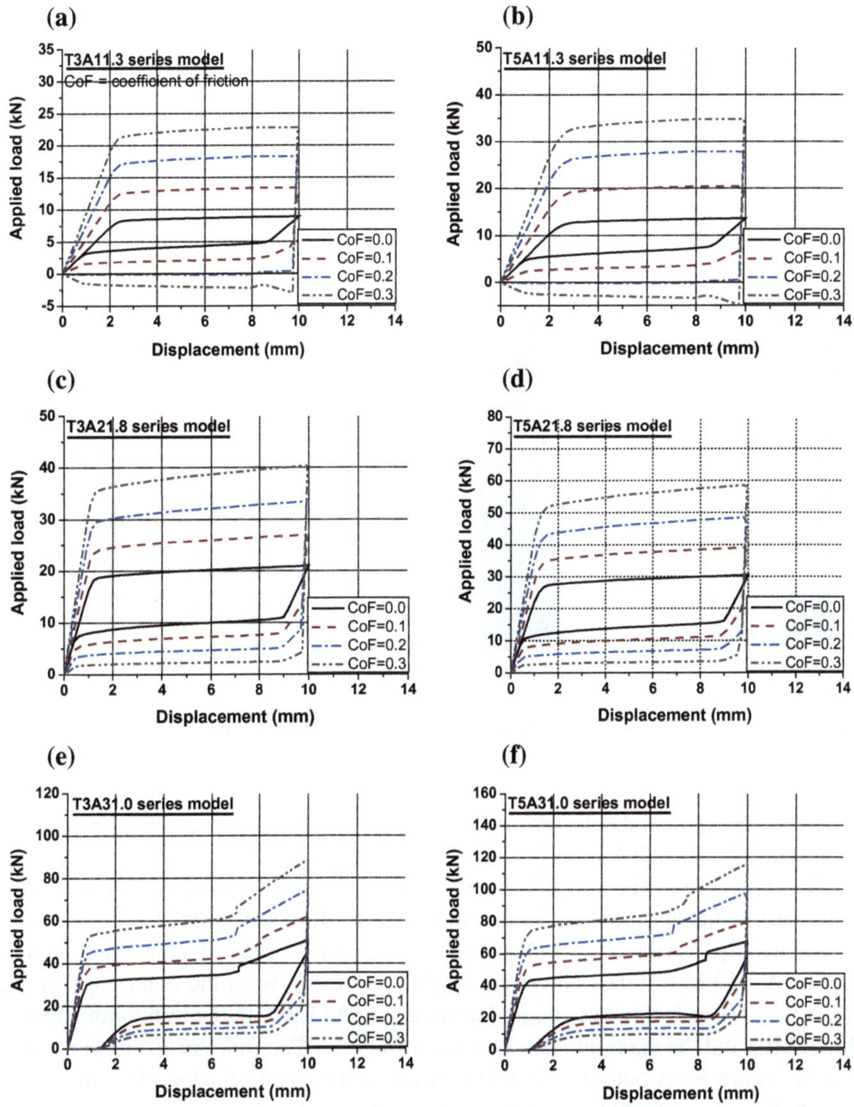

Fig. 2.42 Load–displacement hysteretic responses of SMA ring spring FE models: **a** T3A11.3 series, **b** T5A11.3 series; **c** T3A21.8 series, **d** T5A21.8 series; **e** T3A31.0 series, **f** T5A31.0 series

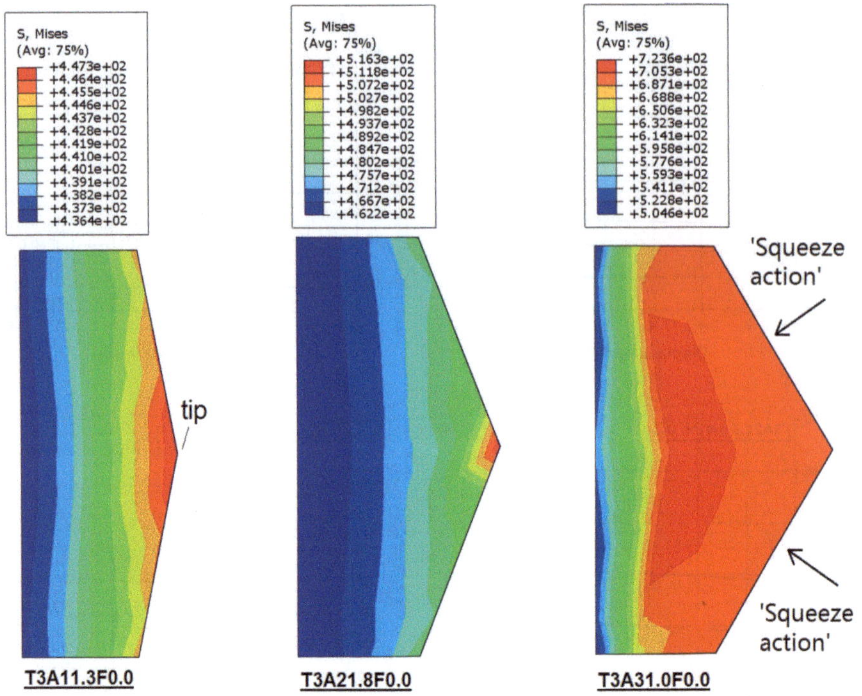

Fig. 2.43 Stress distributions of SMA outer rings when fully compressed

than 1.0 mm residual displacements are induced due to the permanent plastic strain developed at critical locations.

The typical von Mises stress distributions of the fully loaded SMA outer rings with various taper angles are shown in Fig. 2.43. It can be seen that the stress is almost evenly distributed over the ring thickness for the case of small taper angles, e.g. model T3A11.3F0.0. The stress slightly increases from the outer edge of the SMA ring towards the inner edge. For model T3A11.3F0.0, the minimum stress is 436.4 MPa and the maximum stress achieves 447.3 MPa. This stress range indicates that the entire SMA outer ring is within the forward transformation plateau.

Similar stress distribution is exhibited for the model with an increased taper angle of 21.8°. The stress gradient is slightly increased, where the maximum stress is 516.3 MPa which is developed at the tip of the taper face. This indicates that the local area near the tip has experienced martensitic hardening, but is still recoverable. When the taper angle further increases to 31.0°, the stress distribution is significantly changed. The stress gradient is considerably increased, where the peak stress exceeds 700 MPa, indicating the development of plastic stress. This echoes the residual displacements observed in the A31.0 series models, as shown in Fig. 2.42e, f. The much-increased stress distribution can be explained by two reasons: (1) the increase in the taper angle causes more significant expansion of the SMA rings at the

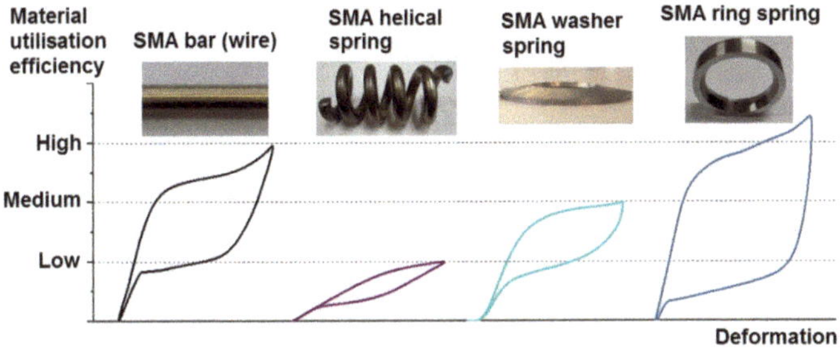

Fig. 2.44 Material utilisation efficiency of various SMA elements

same vertical deformation; and (2) a more evident 'squeeze action', as illustrated in Fig. 2.43, further increases the stress level locally near the tip zone.

We may conclude from the above observation that when the taper angle is appropriately designed, the stress over the cross section of the SMA outer ring can be uniformly distributed. This enables efficient utilisation of the SMA material compared to other possible types of SMA springs such as helical and Belleville washer springs where the maximum stress is only developed at local areas with the remaining areas being less stressed or even unstressed. In fact, if the geometrical configuration of the SMA ring springs is appropriately designed, and a reasonable balance between self-centring tendency and energy dissipation capacity (provided by superelasticity plus friction) is ensured, a higher material utilisation efficiency can be achieved for the SMA ring springs than their bar/wire counterparts. Figure 2.44 demonstrates the typical hysteretic curves indicating the material utilisation efficiencies of various SMA elements. The material utilisation efficiency can be defined as the ratio of the required performance (e.g. load resistance, energy dissipation, etc.) to material consumption.

2.6.3 Analytical Expressions

The numerical study reveals that an SMA ring spring has a clear and straightforward load transfer path. The hysteretic behaviour of an SMA ring spring system can be described via analytical expressions. A viable way is to capture several key points within the hysteretic curve, and then to link these characteristic points using straight lines to form a closed hysteretic loop. As it is reasonable to design SMA devices without inducing plastic stresses, the proposed analytical solution only considers the case where the maximum stress σ_{max} does not exceed the plastic stress σ_p. For simplicity, it may be assumed that the HSS inner rings act rigidly.

Let's consider a simple case where the maximum stress of the SMA does not exceed σ_{Mf}. The characteristic points along a hysteretic loop can be selected based on four stress conditions, namely forward transformation start stress σ_{Ms}, maxi-

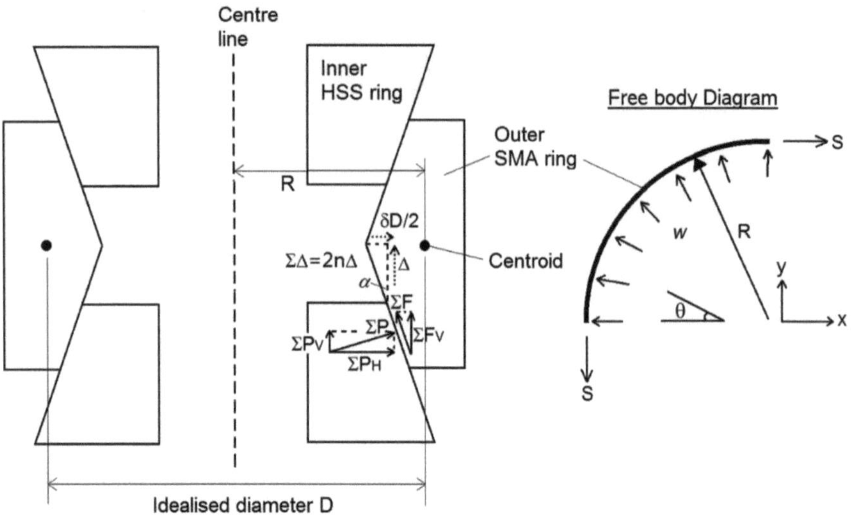

Fig. 2.45 Analytical model and free body diagram for SMA ring springs

mum stresses σ_{max}, stress when reverse transformation is initiated σ_{Rs} (as shown in Fig. 1.25, noting that this stress is different from σ_{As}), and reverse transformation finish stress σ_{Af}. When the maximum stress σ_{max} exceeds σ_{Mf} but is less than σ_p, five characteristic points may be required, which should be σ_{Ms}, σ_{Mf}, σ_{max}, σ_{As} and σ_{Af}. The selected strain–stress pairs are then mapped to the displacement–force pairs for constructing the complete hysteretic curve of the SMA ring spring system using the following procedure.

In order to obtain the relationship between the strain and the vertical displacement $\Sigma\Delta$ of the entire system, the SMA outer ring can be considered as an idealised ring with the hoop axis in line with the centroid of the cross section, as illustrated in Fig. 2.45. Based on this assumption, the initial (non-stressed) diameter D of the idealised ring can be determined. The overall displacement $\Sigma\Delta$ is the sum of the displacement Δ of each half piece of the SMA outer ring, i.e. $\Sigma\Delta = 2n\Delta$, where n is the total number of the SMA outer rings in the considered system. Considering a taper angle of α, tanα is the ratio of the increase in the radius R of the idealised SMA ring over the increase in displacement Δ:

$$\tan\alpha = \frac{\delta R}{\Delta} = \frac{\delta D}{2\Delta} \tag{2.6}$$

where δR and δD are the increases in the ring radius and diameter, respectively. The average strain of the ring can be expressed by the ratio of the increase in the perimeter over the original perimeter, i.e. $\varepsilon = \pi\delta D/\pi D = \delta D/D$, and replace δD with $2\Delta\tan\alpha$ based on Eq. (2.6), the strain of the ring can be expressed by

$$\varepsilon = \frac{\pi \cdot \delta D}{\pi D} = \frac{2\Delta \tan\alpha}{D} \tag{2.7}$$

Similar to the relationship between strain and displacement, the stress condition of the SMA outer ring can also be related to the load applied to the entire system. Assuming a uniform stress distribution over the cross section, which generally holds true based on the FE study, the hoop tensile force S can be expressed by $S = \sigma A$, where A is the cross section area of the SMA ring and σ is the average stress. The SMA outer ring is expanded due to a uniformly distributed load w along the perimeter, and considering a quarter of the ring using the free body diagram shown in Fig. 2.45, the x-axis (or y-axis) component of the sum of w should be equal to the tensile force S in order to achieve equilibrium:

$$\sigma A = S = \int_0^{\pi/2} wR\cos\theta d\theta \tag{2.8}$$

which leads to $w = S/R$, and therefore, the overall horizontal component of the force applied along the perimeter of each SMA ring is

$$2\Sigma P_H = 2\pi Rw = 2\pi R\frac{S}{R} = 2\pi S \tag{2.9}$$

Therefore, $\Sigma P_H = \pi S$ (2.10)

where ΣP_H is the horizontal component of the force applied at one taper face (noting that the expansion of each SMA ring is contributed by the total horizontal component of the force applied at two taper faces). The symbol 'Σ' indicates the sum of the force applied over the entire perimeter of the ring (i.e. from 0° to 360°). The vertical component of the force ΣP_V can be obtained via:

$$\Sigma P_V = \tan\alpha \cdot \Sigma P_H \tag{2.11}$$

and ΣP_V directly contributes to the load carrying capacity of the entire system. If friction exists, the resulting normal contacting force ΣP ($\Sigma P = \Sigma P_H/\cos\alpha$) between the outer and inner rings leads to an overall frictional force of ΣF along the perimeter:

$$\Sigma F = \Sigma P \cdot C_F \tag{2.12}$$

where C_F is the coefficient of friction. The vertical frictional component ΣF_V ($\Sigma F_V = \Sigma F\cos\alpha$) adds to the total resisting load of the system. Therefore, the total resisting load of the system can be expressed by

$$Load = \Sigma P_V \pm \Sigma F_V \tag{2.13}$$

where '+' is for the loading stage, and '−' is for the unloading stage.

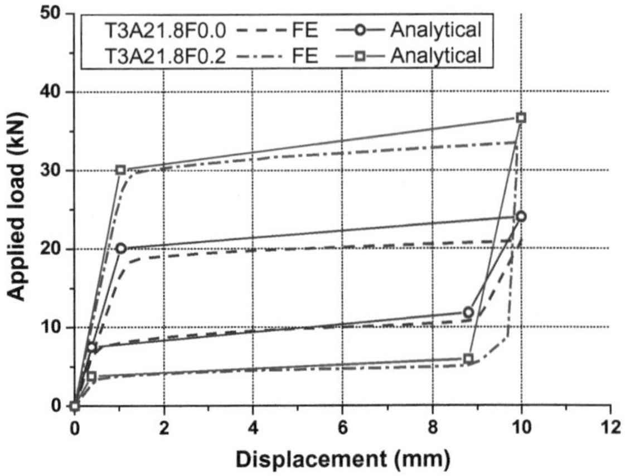

Fig. 2.46 Comparison between numerical and analytical predictions for SMA ring spring systems

Employing the above analytical procedure, the comparisons between the FE and analytical predictions of the load–displacement curves of the typical models are shown in Fig. 2.46. Good agreements are generally observed. The discrepancies may be caused by the assumptions and idealisations made in the analytical model. It is recalled that a uniform stress distribution is assumed over the cross section of the SMA outer ring in the analytical model, but as discussed previously, a slight stress gradient indeed exists. The assumption that the HSS inner rings always behave rigidly can also slightly overestimate the expansion of the SMA rings, leading to a slight overestimation of the load resistance. In addition, only four characteristic load–displacement pairs are considered, which neglects the status between these characteristic points. For example, the quick drop of the load resistance due to friction reversal at the beginning of the unloading stage is not captured. More characteristic points may be selected if a more accurate load–displacement loop shape is required. In any case, the analytical model enables designers to more clearly understand the overall load–deformation behaviour of the SMA ring spring systems, with available SMA material properties provided by material suppliers.

2.6.4 Experimental Investigation

Physical tests were conducted by Wang et al. (2017) to further confirm the feasibility of the SMA ring spring systems. The details of the test specimens are summarised in Table 2.7. Two loading protocols are considered, namely, constant amplitude and incremental amplitudes. For the former case, the specimens were subjected to 30 loading cycles with a constant amplitude of 5 mm; for the latter case, the specimens were loaded with incremental amplitudes from 1 to 4 mm with a 1 mm interval (two cycles per amplitude), and finally by 20 cycles with an amplitude of 5 mm.

Table 2.7 Details of SMA ring spring test specimens (Wang et al. 2017)

Specimen code	D_e (mm)	H (mm)	t (mm)	α (°)	Material	Grease condition
SMA-T3-G-CA	40	10	3	21.8	SMA	Greased
SMA-T3-G-IA	40	10	3	21.8	SMA	Greased
SMA-T5-G-IA	44	10	5	21.8	SMA	Greased
SMA-T3-NG-CA	40	10	3	21.8	SMA	Non-greased
ALU-T3-G-CA	40	10	3	21.8	Aluminium	Greased

The SMA outer rings were made of commercial Ti-55.9wt%Ni SMA. The alloy was processed by hot rolling at temperatures between 800 and 900 °C, and was then forged and straightened to a final required diameter. The large-diameter SMA bars were cut into segments which were then machined into the desired outer ring shape. The SMA rings were then annealed at 400 °C for 30 min and water quenched (Wang et al. 2016). For comparison purpose, an additional aluminium outer ring was produced and tested to highlight the unique superelastic property of SMA, and to clearly demonstrate how this effect significantly changes the load–deformation characteristics of the ring spring system. The specimen code in Table 2.7 starts with the material type (SMA = SMA or ALU = aluminium), followed by the ring thickness (T3 or T5) and the grease condition (G = greased or NG = non-greased), and ends with the loading protocol (CA = constant amplitude or IA = incremental amplitude). The testing device designed to facilitate the cyclic tests of SMA ring springs is shown in Fig. 2.47.

The flag-shaped compressive load–displacement responses of the individual SMA ring spring specimens are shown in Fig. 2.48. The yield loads at which the forward transformation plateau starts range between 20 and 30 kN, and the maximum load resistance ranges between 40 and 70 kN at the maximum displacement. Upon unloading, the load resistance first drops quickly due to the effect of reversal friction, and then a reverse transformation plateau is developed, promoting deformation recovery of the system.

Similar to the other types of SMA elements, the SMA rings show degradation effects, although the hysteretic response tends to be stabilised after a few loading cycles. Increasing the SMA ring thickness could effectively increase the strength. In addition, the hysteretic loops are widened in the absence of grease, which causes an increase in the maximum load resistance and energy dissipation but at the cost of compromised self-centring capability. Residual deformation accumulates as a result of material degradation plus possible abrasion over the contacting faces. Because of the abrasion, the self-centring capability of specimen SMA-T3-NG-CA is gradually lost with repeated loading cycles. This observation underlines the importance of maintaining a balance between the load resistance/energy dissipation and self-centring capability when determining the friction condition of the taper face.

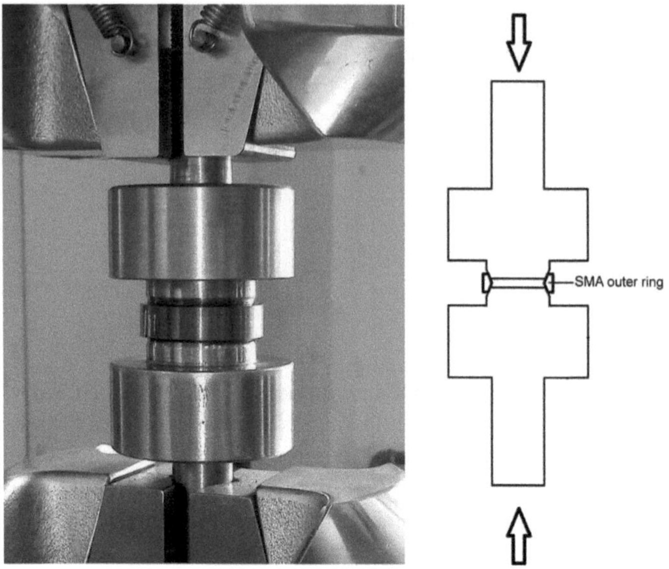

Fig. 2.47 Test setup for SMA ring springs

On the other hand, the aluminium ring exhibits a much less favourable hysteretic response, as shown in Fig. 2.49. As plastic strain is developed, very limited deformation recoverability is available during the unloading process, and as a result, significant residual deformation is induced immediately after the first loading cycle. In the subsequent loading cycles, the deformed aluminium ring undergoes limited elastic deformation, where the friction action attributes to the minor hysteretic loop.

Figure 2.50 further shows the accumulated residual displacements of the SMA ring spring specimens as a function of cycle number. The results confirm that overly large friction, due to the absence of grease, hinders the self-centring capability, whereas for the greased specimens, the maximum accumulated residual deformation is well controlled, corresponding a recovery rate of at least 76%. Similar to the SMA Belleville washers, the SMA rings may be trained by several loading cycles prior to their long-term service. A preload (pre-deformation) can also be applied to enhance their initial stiffness and the self-centring potential. As can be deduced from the limited test results, a pre-deformation of 20% of the maximum available deformation is adequate for suppressing the residual deformation of appropriately greased SMA ring springs.

Figure 2.51 shows that the maximum EVD observed in the SMA ring spring specimens could exceed 9%, while a stabilised value is kept between 6 and 7%. Due to the material degradation effect, a minor decrease in the EVD is observed with increasing cycle numbers. The non-greased specimen shows higher EVDs, which is clearly attributed to the increased friction effect.

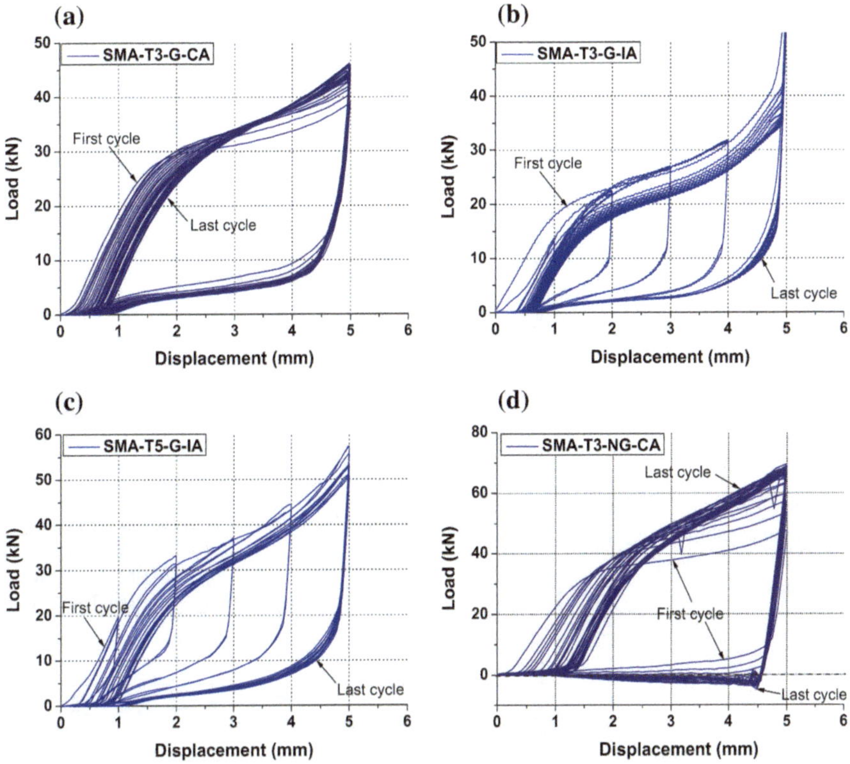

Fig. 2.48 Load–deformation responses of individual SMA ring spring specimens: **a** SMA-T3-G-CA, **b** SMA-T3-G-IA, **c** SMA-T5-G-IA, **d** SMA-T3-NG-CA

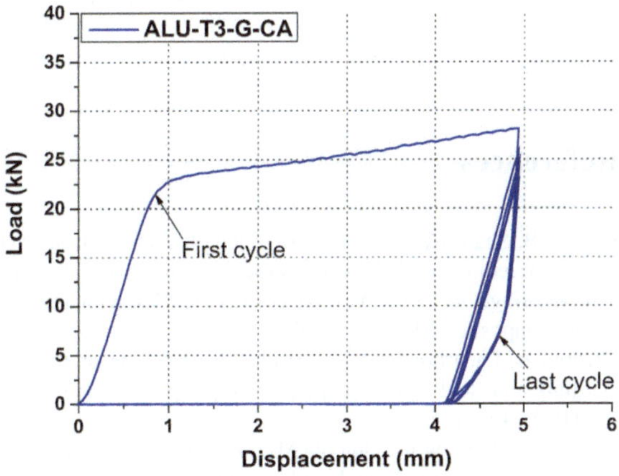

Fig. 2.49 Load–deformation responses of individual aluminium ring spring specimen

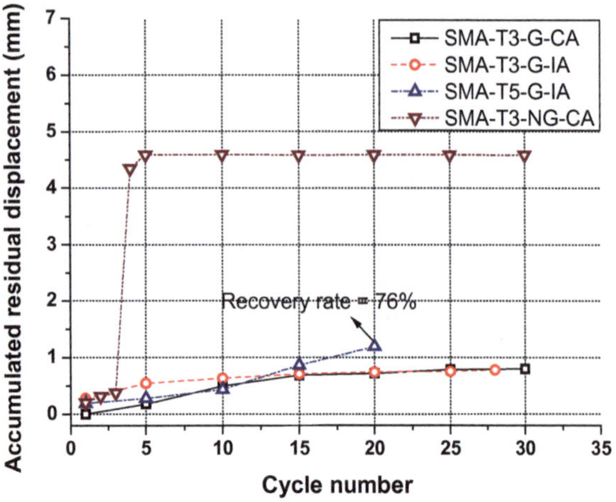

Fig. 2.50 Self-centring capability of individual SMA ring spring specimens

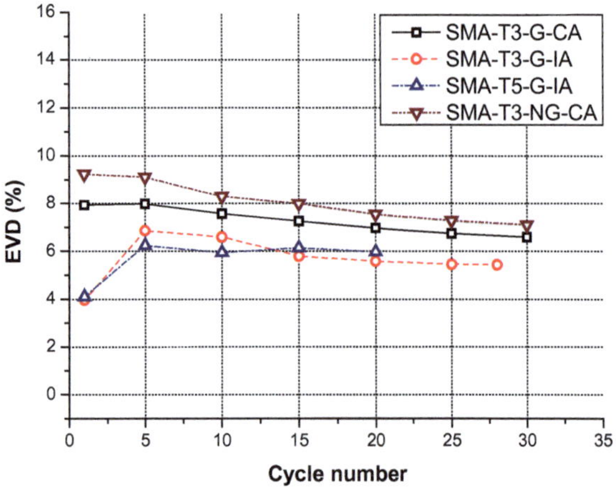

Fig. 2.51 Energy dissipation capability of individual SMA ring spring specimens

References

Biggs DB, Shaw JA (2016) Experimental characterization of shape memory alloy actuator cables. In: Proceedings of SPIE smart structures and materials + nondestructive evaluation and health monitoring, Las Vegas, Nevada, USA, 2016; Behavior and mechanics of multifunctional materials and composites 2016, vol 9800

Carboni B, Lacarbonara W, Auricchio F (2014) Hysteresis of multiconfiguration assemblies of nitinol and steel strands: experiments and phenomenological identification. J Eng Mech-ASCE 141(3):04014135

Churchill CB, Shaw JA, Iadicola MA (2009) Tips and tricks for characterizing shape memory alloy wire: part 2—fundamental isothermal responses. Exp Tech 33(1):51–62

DesRoches R, McCormick J, Delemont MA (2004) Cyclical properties of superelastic shape memory alloys. J Struct Eng-ASCE 130(1):38–46

Dolce M, Cardone D (2001a) Mechanical behaviour of SMA elements for seismic applications—part 2 austenite NiTi wires subjected to tension. Int J Mech Sci 43(11):2657–2677

Dolce M, Cardone D (2001b) Mechanical behaviour of SMA elements for seismic applications—part 1 martensile and austenite NiTi bars subjected to torsion. Int J Mech Sci 43(11):2631–2656

Dolce M, Marnetto R (1999) Seismic devices based on shape memory alloys. Manside Project, II105–134, Italian Dept for National Technical Services, Rome

Fang C, Yam MCH, Lam ACC, Xie LK (2014a) Cyclic performance of extended end-plate connections equipped with shape memory alloy bolts. J Constr Steel Res 94:122–136

Fang C, Vemury CM, Yam MCH (2014b) A numerical study of cyclic behaviour of NiTi shape memory alloy Belleville washers. In: Proceedings of the international conference on advances in civil, structural and mechanical engineering, Birmingham, 2014

Fang C, Yam MCH, Ma HW, Chung KF (2015a) Tests on superelastic Ni-Ti SMA bars under cyclic tension and direct-shear: towards practical recentring connections. Mater Struct 48(4):1013–1030

Fang C, Yam MCH, Lam ACC, Zhang YY (2015b) Feasibility study of shape memory alloy ring spring systems for self-centring seismic resisting devices. Smart Mater Struct 24(7):075024

Fang C, Zhou XY, Osofero AI, Shu Z, Corradi M (2016) Superelastic SMA Belleville washers for seismic resisting applications: experimental study and modelling strategy. Smart Mater Struct 25(10):105013

Fang C, Wang W, He C, Chen YY (2017) Self-centring behaviour of steel and steel-concrete composite connections equipped with NiTi SMA bolts. Eng Struct 150:390–408

Fang C, Zheng Y, Chen J, Yam MCH, Wang W (2019) Superelastic NiTi SMA cables: thermal-mechanical behavior, hysteretic modelling and seismic application. Eng Struct 183:533–549

Fromm E, Kleiner W (2003) Handbook for disc springs. Schnorr Corporation, Hela Werbung, Heilbronn

Fugazza D (2005) Experimental investigation on the cyclic properties of superelastic NiTi shape-memory alloy wires and bars. Rose School, European school for advanced studies in reduction of seismic risk, Pavia

Jee KK, Han JH, Kim YB, Jang WY (2008) New method for improving properties of SMA coil springs. Eur Phys J-Spec Top 158(1):261–266

Labrecque C, Braunovic M, Terriault P, Trochu F, Schetky M (1996) Experimental and theoretical evaluation of the behavior of a shape memory alloy Belleville washer under different operating conditions. In: Proceedings of the 42rd IEEE Holm conference on electrical contacts, joint with the 18th international conference on electrical contacts, Chicago, 1996

Liang C, Rogers CA (1997) Design of shape memory alloy springs with applications in vibration control. J Intel Mat Syst Str 8(4):314–322

Lim TJ, McDowell DL (1995) Path dependence of shape memory alloys during cyclic loading. J Intel Mat Syst Str 6(6):817–830

Maletta C, Filice L, Furgiuele F (2013) NiTi Belleville washers: design, manufacturing and testing. J Intel Mat Syst Str 24(6):695–703

Mas B, Biggs D, Vieito I, Cladera A, Shaw J, Martínez-Abella F (2017) Superelastic shape memory alloy cables for reinforced concrete applications. Constr Build Mater 148:307–320

Mirzaeifar R, DesRoches R, Yavari A (2011) A combined analytical, numerical, and experimental study of shape-memory-alloy helical springs. Int J Solids Struct 48(3–4):611–624

Morgan NB, Broadley M (2004) Taking the art out of smart! - forming processes and durability issues for the application of NiTi shape memory alloys in medical devices. In: Proceedings for the materials and processes for medical devices conference, Anaheim, CA, 2004

Ozbulut OE, Daghash S, Sherif MM (2015) Shape memory alloy cables for structural applications. J Mater Civil Eng 28(4):04015176

Piedboeuf MC, Gauvin R, Thomas M (1998) Damping behaviour of shape memory alloys: strain amplitude, frequency and temperature effects. J Sound Vib 214(5):885–901

Qiu CX (2016) Seismic-resisting self-centering structures with superelastic shape memory alloy damping devices. PhD thesis, The Hong Kong Polytechnic University

Reedlunn B, Daly S, Shaw J (2013) Superelastic shape memory alloy cables: part I—isothermal tension experiments. Int J Solids Struct 50(20–21):3009–3026

Savi M, Pacheco PMCL, Garcia MS, Aguiar RAA, Souza LFGD, Hora RBD (2015) Nonlinear geometric influence on the mechanical behavior of shape memory alloy helical springs. Smart Mater Struct 24(3):035012

Sgambitterra E, Maletta C, Furgiuele F (2014) Modeling and simulation of the thermo-mechanical response of NiTi-based Belleville springs. Remote Sens Environ 113(2):445–457

Sherif MM, Ozbulut OE (2018) Tensile and superelastic fatigue characterization of NiTi shape memory cables. Smart Mater Struct 27(1):015007

Shigley JE, Mischke CR, Budynas RG (1989) Mechanical engineering design. McGraw-Hill, New York

Speicher MS, Hodgson DE, DesRoches R, Leon RT (2009) Shape memory alloy tension/compression device for seismic retrofit of buildings. J Mater Eng Perform 18(5):746–753

Speicher MS, DesRoches R, Leon RT (2011) Experimental results of a NiTi shape memory alloy (SMA)-based recentering beam-column connection. Eng Struct 33(9):2448–2457

Sun SS, Lv M (2016) Dynamic response of a high-rise chemical tower controlled by pseudo-elastic shape memory alloy cables. J Intel Mat Syst Str 27(10):1412–1422

Sutton MA, Orteu JJ, Schreier HW (2009) Image correlation for shape, motion and deformation measurements: basic concepts, theory and applications. Springer, Boston

Tamai H, Kitagawa Y (2002) Pseudoelastic behavior of shape memory alloy wire and its application to seismic resistance member for building. Comp Mater Sci 25(1–2):218–227

Tamai H, Miura K, Kitagawa Y, Fukuta T (2003) Application of SMA rod to exposed-type column base in smart structural system. In: Liu SC (ed) Proceedings of SPIE 5057: smart structures and materials 2003: smart systems and nondestructive evaluation for civil infrastructures, 2003

Tobushi H, Shimeno Y, Hachisuka T, Tanaka K (1998) Influence of strain rate on superelastic properties of TiNi shape memory alloy. Mech Mater 30(2):141–150

Wang W, Fang C, Liu J (2016) Large size superelastic SMA bars: heat treatment strategy, mechanical property and seismic application. Smart Mater Struct 25(7):075001

Wang W, Fang C, Yang X, Chen YY, Ricles J, Sause R (2017) Innovative use of a shape memory alloy ring spring system for self-centering connections. Eng Struct 153:503–515

Wang W, Fang C, Zhang A, Liu XS (2019) Manufacturing and performance of a novel self-centring damper with shape memory alloy ring springs for seismic resilience. Struct Control Hlth. https://doi.org/10.1002/stc.2337

Wolons D, Gandhi F, Malovrh B (1998) An experimental investigation of the pseudoelastic hysteresis damping characteristics of shape memory alloy wires. J Intel Mat Syst Struct 9(2):116–126

Wu MH (2002) Fabrication of nitinol materials and components. Mater Sci Forum 394–395:285–292

Zheng Y, Dong Y, Li Y (2018) Resilience and life-cycle performance of smart bridges with shape memory alloy (SMA)-cable-based bearings. Constr Build Mater 158:389–400

Chapter 3
Steel Beam-to-Column Connections with SMA Elements

Abstract Having gained the necessary knowledge on the behaviour of various SMA elements, this chapter focuses on a series of novel SMA-based self-centring steel beam-to-column connections. Different SMA elements, including SMA bolts, SMA Belleville washer springs and SMA ring springs, are considered as kernel elements for these connections. The combined use of different types of SMA elements is also attempted. The influence of the slab system on the behaviour of self-centring connections is discussed, and the issue of frame expansion is particularly raised and addressed. The available experimental studies on these novel connections are presented, enabling a comprehensive understanding of their key performances such as moment–rotation response, ductility, self-centring capability, energy dissipation and possible failure modes. The design recommendations for some typical connections are also presented in this chapter.

3.1 Introduction

Beam-to-column connections, hereafter simplified as connections, play a critical role in ensuring structural safety against the seismic hazard. Conventional seismic design code requires the connections to be capable of withstanding design earthquakes without experiencing failure (e.g. weld fracture). In the early days, special moment-resisting frames (SMRFs) with fully restrained welded steel connections were recognised as ideal seismic-resistant systems. As the stable yielding behaviour of the beam sections encourages adequate energy dissipation against collapse, engineers at that time believed that these fully restrained connections with welded beam flange and welded or bolted beam web can reliably make the main inelastic deformation to occur at the adjacent steel beams rather than at the connection zone. Partly as a result of this belief, many steel SMRFs with welded connections were constructed from the 1960s, especially in the USA. However, brittle fractures of a large number of the welded connections were observed after the Northridge earthquake in 1994 and later in the Kobe earthquake in 1995, which initiated a worldwide controversy on this design concept (FEMA 2000a).

C. Fang and W. Wang, *Shape Memory Alloys for Seismic Resilience*, https://doi.org/10.1007/978-981-13-7040-3_3

After these two major events, people began to re-evaluate the inelastic behaviour of fully restrained welded connections. Investigations revealed that weld discontinuities, potentially resulting from the use of low-toughness welding electrodes with inappropriate welding processes, mainly attributed to the poor seismic performance of these connections (Youssef et al. 1995; FEMA 2000a). Meanwhile, alternative connection types were explored, and it was concluded that appropriately designed bolted connections could offer good seismic performance with sufficient strength, ductility, and rigidity. More flexible pre-qualified semi-rigid connections might also be employed (FEMA 2000b).

The improved connection types are often called 'post-Northridge' connections. Newly built structures with 'post-Northridge' connections are expected to have controllable yielding behaviour against collapse under strong seismic excitations. However, the ductility demand of the structures is provided at the cost of significant inelastic/permanent deformations of the structural members, and the associated damage could be costly, time-consuming and difficult to repair. Some studies suggested that repairing damaged structures becomes uneconomical if the residual inter-storey drift is greater than 0.5% after earthquakes (McCormick et al. 2008). These structures, although neither collapsed nor severely damaged, may need to be demolished due to the excessively large residual deformations. This places a great challenge on the affected communities to restore essential services and to initiate the process of recovery. Even if the damaged structures are repairable, real estate owners are often faced with a dilemma of whether or not to repair them, as sometimes the cost of the repair work can be comparable to or even higher than that of building a new structure. In any case, the downtime inevitably puts further pressure on launching recovery efforts. Some critical structures, such as hospitals, government headquarters, nuclear plants and other lifeline facilities, cannot even afford a temporary occupancy suspension.

Driven by the next-generation performance-based design framework, which tends to considerably broaden the seismic design targets, along with a growing appeal for more sustainable and resilient cities and infrastructures, clients and real estate owners are no longer willing to compromise their properties with severe or irreparable post-earthquake damages, and people are now more interested in pursuing resilient communities and cities which are able to rebound quickly from a disaster (e.g. earthquake) that has caused structural collapse, casualties, infrastructure disruption and business interruption (NAS 2011).

From the structural engineering standpoint, one possible solution to speed up the recovery process is to incorporate the self-centring concept in the design of structural systems. The core intention of the self-centring structural design, under an idealised situation, is to eliminate the post-earthquake permanent deformation of the structures, hence minimising the repair work. Towards this goal, structural engineers and researchers have been exploring new self-centring design concepts, among which endowing connections with self-centring capabilities has been considered as one of the most feasible strategies. In the field of steel structures, the earliest attempt was to incorporate post-tensioned (PT) cables/tendons into the connections to provide the self-centring mechanism (Christopoulos et al. 2002; Ricles et al. 2001; Garlock et al. 2005, 2008; Lin et al. 2013a, b; Zhang et al. 2016), as typically illustrated in Fig. 3.1.

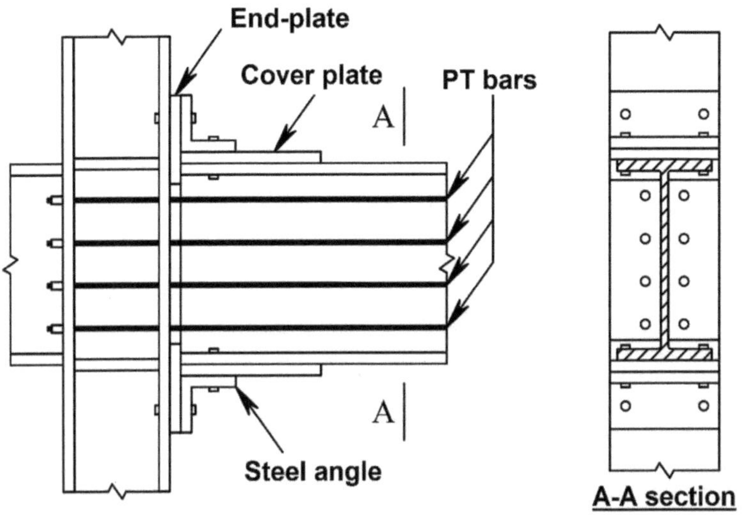

Fig. 3.1 Typical PT-based self-centring connections

As an alternative to the 'structural-based' self-centring strategy, a 'material-based' approach using superelastic SMAs is considered as an encouraging solution. While both solutions have their own advantages and deserve long-term research effort, this chapter only focuses on self-centring connections using the SMA-based strategy.

3.2 Connections with SMA Bolts

The basic SMA-based connection type is the one that is equipped with SMA bolts. The initial study can be traced back to the early 2000s, where Ocel et al. (2004) first conducted two proof-of-concept tests on shear tab steel connections employing martensitic SMA bolts. The self-centring capability was realised by utilising the SME of the bolts via heating. The connection specimens exhibited good energy dissipation, and more than half of the residual rotation was recovered after reheating. Spontaneous self-centring capabilities were observed by Penar (2005) who carried out follow-up tests on connections with similar configurations but with superelastic SMA bolts. Abolmaali et al. (2006) observed early bolt fracture which was due to the insufficient net threaded-to-shank diameter ratio of the SMA bolts, as discussed in Chap. 2. The experimental data pool was further broadened by Speicher et al. (2011), who undertook four more tests on shear tab connections with different types of SMA bolts. It was found that the permanent deformation of the connection with martensitic SMA bolts could be recovered by 75% via reheating. For the connections with superelastic SMA bolts, the residual rotation after a maximum drift of 5% was between 0.5 and 1.0%, and the EVD could achieve up to 13%. Concerning the issues

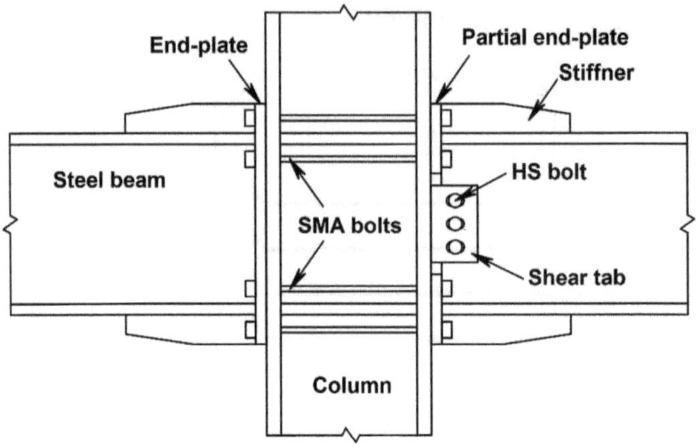

Fig. 3.2 Illustration of practical SMA-bolted connections

of installation complexity, Ma et al. (2007) proposed an easy-to-install self-centring connection type, where the conventional HS bolts in extended end-plate connections are replaced by superelastic SMA bolts which provide ductility and energy dissipation capability. This concept was later experimentally verified (Fang et al. 2014), and the concept was also used for developing high-performance I-beam-to-tubular column connections (Wang et al. 2015a).

Clearly, many research groups have successfully completed physical tests on SMA-bolted connections, and these specimens generally showed satisfactory self-centring capability, energy dissipation and ductility. Shear tab and end-plate configurations are two most commonly considered connection types that incorporate SMA bolts, although the load carrying mechanism for the two cases is similar. The combined use of partially extended end-plates and shear tabs may be a good option, as illustrated in Fig. 3.2. For practical application, the superelastic form of SMA bolts is more preferred than the martensitic form because the former needs no heating to accomplish self-centring.

This section presents the typical testing procedure and results for SMA-bolted connections. In addition, some practical issues which have been identified from the existing investigations are also mentioned.

3.2.1 Basic Connection Behaviour

The behaviour of SMA-bolted connections largely depends on the bolt dimension and layout. The strength and stiffness, which are two fundamental properties of beam-to-column connections, are increased with increases in the SMA bolt diameter and the distance between the external bolt rows, i.e. lever arm for bending. Given the fact that the expected connection deformation is mainly provided by the SMA bolts, simple design equations (CEN 2005) can be employed to estimate the initial stiffness $K_{c,ini}$ and yield moment $M_{j,Rd}$ of SMA-bolted connections:

$$K_{c,ini} = \sum_i k_i h_i^2 = \sum_i \frac{\sigma_{Ms} E A_b}{(\sigma_{Ms} - \sigma_{pre}) L_p} h_i^2 \tag{3.1}$$

$$M_{j,Rd} = \sum_i \sigma_{Ms} A_b \frac{h_i^2}{h_1} \tag{3.2}$$

where k_i is the 'effective' stiffness of each SMA bolt with the consideration of preload; E is the elastic modulus of SMA material; A_b is the cross-sectional area of each SMA bolt; σ_{Ms} is the forward transformation start stress; σ_{pre} is the bolt prestress, L_p is the parallel/working length of the SMA bolts; and h_i is the distance of the ith bolt row to the rotation centre which might be assumed to be near the bottom stiffener. The connection yield moment $M_{j,Rd}$ may be conservatively determined by the moment when the furthest bolt row first reaches σ_{Ms} [i.e. Eq. (3.2)]. However, if the utilisation of the connection overstrength beyond the first bolt 'yielding' is allowed, and assuming an elastic-perfectly plastic response of the SMA bolts, the maximum design moment resistance might be increased to

$$M_{j,Rd} = \sum_i \sigma_{Ms} A_b h_i \tag{3.3}$$

An important issue for SMA bolt design is the determination of the net threaded-to-shank bolt diameter ratio, which, as discussed in Chap. 2, can greatly influence the fracture resistance and ductility of the bolts. In addition, an appropriate length-to-diameter ratio of the bolt is also desired. If this ratio is too small, i.e. the SMA bolt is too stocky, the ductility is decreased, and the bolt tends to be subjected to an unwanted combination of axial tension and bending/shear when the connection rotates. A sufficient length-to-diameter ratio means that the SMA bolt is predominantly subjected to a uniaxial loading state, which is a preferable load resisting mechanism. Nevertheless, an overly large length-to-diameter ratio should also be avoided as this will cause an unfavourable twisting effect when the bolts are preloaded with a wrench.

Preload/prestrain level is another important factor which determines the initial stiffness and self-centring capacity of the connection. An appropriate bolt preload level ensures a reasonable recovery tendency and benefits the initial stiffness of the connections. If no or insufficient bolt preload is applied, the connection would have an inadequate decompression moment resistance, leading to low initial stiffness. The connection stiffness can be further compromised due to the TIF degradation. Of course, an excessively large bolt preload, e.g. ε_{pre} being much larger than ε_{Ms}, is unnecessary as this would reduce the available deformability of the SMA bolts and may cause more pronounced TIF degradation because a high strain level is constantly maintained.

Where applicable, special attention should also be paid to the end-plate design. For conventional end-plate connections, i.e. those with HS bolts, relatively 'thin' end-plates could offer good ductility, provided that the detrimental 'prying action' does not cause premature failure of the bolts. For SMA-bolted connections, however, the end-plate should be designed as a 'thick plate' because the inelastic deformation of a thin end-plate would cause unwanted biaxial bending of the SMA bolts (Yam et al. 2015). The bending action can be detrimental to the self-centring and energy dissipation capabilities of the connection. More importantly, the non-uniform cross-sectional stress distribution of the SMA bolts induced by the bending action makes the threaded area more susceptible to fracture. Therefore, the elastic behaviour of the end-plate should be ensured to encourage a more reliable self-centring mechanism of the SMA connections. Based on the yield line theory, the required end-plate thickness $t_{pReq'd}$ can be calculated by (Murray and Sumner 2003)

$$t_{pReq'd} = \sqrt{\frac{1.11\phi M_{np}}{\phi_b f_{yp} Y_p}} \tag{3.4}$$

where $\phi = 0.75$; $\phi_b = 0.9$; f_{yp} is the yield strength of the end-plate; M_{np} is the bolt moment strength, which is obtained from $M_{np} = \Sigma P_t h_i$ where P_t is the bolt tensile strength; and Y_p is a yield line mechanism parameter, which is given in detail in Murray and Sumner (2003) for varied connection types. For the SMA bolts, the martensitic transformation finish stress σ_{Mf} may be considered to be the expected bolt tensile force. This assumption is reasonable if one chooses not to utilise the reserve of the SMA tensile strength beyond σ_{Mf}.

3.2.2 *Proof-of-Concept Experimental Investigation*

While the load transfer mechanism of SMA-bolted connections is simple and straightforward, experimental investigations are necessary to enable a more comprehensive understanding of the connection behaviour under seismic loading. Either a single-sided or double-sided substructural assembly can be used for examining the cyclic behaviour of the SMA connections. Figure 3.3 illustrates a typical single-sided substructural assembly that consists of a cantilever steel beam, a steel column, and the associated SMA-bolted connection zone (Fang et al. 2014). The column can be oriented either vertically or horizontally and should be fixed to a supporting frame/strong floor via rigid connections at both column ends. A double action hydraulic actuator is used to apply the required cyclic load at the beam tip.

For any connection test, the two most important measurements are connection moment and rotation. The former can be easily obtained through multiplying the beam tip load by the beam arm length. For the definition of connection rotation, a 'drift ratio', or sometimes simplified as 'drift', is defined as the total rotation of the substructure, i.e. the beam tip deflection divided by the arm length of the beam. A 'plastic rotation' (θ_p) is that obtained by deducting the elastic deformation of the system from the total rotation, as expressed by (Stojadinović et al. 2000)

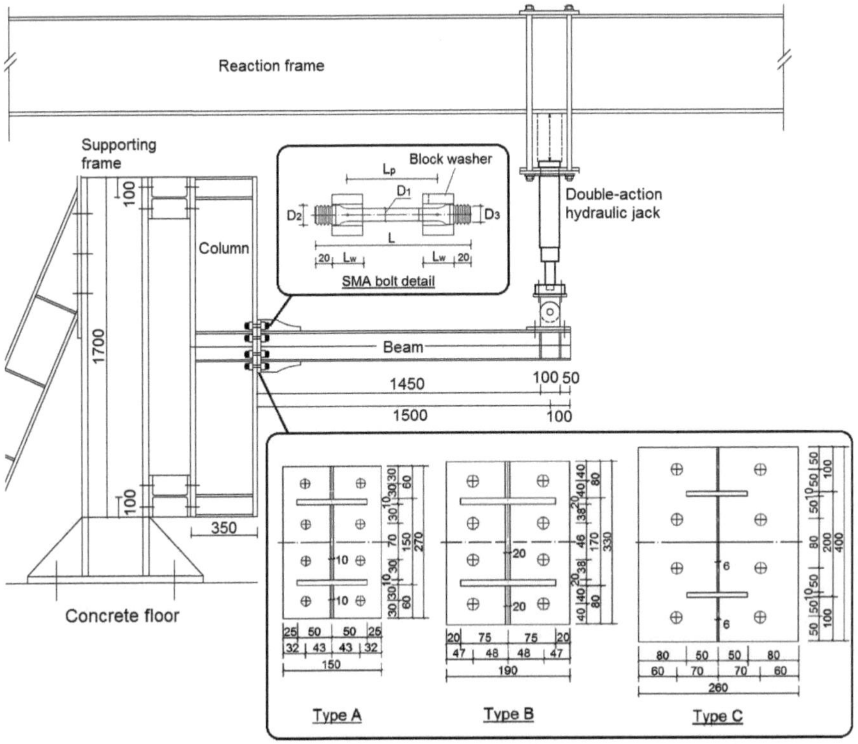

Fig. 3.3 Typical single-sided test setup for SMA connections (Fang et al. 2014)

$$\theta_p = \frac{\Delta_b - P/K_e}{L_{arm}} \tag{3.5}$$

where Δ_b is the beam tip displacement, P is the applied load, K_e is the elastic stiffness of the system and L_{arm} is the arm length. Under certain circumstances, a 'concentrated connection rotation' needs to be defined which is the local rotation contributed by the connection only.

Table 3.1 gives the details of the eight proof-of-concept connection specimens conducted by Fang et al. (2014), where the symbols for the key dimensions of the bolts are given in Fig. 3.3. In the experimental programme, seven tests were performed on extended end-plate steel connections equipped with superelastic SMA bolts, while the remaining specimen was a conventional extended end-plate steel connection with HS bolts (hereafter named as conventional connection). The main test parameters for the SMA connections were bolt length and bolt diameter. Each connection is designated with a specimen code according to the bolt details, as shown in Table 3.1. For example, 'SMA-D10-240' represents the connection using the SMA bolts with a nominal diameter of 10 mm and a nominal total bolt length (including the length of the threaded area) of 240 mm.

Table 3.1 Geometric details of connection specimens (Fang et al. 2014)

Specimen code	Shank diameter D_1	Threaded gross diameter D_2	Threaded net diameter D_3	Bolt layout type and beam section
SMA-D10-190	9.9	12.4	10.1	Type A (150 × 100 × 10 × 10)
SMA-D10-240	9.9	12.4	10.0	Type A (150 × 100 × 10 × 10)
SMA-D10-240-d	9.9	12.4	10.1	Type B (170 × 150 × 20 × 20)
SMA-D10-290	9.9	12.4	10.1	Type A (150 × 100 × 10 × 10)
SMA-D16-190	15.9	18.9	15.6	Type A (150 × 100 × 10 × 10)
SMA-D16-240	15.8	18.9	15.4	Type B (170 × 150 × 20 × 20)
SMA-D16-290	15.9	18.9	15.5	Type B (170 × 150 × 20 × 20)
HS	M20 grade 10.9			Type C (200 × 100 × 9.8 × 6)

Note All units are in mm

To ensure adequate initial stiffness and reliable self-centring capability, a preload corresponding to a stress of $0.65\sigma_{Ms}$ was applied to each SMA bolt. For the conventional connection, the HS bolts were installed by the turn-of-nut method with 1/2 turn after a snug-tightened condition. After finishing the bolt preload process, a cyclic load was applied using the SAC loading protocol (SAC 1997). This loading protocol uses drift as the governing parameter, and employs the following standard loading sequence: 0.375% drift for six cycles, 0.50% drift for six cycles, 0.75% drift for six cycles, 1% drift for four cycles, 1.5% drift for two cycles, 2% drift for two cycles, and then continuous with a 1% drift interval until 4% drift, or to failure of the connection.

The typical deformation mode of the SMA connection specimens is shown in Fig. 3.4. As anticipated, the main deformation occurs in the SMA bolts, and the end-plate behaves in a thick plate manner. The governing failure mode is bolt fracture, which is caused by insufficient net threaded-to-shank diameter ratio. Figure 3.5 shows the moment–plastic rotation responses of the eight test specimens. In general, all the seven SMA connections show recognisable flag-shaped responses. Different bolt layouts and bolt geometric dimensions result in varied shapes of the hysteretic loop. The friction between the end-plate and the column flange face provides sufficient shear resistance with no slippage observed. The friction is enabled by the bolt preload plus the firm contact of the two plates in the centre of compressive when the connection rotates. Caution should be exercised that this shear resisting mechanism is not reliable, although it is accepted for proof-of-concept investigation purpose. An

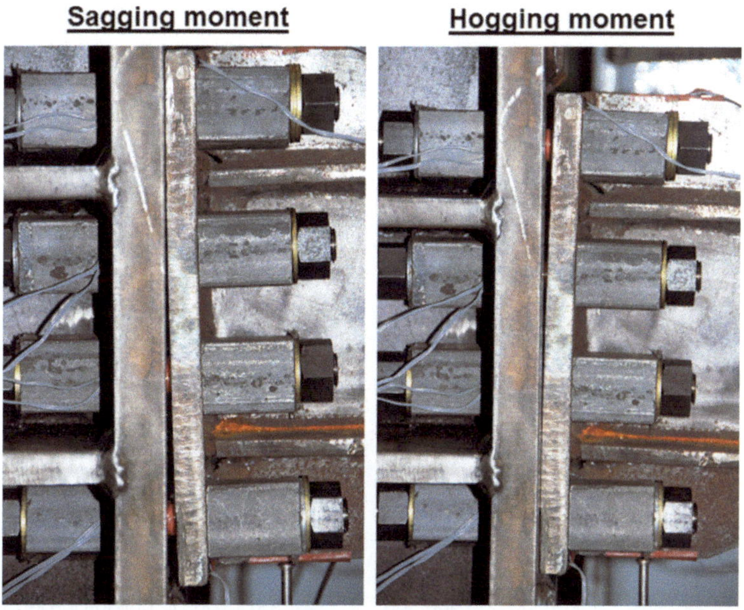

Fig. 3.4 Deformation mode of SMA-bolted end-plate connections

additional shear tab is recommended for practical use. On the other hand, the conventional connection specimen shows a full hysteretic moment–rotation response, but the stable hysteretic curve is accompanied by considerable residual rotations upon unloading.

Beam-to-column connections are often classified in terms of their strength and stiffness such that an appropriate design model/assumption can be made for global analysis (CEN 2005). This also applies to SMA connections. For strength, a connection is classified as nominally pinned, partial-strength, or full-strength through comparisons against the plastic moment resistance ($M_{b,pl}$) of the connected beam. A connection is nominally pinned if the moment capacity (M_{max}) does not lead to significant moments which might adversely affect the members or the whole structure (i.e. $M_{max} < 0.25\ M_{b,pl}$), while it is a full-strength connection if its strength is larger than the design plastic moment resistance of the connected beam (i.e. $M_{max} > M_{b,pl}$). A connection which has non-negligible moment capacity but the strength is less than the plastic moment resistance of the connected member, is considered as a partial-strength connection (i.e. $0.25\ M_{b,pl} < M_{max} < M_{b,pl}$).

On the other hand, a connection is classified in terms of stiffness as nominally pinned, semi-rigid, or rigid through comparisons with EI/L_b, where E, I and L_b are the modulus of elasticity, second moment of area, and span of the connected member (e.g. connected beam), respectively. A connection is nominally pinned if the initial rotational stiffness $K_{c,ini}$ is less than $0.5EI/L_b$, while it is deemed as rigid, in sway frames (e.g. MRFs), if the initial rotational stiffness is greater than $25EI/L_b$.

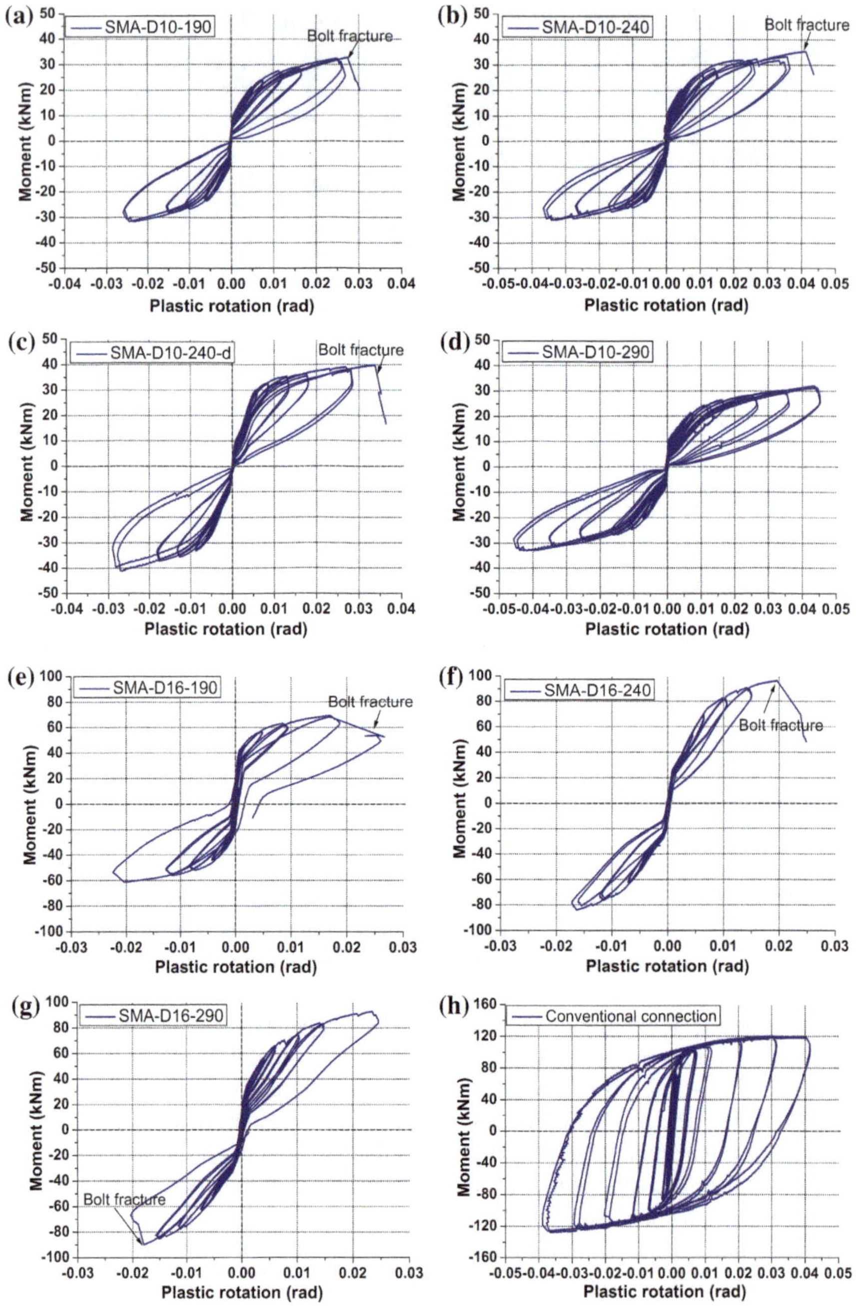

Fig. 3.5 Moment-plastic rotation responses of SMA-bolted connections: **a** SMA-D10-190, **b** SMA-D10-240, **c** SMA-D10-240-d, **d** SMA-D10-290, **e** SMA-D16-190, **f** SMA-D16-240, **g** SMA-D16-290, **h** conventional connection

Table 3.2 Stiffness and strength of SMA-bolted connections

Specimen	Stiffness			Strength		
	$K_{c,ini}$	$K_{c,ini}/(EI/L)$	Classification	M_{max}	$M_{max}/M_{b,pl}$	Classification
SMA-D10-190	11,295	15.2	Semi-rigid	32.8	0.65	Partial-strength
SMA-D10-240	11,154	15.0	Semi-rigid	32.1	0.63	Partial-strength
SMA-D10-240-d	7955	3.0	Semi-rigid	40.4	0.28	Partial-strength
SMA-D10-290	8714	11.7	Semi-rigid	33.6	0.66	Partial-strength
SMA-D16-190	29,411	39.6	Rigid	69.2	1.36	Full-strength
SMA-D16-240	16,926	6.3	Semi-rigid	90.1	0.65	Partial-strength
SMA-D16-290	18,818	7.0	Semi-rigid	91.5	0.66	Partial-strength

Note 1 the unit for $K_{c,ini}$ is kNm/rad, and the unit for M_{max} is kNm
Note 2 the beam length is assumed to be 3.2 m for stiffness classification

A connection with the initial rotational stiffness that falls between the two thresholds is classified as semi-rigid. The above terminologies may slightly vary in different standards, but the rationale behind the different classification methods is similar. Considering these definitions, the classifications of the SMA connection specimens are summarised in Table 3.2. Incidentally, the steel beams designed for the test specimens were deliberately oversized to make sure that the inelastic deformation majorly concentrates on the connection zone. This is the main reason that most of the specimens are semi-rigid and partial-strength connections.

A critical lesson drawn from this proof-of-concept experimental programme is the occurrence of early bolt fracture which significantly compromises the connection ductility. According to Eurocode 8 (CEN 2004), connections should satisfy a plastic hinge rotational capacity of at least 0.035 radians and 0.025 radians for the Ductility Class High (DCH) and Ductility Class Medium (DCM) criteria, respectively. AISC (2010) requires that special moment frames (SMFs) and intermediate moment frames (IMFs) should provide inter-storey drifts of more than 4% and 2%, respectively. Based on these restrictions, only specimens SMA-D10-240 and SMA-D10-290 satisfy the DCH or SMF criterion, while the other SMA connection specimens only meet the DCM or IMF requirement. The bolt fracture strain, with the prestrain inclusive, ranges between 4.8 and 5.6% for the 10 mm-diameter bolts, and between 3.3 and 3.8% for the 16 mm-diameter bolts. As shown in Table 3.1, the average net threaded-to-shank diameter ratios are only 1.02 and 0.97 for the 10 mm and 16 mm SMA bolts, respectively. The low net threaded-to-shank diameter ratio, coupled with potential material imperfection and machining deficiency, is the main reason causing the premature fracture of the threaded area. As mentioned in Chap. 2, increasing the diameter of the threaded part of the SMA bolts can effectively enhance the connection ductility.

While the proof-of-concept SMA connection specimens exhibited unsatisfactory ductility, the SMA bolts did provide excellent self-centring capability prior to fracture. The residual deformations of the specimens are illustrated in Fig. 3.6 together with the recovery rate. For all the SMA connections, the recovery rate prior to bolt fracture is more than 90%. On the other hand, the conventional connection exhibits no self-centring capability, although much-increased energy dissipation is offered. As shown in Fig. 3.7, the conventional connection provides a maximum EVD of more than 40%, while an EVD of up to 16% is observed for the SMA connections. For the former case, the energy is largely dissipated via the plastic hinge of the steel beam.

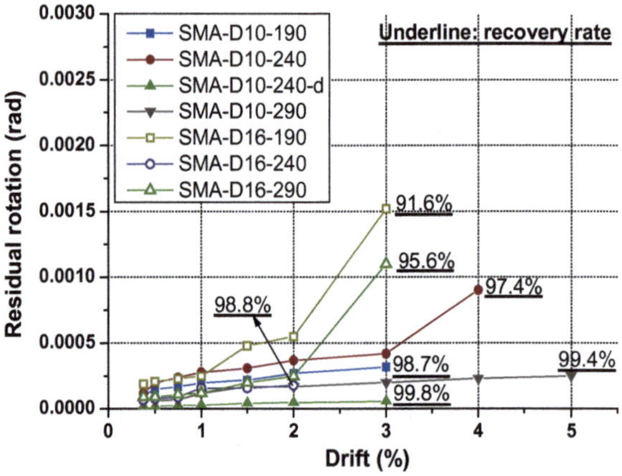

Fig. 3.6 Self-centring capability of SMA-bolted connections

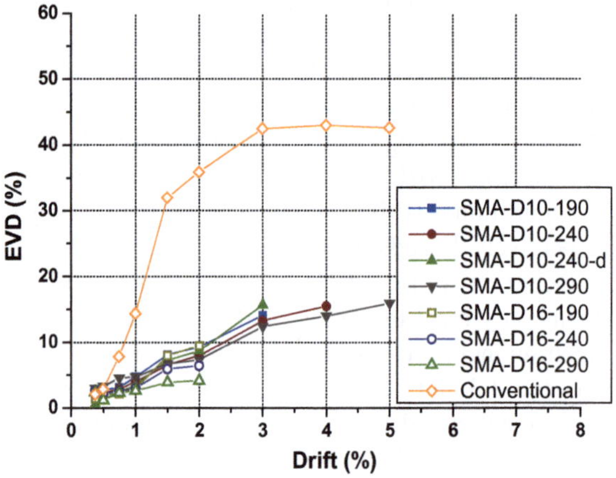

Fig. 3.7 Energy dissipation of connection specimens

The main findings and conclusions obtained from the proof-of-concept experimental study conducted by Fang et al. (2014) on SMA-bolted connections can be summarised as follows:

(1) The feasibility of SMA-bolted connections is confirmed, and the SMA bolts are shown to display anticipated functions. The behaviour of the SMA-bolted connection specimens is in line with the initial expectations, where the main source of recoverable deformation is offered by the SMA bolts, and the remaining structural components mainly stay elastic.
(2) Although the wire form of SMA has been found to be quite ductile (Lagoudas 2008), the ductility of the connection specimens observed from the experimental study seems to be less satisfactory. A sufficient net threaded-to-shank diameter ratio of the SMA bolts should be guaranteed in practical design.
(3) The connection specimens with the 10 mm-diameter SMA bolts seem to exhibit better performance in terms of ductility and hysteretic stability than the specimens with the 16 mm-diameter bolts. Size effect could be a reason behind this difference. Another possible reason is the stress state, which differs in SMA bolts with different length-to-diameter ratios.
(4) Apart from the length-to-diameter ratio, the selection of the bolt dimension should cater to the basic design requirements regarding strength and ductility. A larger bolt diameter and a longer bolt normally lead to increases in the connection moment resistance and ductility, respectively. With the same bolt diameter and length, pursuing higher moment resistance and higher ductility at the same time via only adjusting the bolt row positions could be difficult, and thus a trade-off is often needed.

3.2.3 Effect of Floor Slabs

Composite flooring systems, typically consisting of steel beams, cast-in-place concrete with reinforcement, shear studs, and metal decks, are nowadays mainstream structural forms for residential and commercial buildings. However, self-centring connections should be carefully used in the composite flooring systems since the basic principle of self-centring connection design lies in the gap opening mechanism where the restoring forces are provided by either PT bars/cables or SMA bolts. During earthquakes, the connections are cyclically deformed with repeated gap opening, which causes frame expansion (also known as beam growth) of the entire structural system (Fig. 3.8). This raises an incompatibility issue between the gravity frames (including the floor diaphragm) and the lateral resisting system. Slab integrity is significantly compromised when the frame tries to expand, and in addition, the presence of the slab may affect the self-centring capability of the connections themselves. Extra bending demands are also exerted to the columns which tend to be pushed apart due to the frame expansion effect, and as a result, more significant yielding could occur in the structural members.

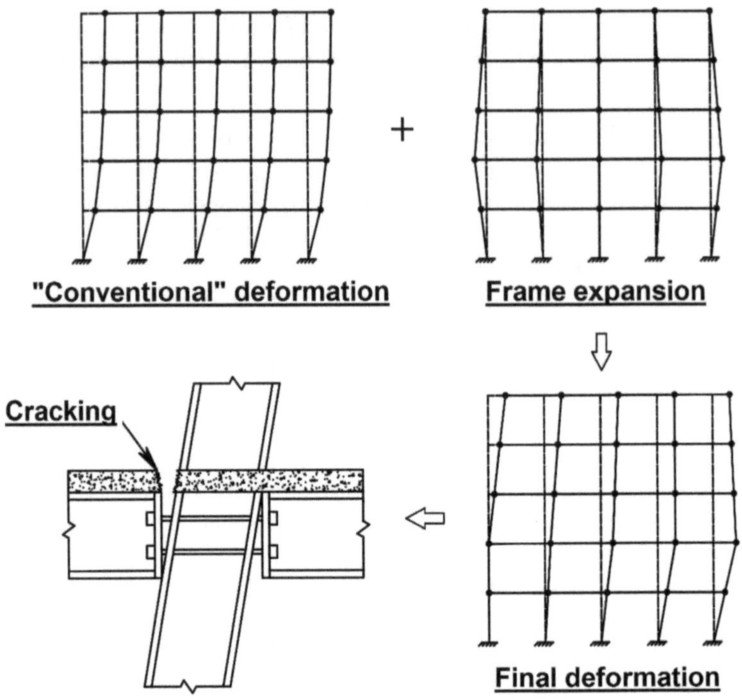

Fig. 3.8 Illustration of frame expansion

The cyclic behaviour of end-plate type composite connections equipped with SMA bolts was examined by Fang et al. (2017). Figure 3.9 illustrates the test specimens used for demonstrating the typical deformation and damage pattern of SMA-based composite connections. The specimen configuration and test setup are similar to those described in Sect. 3.2.2. The SMA bolts were more conveniently installed through the full depth of the column. Both 'bare steel' and composite connection specimens were included in the test programme for comparison purpose. The slab system consists of a continuous 1 mm-thick ribbed metal deck, with the rib direction (metal deck orientation) perpendicular to the beam axis. The metal deck acted as permanent formwork for wet concrete pouring. Both anti-crack meshes and longitudinal reinforcement were used in the concrete slab. A series of headed shear studs were welded to the beam top flange to provide the necessary composite action.

As shown in Fig. 3.10, the deformation mode of the SMA-bolted bare steel connection specimens is similar to that observed in the proof-of-concept connections discussed in Sect. 3.2.2, but a significantly improved ductility is exhibited, owing to the proper design of the bolt geometric detailing with an increased net threaded-to-shank bolt diameter ratio of 1.25.

On the other hand, the composite connections experience severe damage to the concrete slab during gap opening, as shown in Fig. 3.11. Minor cracks near the

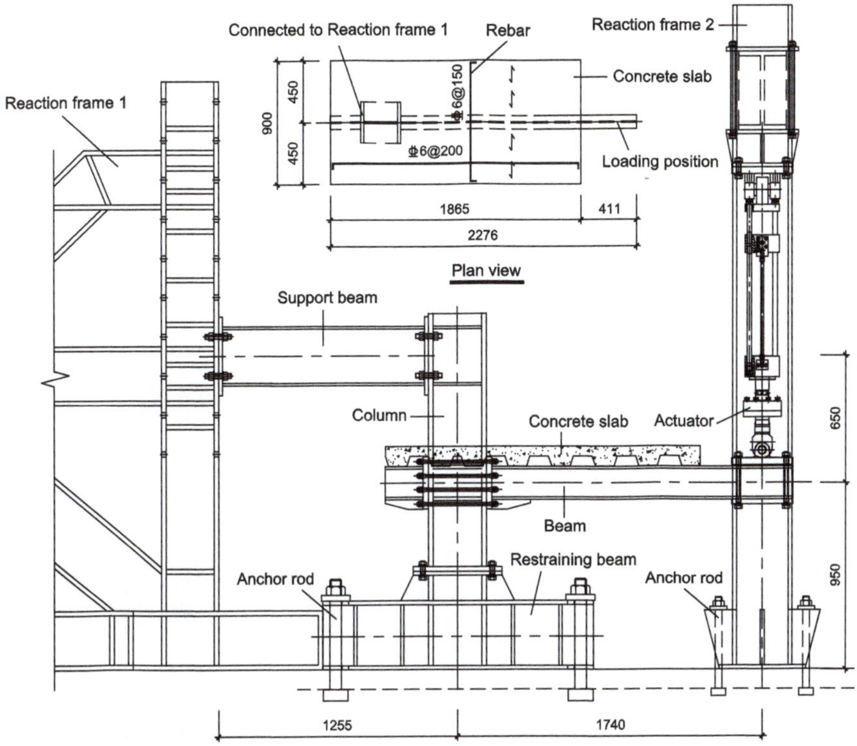

Fig. 3.9 Test for composite connections with SMA bolts (Fang et al. 2017)

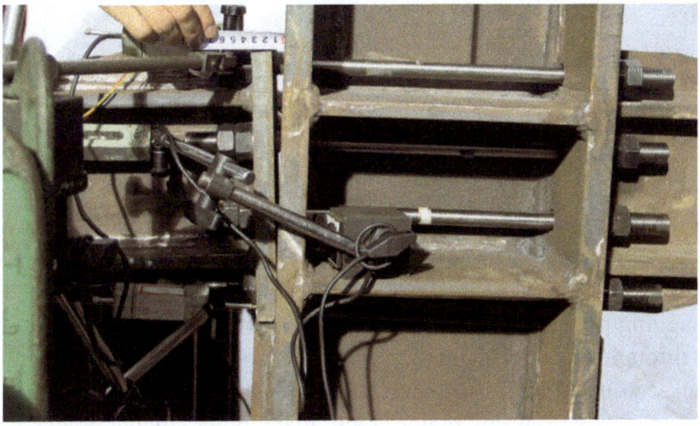

Fig. 3.10 Deformation mode of steel connections with SMA bolts

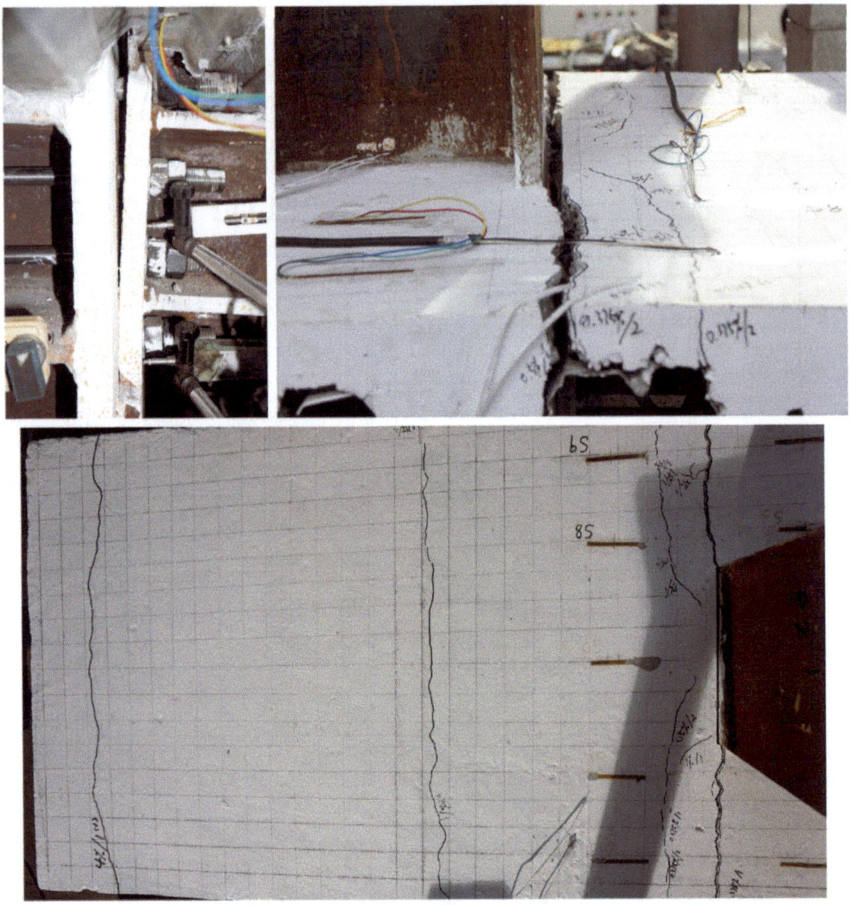

Fig. 3.11 Deformation and damage condition of SMA-bolted composite connections

beam-to-column interface occurred immediately after undergoing the 0.375% drift under the hogging/negative moment (i.e. concrete under tension). The cracks quickly propagated towards the two edges of the slab, although they were reclosed under the reversed sagging/positive moment. At 0.5–0.75% drifts, a major crack was developed which cut through the entire cross section of the slab. With increasing amplitudes, the major crack continued to be widened, and the damage to the concrete slab became concentrative. Apart from this major crack, many other smaller cracks were developed at other locations of the slab, but these cracks were not significantly spread. In addition to the significant concrete cracking, evident deformation of the metal deck, together with debonding between the concrete and the deck, was observed. Complete fracture of the longitudinal reinforcement occurred at 3–4% drifts.

The good news is that no inelastic deformation was induced in the main steel members, including the beam, column, end-plate, and most part of the stiffeners. In addition, no significant crushing of the concrete is observed when the slab is under compression. More importantly, no bolt fracture was observed throughout the entire testing process.

The measured moment–drift responses of the SMA-bolted bare steel and composite connection specimens are shown in Fig. 3.12. The presence of the slab does affect the hysteretic behaviour of the connections. Under sagging moment, i.e. slab in compression, the presence of the slab increases the moment resistance of the connection. Under hogging moment, i.e. slab in tension, the initial stiffness of the composite connection is evidently increased prior to first cracking of the concrete. As the reinforcement keeps contributing to the moment resistance when the major crack first forms, the peak moment is higher than that of the bare steel connections prior to 3% drift. The successive fractures of the longitudinal reinforcement are responsible for the observed sudden drop of the moment resistance, and once the reinforcement completely fractures, the peak moments of the two types of connections become similar.

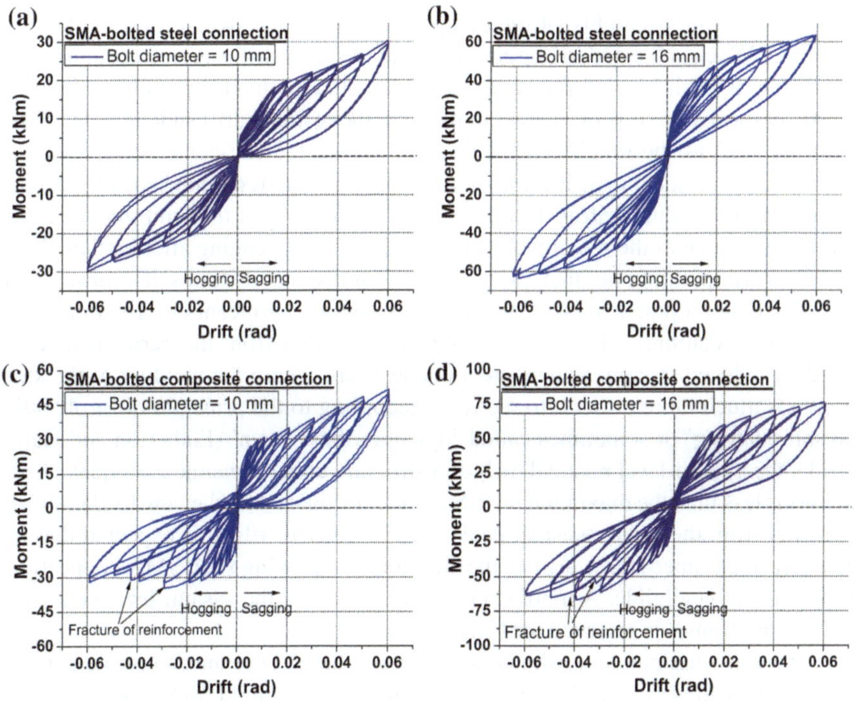

Fig. 3.12 Moment–drift responses of SMA-bolted steel and composite connection specimens: **a** steel connection with 10 mm bolt diameter, **b** steel connection with 16 mm bolt diameter, **c** composite connection with 10 mm bolt diameter, **d** composite connection with 16 mm bolt diameter

Fig. 3.13 Residual drift of connection specimens

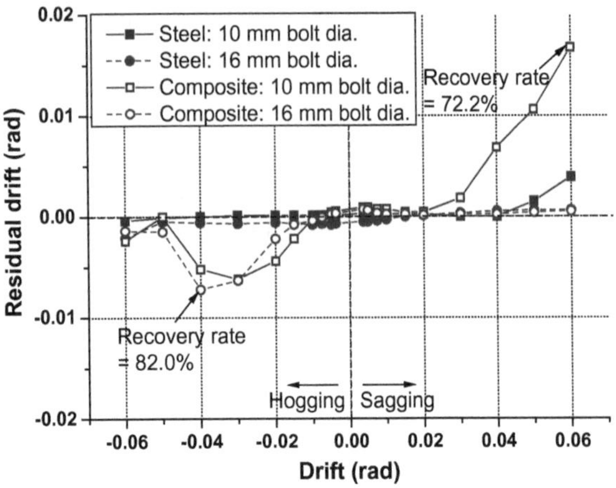

The self-centring capability of the specimens can be evaluated in more detail by monitoring the residual drift, as shown in Fig. 3.13. The bare steel connections generally show very small residual drifts, with a recovery rate of more than 90%. The self-centring capability of the composite connections depends on the loading direction. When subjected to hogging moment, residual drift starts to accumulate at 2% drift as a result of reinforcement yielding together with other possible sources of plastic deformation. It is of interest to find that the residual drift suddenly decreases after 4% drift as the reinforcement fractures. Once the self-centring capability is regained, the residual drift is kept small throughout the remaining loading cycles. Although the residual drift is decreased after the fracture of the reinforcement, this failure mode makes the repair work of the slab very challenging after earthquakes.

It is worth pointing out that the occurrence of residual drift in self-centring connections is not a practical issue as the corresponding permeant deformation could be easily repositioned. This is fundamentally different from the permanent residual drift induced in conventional steel connections, where the yielding of the steel beams is much more difficult to recover. According to the principle of 'probabilistic self-centring' or sometimes called 'dynamic shakedown' (Eatherton and Hajjar 2011), the system-level residual drift of a structure under dynamic earthquake excitations is less than the maximum possible static residual drift when the load is slowly removed from the same transient peak drift. In other words, it is expected that the actual residual inter-storey drift of the structures employing such SMA connections is practically negligible. A more detailed discussion on the system-level behaviour of building frames is reserved for Chap. 6.

For energy dissipation, as shown in Fig. 3.14, the EVD of the composite connections could achieve 12%, which is higher than the upper limit EVD value of 10% for the bare steel connections. This implies that the inelastic deformation and damage behaviour of the concrete slab contribute to the energy dissipation of the connection to a certain extent.

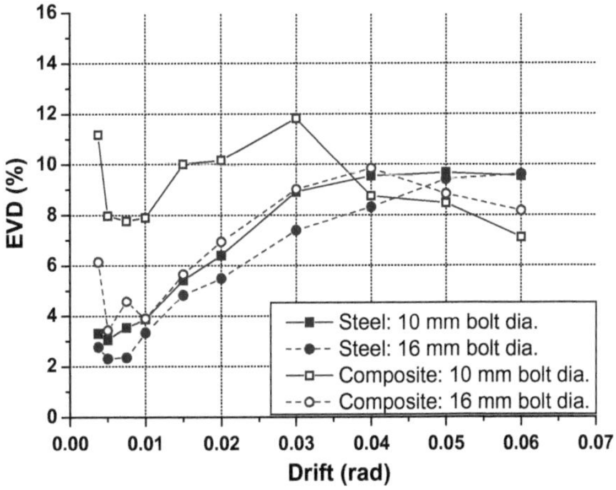

Fig. 3.14 Energy dissipation of connection specimens

The experimental investigation reveals that SMA-bolted composite connections with 'conventional' slab detailing can have satisfactory self-centring and ductility performance, but have to suffer evident slab cracking due to gap opening. The extensive damage to the floor diaphragm not only makes the post-earthquake repair work difficult but also affects the transformation of the lateral inertia force. In addition, the compromised integrity of the floor diaphragm increases the risks of collapse and fire spread during and after earthquakes. Many researchers have already been aware of this issue and attempted to mitigate the frame expansion effect by introducing new types of connections or flooring systems. The simplest way to avoid extensive slab damage is to physically separate the slab system, including the reinforcement and the steel deck, from the column. It is also possible to design composite action only at one rigid bay and to allow sliding in the other bays (Chou and Chen 2011a, b). However, these approaches inevitably complicate the construction detailing, and lose the benefit of the composite action at the connection zone. Specially designed collector beams may be used in the floor diaphragm to transfer the inertia forces to the self-centring frame and to accommodate frame expansion (Garlock and Li 2008), but this method is quite demanding in terms of design and construction, and in addition, tends to cause secondary damage (Khoo et al. 2013).

If complete elimination of the gap opening for the mainframe connections is required, prefabricated self-centring modular panels, as illustrated in Fig. 3.15, may be a good choice. These panels are embedded within frame bays and act like shear wall panels (Wang et al. 2017a, 2018). In this case, gap opening only occurs in the joints of the self-centring modular panels, whereas the mainframe connections can be designed as normal ones. In fact, the idea of self-centring modular panels is more suited to a less common structural system called beam-through frames, where the beams are continuous and have relatively large sizes, whereas the slender columns are constructed floor by floor.

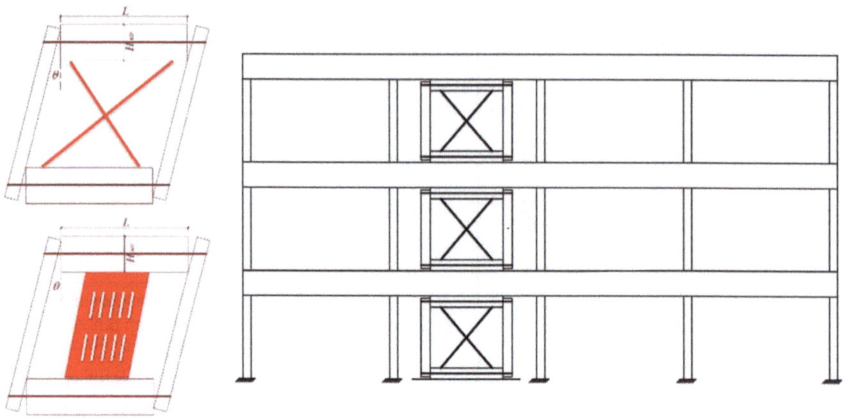

Fig. 3.15 Prefabricated self-centring modular panels

Another effective strategy to minimise connection–slab interaction is to shift the joint rotation centre to the top beam flange, as illustrated in Fig. 3.16. Employing this principle, a sliding-hinge joint connection was proposed by MacRae (2010), although self-centring capability was not the main objective when this type of connection was initially proposed. An amended version, which utilises friction springs (conventional ring springs) to offer self-centring capability, was later developed and experimentally verified (Khoo et al 2012, 2013). For such connections, energy dissipation is mainly offered by friction between the inner and outer HSS rings, and extra friction pads can be used to provide more energy dissipation. Dowden et al. (2016) and Dowden and Bruneau (2016) developed a new class of PT connections, namely, NewZ-BREAKSS connections. Compared with the 'conventional' PT connections which deform via gap opening, the new connections are free from frame expansion thanks to the special detailing of the beam ends. More recently, a shop-fabricated self-centring coupled beam has been proposed to eliminate frame expansion (Maurya and Eatherton 2016; Huang et al. 2017). The novel idea was successfully verified through reduced-scale tests. The coupled beam system shows a satisfactory flag-shaped hysteretic response, but a potential drawback is that it takes up architectural spaces beneath the main girder along the entire beam length.

Most of the existing studies, which try to deal with frame expansion, rely on the PT solution. Building on these design concepts, more efficient SMA-based self-centring connections free from frame expansion may be developed. The detailed solution is presented later in Sect. 3.4.

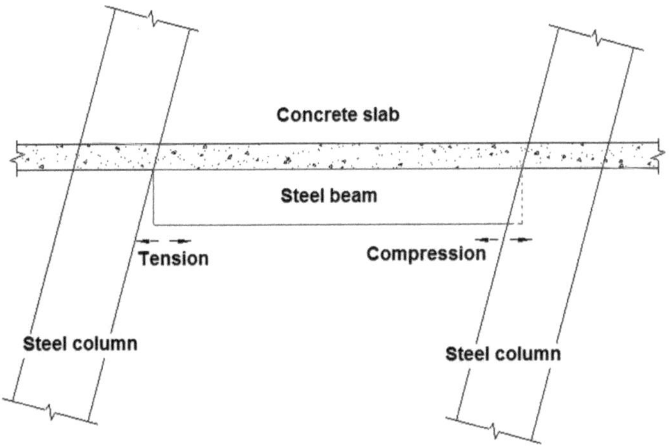

Fig. 3.16 Illustration of connection design for avoiding frame expansion

3.3 Hybrid Connections

This section proceeds with the discussion of more practical SMA connection solutions. One problem for the aforementioned proof-of-concept end-plate connections is the inadequacy of the shear resistance. An additional shear tab with slotted bolt holes may be used, but particular attention should be paid to the preload applied to the shear bolts. An appropriate level of supplemental friction resulting from the preloaded shear bolts (HS bolts) can be beneficial for energy dissipation, but overly large friction hinders the recoverability of the connection. Supplementary energy dissipation can be alternatively provided by low yield strength metals or other special SMA components. To address the shear resistance issue and to further improve the energy dissipation property, some novel solutions for SMA connections are described. It is worth pointing out that the development of these hybrid connections is not intended to address the frame expansion issue, which is elaborated later in Sect. 3.4.

3.3.1 Hybrid Connections with SMA Washers and Bolts

With the aim of increasing the shear resistance of SMA-based end-plate connections while ensuring ease of construction, a new type of hybrid self-centring connection incorporating combined SMA bolts, HS bolts, and SMA Belleville washers (hereafter abbreviated as SMA-WB connections) has been proposed by Fang et al. (2018a). As schematically illustrated in Fig. 3.17, the SMA-WB connection includes a series of external SMA bolts and internal HS bolts. The external SMA bolts are supposed to provide moment resistance, energy dissipation and self-centring capability. The shear resistance of the connection is mainly provided by the internal HS bolts, which

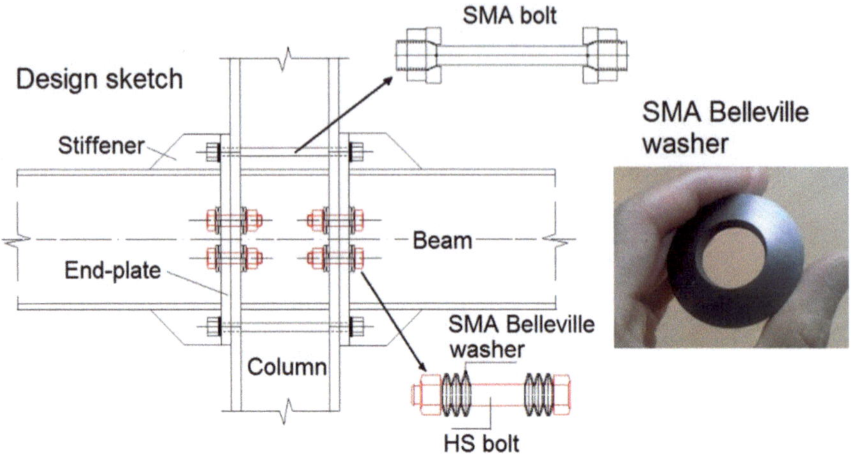

Fig. 3.17 Illustration of SMA-WB connections

are used in conjunction with the SMA Belleville washer sets. The SMA Belleville washers offer a certain level of rotational flexibility for the HS bolts and further encourage self-centring and energy dissipation capabilities. With this arrangement, the deformation demand of the internal HS bolt rows is mainly provided by the SMA Belleville washers, and hence, the HS bolts are unlikely to fail. The installation procedure of the SMA-WB connections is similar to that used in conventional end-plate connections.

The cyclic behaviour of the SMA-WB connections is mainly determined by the behaviour and arrangement of the SMA bolts and washers. The preload applied to these elements is also influential. According to the 'component-based' design framework stipulated in Eurocode 3 (CEN 2005), the design yield moment resistance of a SMA-WB connection is contributed by the moment of forces provided by the SMA bolt and washer rows in relation to the rotation centre. Assuming a V-shaped gap opening mode, and allowing a fully 'yielding' behaviour of the SMA elements, the design yield moment resistance $M_{j,Rd}$ can be calculated by

$$M_{j,Rd} = \sum_{i} F_{b,y}h_i + \sum_{j} F_{w,y}h_j \tag{3.6}$$

where $F_{b,y}$ is the 'yield' resistance of each SMA bolt, $F_{w,y}$ is the 'yield' resistance of each SMA washer set and h_i and h_j are the distances from the SMA bolts and SMA washers to the rotation centre of the connection, respectively.

In Eq. (3.6), the yield resistance of the SMA bolts is easily calculable by multiplying the design forward transformation stress σ_{Ms} by the shank area A_b. For the SMA washers, the calculation method for yield resistance has been discussed in Chap. 2, and more details can be found in the handbook (Fromm and Kleiner 2003).

Employing the principle of the component-based design approach, the initial stiffness $K_{c,ini}$ of the SMA-WB connection can be calculated by

$$K_{c,ini} = \sum_i \frac{F_{b,y} E A_b}{\left(F_{b,y} - F_{b,pre}\right) L_p} h_i^2 + \sum_j \frac{F_{w,y} k_w}{\left(F_{w,y} - F_{w,pre}\right)} h_j^2 \quad (3.7)$$

where L_p is the working length of the SMA bolts and k_w is the stiffness of the considered SMA washer group. The stiffness of each individual SMA washer can be calculated by the derivation of Eq. (2.5), and when the washers are stacked with various combinations, the value of k_w should be adjusted accordingly. Equation (3.7) enables the consideration of preload, where $F_{b,pre}$ and $F_{w,pre}$ are the preloads applied to the SMA bolts and SMA washers, respectively. The equation reflects the fact that the inclusion of the preload tends to increase the design stiffness of the connection, and when the preload equals to the yield resistance, the design stiffness can be considered as infinite prior to decompression.

Figures 3.18 and 3.19 show the test setup and deformation mode of a representative small-scale test specimen designed to verify the concept of hybrid SMA-WB connections (Fang et al. 2018a). The basic information of the test specimens is provided in Table 3.3, where each specimen is designated with a specimen code according to the bolt size and SMA washer arrangement. The SMA washers used for the tests

Fig. 3.18 Test setup for examining SMA-WB connection specimens

Fig. 3.19 Deformation of typical SMA-WB connection specimen

Table 3.3 Details of SMA-WB connection specimens (Fang et al. 2018a)

Specimen designation	Shank diameter D_1	Threaded net diameter D_3	No. of SMA washer per HS bolt	Washer arrangement
D8L290W8	8.0	10.0	4	×4 in series
D8L240W16	8.0	10.0	8	×2 in parallel × 4 in series
D12L290W8	12.0	15.5	4	×4 in series
D8L240	8.0	10.0	0	N/A

are identical to those presented in Sect. 2.5. 'W8' and 'W16', which indicate washer arrangement, mean the total number of washers employed for each specimen. For all the specimens, a prestrain of 0.5% was applied to the external SMA bolts, and the HS bolts were also reasonably tightened with a 1 mm total precompression deformation.

As can be seen in Fig. 3.19, the SMA bolts elongated and concurrently the SMA washers were compressed when the connections deform. Again, the end-plate behaved in a 'thick plate' fashion, enabling a V-shape gap opening mode with insignificant bending of the end-plate itself. The other structural members stayed elastic. The measured moment–drift curves of the specimens given in Fig. 3.20 show typical flag-shaped hysteretic curves with negligible residual rotation. The shape of the hysteretic curves and the maximum moment (M_{max}) are influenced by the arrangements of the SMA bolts and washers, as well as the size of the SMA bolts. We can see that all the specimens satisfy the DCH and SMF criteria in terms of ductility.

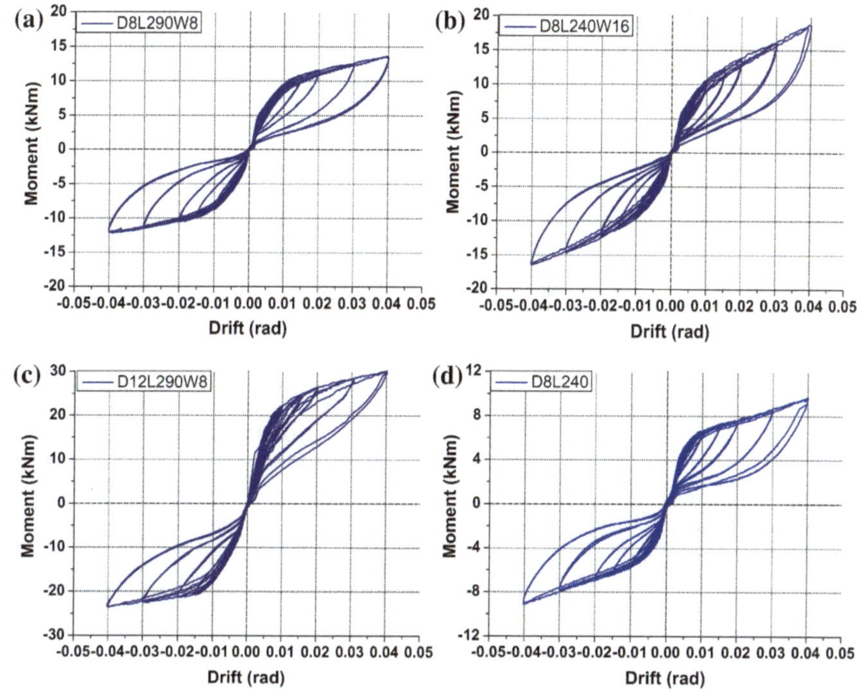

Fig. 3.20 Moment–drift responses of SMA-WB connection specimens: **a** D8L290W8, **b** D8L240W16, **c** D12L290W8, **d** D8L240

The good recoverability of the SMA bolts and washers allows the connection specimens to exhibit excellent self-centring capability. As no residual deformation is observed, the self-centring tendency can be further evaluated by checking the moment when the reverse transformation completes. This moment is called restoring moment M_{rst}, as illustrated in Fig. 3.21. A positive value means a fully self-centred behaviour with no residual rotation, whereas a negative value indicates that residual rotation is developed. In practical design, M_{rst} should, on one hand, be reasonably low to ensure sufficient energy dissipation, but on the other hand, be suitably high to allow an adequate self-centring tendency. As shown in Fig. 3.21, the M_{rst} of the specimens decrease with increasing amplitudes, but the values are always above zero. At 4% drift, the M_{rst}/M_{max} ratio of the specimens ranges from approximately 5 to 12%. For the energy dissipation characteristic, as shown in Fig. 3.22, the maximum EVD could achieve 13% for these hybrid connection specimens.

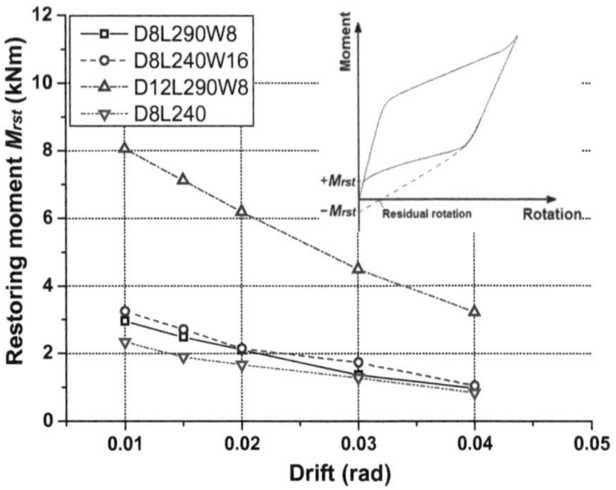

Fig. 3.21 Self-centring capability of SMA-WB connection specimens

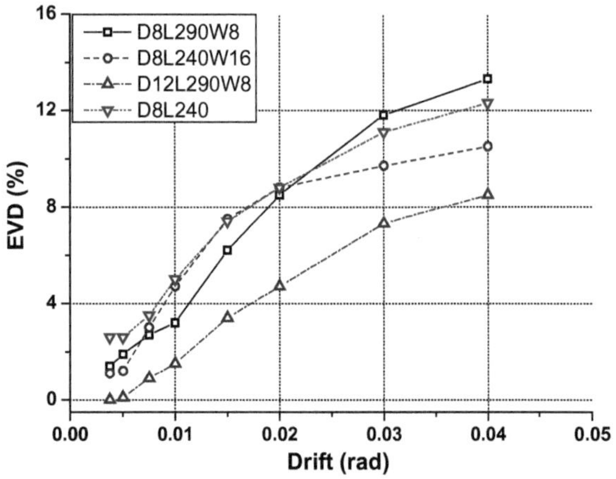

Fig. 3.22 Energy dissipation of SMA-WB connection specimens

3.3.2 Hybrid Connections with SMA Bolts and Steel Angles

Although the potential of SMA elements for enhancing structural seismic resilience has appeared, SMA elements are still quite costly, and hence, an overconsumption of the SMA material may be uneconomical and impractical. Mild steel angles, which are much cheaper, could be alternative supplemental components helping increase the stiffness/strength of the connections, whilst offering extra energy dissipation during earthquakes through material yielding. Moreover, the steel angles could provide additional shear resistance for the connection. Although the use of the steel angles could inevitably affect the self-centring performance of individual connections (Wang et al. 2015b), an appropriate combination of the steel angles and SMA elements can increase the design flexibility, in a sense that a desirable trade-off between performance and cost can be achieved. This concept conforms to the performance-based design framework that allows variation of seismic performance targets. It is also worth mentioning that the steel angles could be conveniently replaced after earthquakes where and when necessary.

Figure 3.23 shows the typical design of hybrid connections with SMA bolts and steel angles, together with a non-hybrid connection equipped with SMA bolts only (Wang et al. 2017b). The angles can be placed either outside or inside the beam flanges. An additional shear tab with slotted bolt holes should be used to provide sufficient shear resistance and to allow certain rotational flexibility. The connection behaviour is closely related to the combination of the SMA bolts and steel angles. For instance, using large size angles or placing the angles outside the beam flanges can help improve the energy dissipation but tends to decrease the self-centring capability. In contrast, by considering a smaller fraction of the moment resistance contributed by the steel angles, e.g. using smaller size angles or placing them inside the beam flanges, the self-centring capability offered by the SMA bolts would be less significantly affected whereas the increase in energy dissipation is also less remarkable.

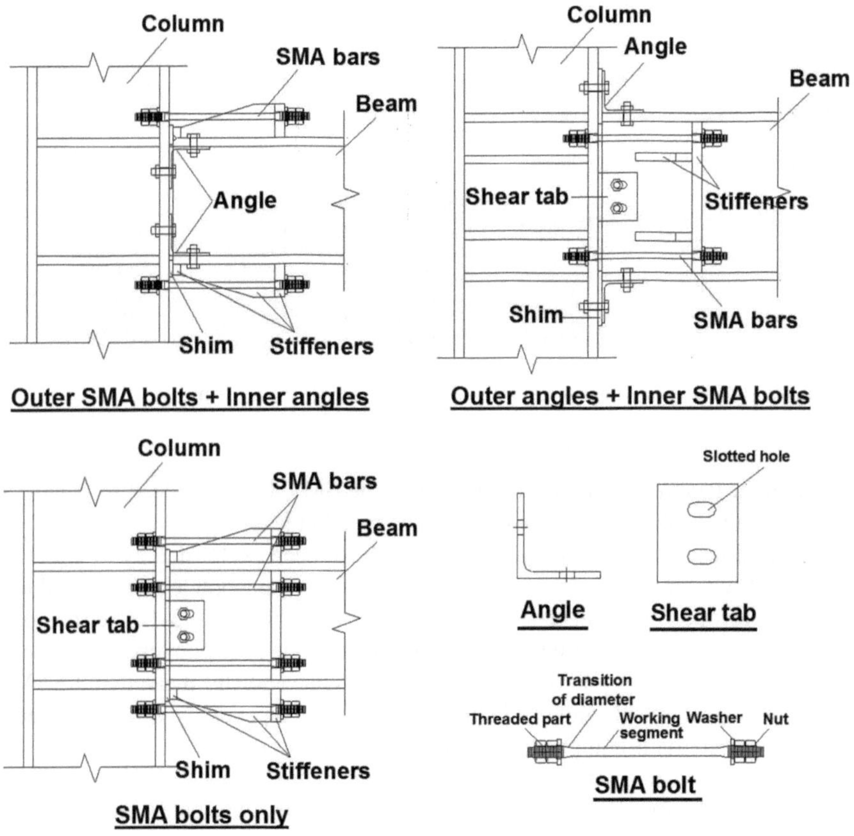

Fig. 3.23 Illustration of hybrid connections with SMA bolts and steel angles

Table 3.4 Details of hybrid connections with SMA bolts and steel angles (Wang et al. 2017b)

Specimen code	Total bolt length L (mm)	Bolt shank length L_p (mm)	Bolt prestrain (%)	Angle thickness t (mm)	Type
Out-P0.60-L6	425	265	0.60	6	Out
Out-P1.25-L6	425	265	1.25	6	Out
Out-P2.50-L6	425	265	2.50	6	Out
Out-P1.25-L6-S	345	185	1.25	6	Out
Out-P1.25-L8	425	265	1.25	8	Out
All-P1.25	425	265	1.25	–	All
Out-P1.25-L6-H	510	350	1.25	6	Out
In-P1.25-L6-H	395	235	1.25	6	In

Fig. 3.24 Hybrid connections with SMA bolts located outside the beam flanges

In addition, increasing the prestrain level of the SMA bolts can lead to enhanced self-centring capability, but may make the connection less ductile.

Physical tests on such hybrid connections have been performed by Wang et al. (2017b). The main test parameters were SMA bolt prestrain, bolt length, angle thickness, and layout of the bolts and angles, as summarised in Table 3.4. The specimen code includes the connection type (Out, In, and All), the prestrain level (in percentage) of the SMA bolts, and the angle thickness. 'Out' and 'In' mean that the SMA bolts are located outside and inside the beam flanges, respectively, and 'All'' means that the connection is equipped with SMA bolts only.

Figures 3.24 and 3.25 show the typical deformation modes of the connection specimens. Anticipated deformation modes and yielding sequences are displayed. The main inelastic deformation of the hybrid connections is provided by the SMA bolts and steel angles, and the steel column and beam can remain elastic.

Fig. 3.25 Hybrid connections with SMA bolts located inside the beam flanges

The moment–drift hysteretic curves of the specimens are given in Fig. 3.26. The residual deformation varies with different SMA bolt and steel angle layouts. Pronounced hardening is exhibited after the yield moment, which is related to the unique behaviour of the SMA bolts and the geometric nonlinearity of the steel angles developed at large deformations, i.e. catenary action. Compared with the connection with SMA bolts only, the self-centring capability of the hybrid connections is decreased to a certain extent, which is caused by the yielding of the steel angles.

It can be more clearly observed in Fig. 3.27 that the residual drift is negligible for all the specimens at 1.5% drift. Specimen Out-P1.25-L8 shows a relatively large residual drift because of the increased angle thickness. With increasing amplitudes, the connections with combined SMA bolts and steel angles start to exhibit evident residual drifts. A maximum residual drift of 2.29%, corresponding to a recovery rate of 54.3%, is observed in specimen Out-P1.25-L6-S at 5% drift. The connection specimen with no steel angles shows negligible residual drift throughout the entire loading process.

The change of the residual drift shown in Fig. 3.27 also reveals the effects of several parameters on the connection self-centring behaviour. First, an increase in the size of the steel angle decreases the self-centring capability of the connection. The same tendency is observed when the SMA bolts are placed inside the beam flanges, where a less effective recovery moment is offered by the internal SMA bolt group. Moreover, an increase in bolt prestrain could, on one hand, facilitate the self-centring capability due to enhanced restoring action, but on the other hand, may be detrimental to the connection self-centring performance at late stages due to more evident TIF degradation. A low prestrain leads to decreased initial stiffness and more rapid loss of the prestrain due to TIF degradation. These explain the 'optimal' self-centring capability observed in the specimen with a moderate prestrain level of 1.25%. For the same reason, using shorter SMA bolts would increase the strain demand, which is responsible for the larger residual drift beyond 3% drift.

For energy dissipation, as shown in Fig. 3.28, the EVD increases with increasing amplitudes and achieves a peak at 3% drift. The EVD is partially contributed by the

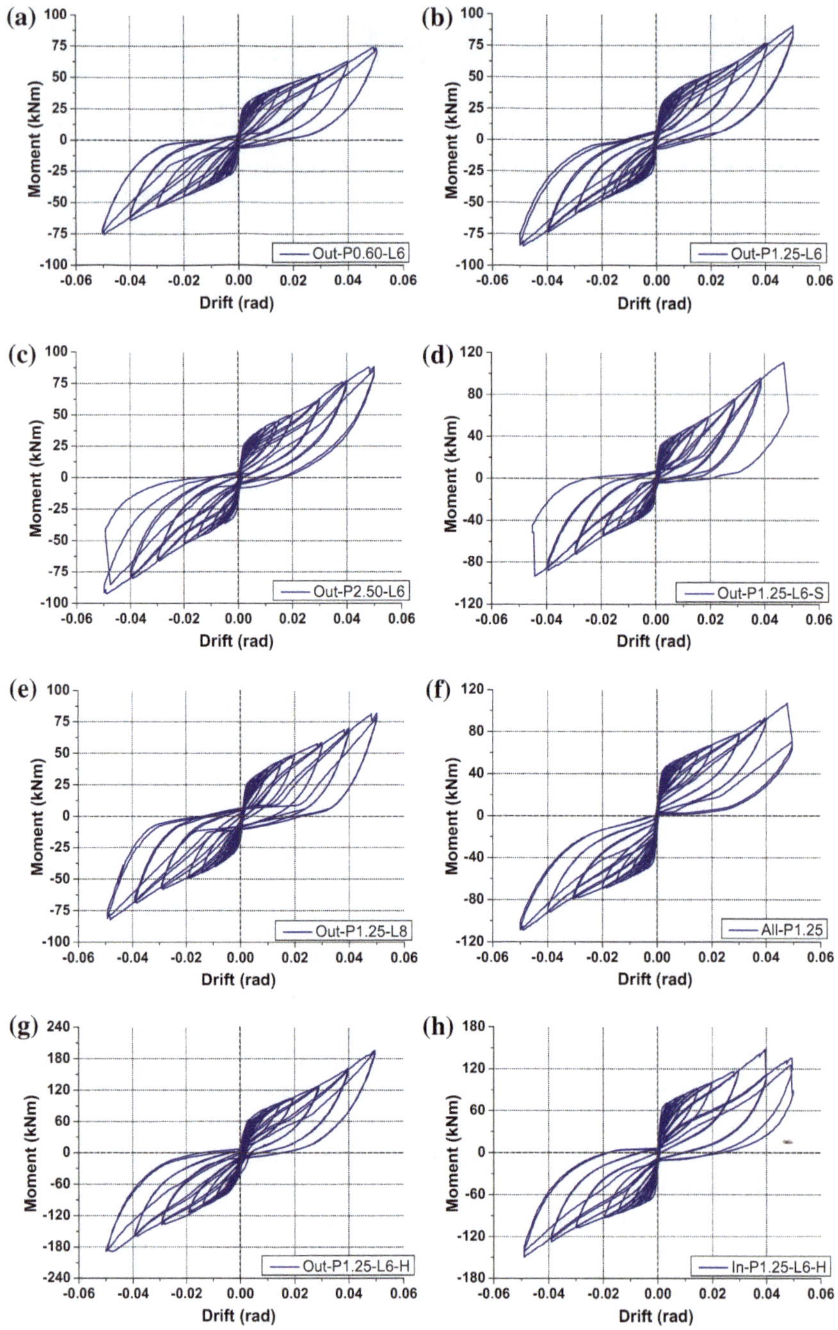

Fig. 3.26 Moment–drift responses of hybrid connections with SMA bolts and steel angles: **a** Out-P0.60-L6, **b** Out-P1.25-L6, **c** Out-P2.50-L6, **d** Out-P1.25-L6-S, **e** Out-P1.25-L8, **f** All-P1.25, **g** Out-P1.25-L6-H, **h** In-P1.25-L6-H

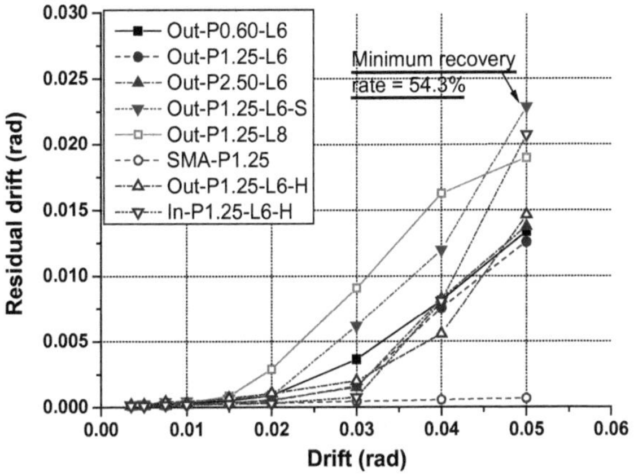

Fig. 3.27 Residual drifts of hybrid connections with SMA bolts and steel angles

yielding of the steel angles, in addition to the energy dissipation provided by the SMA bolts. A maximum EVD of up to 15% is recorded. A decrease in EVD beyond 3% drift is caused by the degradation effect of the SMA bolts. All the specimens generally follow a similar trend in terms of the EVD, although the values vary. Clearly, thicker angles increase the EVD, confirming the positive role played by the steel angles in providing extra energy dissipation.

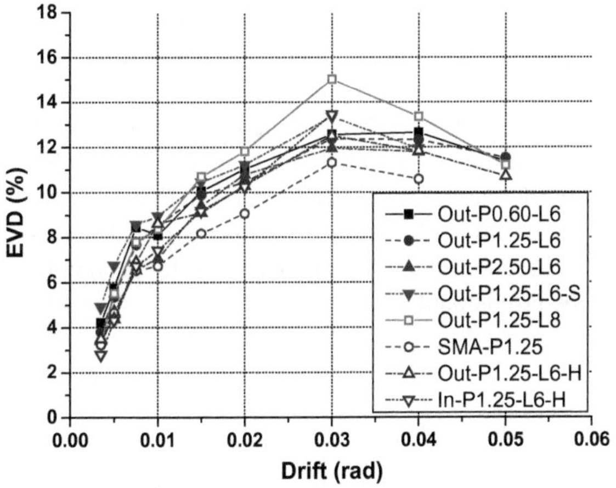

Fig. 3.28 Energy dissipation of hybrid connections with SMA bolts and steel angles

3.3.3 Design of Hybrid Connections

While different combinations of SMA element can be employed in various hybrid connections, the key design principles for these connections are generally consistent. The component-based design framework is particularly suited to hybrid connections.

Taking the connections with SMA bolts and steel angles, for example the study conducted by Wang et al. (2017b) clearly illustrates the dependence of the connection behaviour on the SMA bolt prestrain, SMA bolt length, steel angle dimension and layout of the basic elements. These influential factors should be taken into account in the connection design. The following procedures are an illustrative example of how the hybrid connections with SMA bolts and steel angles can be designed, and how the uncertainty related to the SMA material property could be considered in the design.

- *Step 1: Determination of performance target.* The performance target is essentially the required connection strength/stiffness and deformation demand (i.e. target inter-storey drift). The former is often determined based on the available design information of the steel beams and columns, and the latter can be determined based on the ductility requirements of the design structures. Typical ductility requirements for steel building frames are 4% inter-storey drift for SMFs and 2% inter-storey drift for IMFs (AISC 2010), as mentioned before. While the most 'strict' objective is to design the connections with a negligible residual rotation when unloaded from the target inter-storey drift, this criterion can be relaxed with the permission of certain residual deformation (Eatherton and Hajjar 2011).
- *Step 2: Determination of SMA properties.* After receiving the appropriately annealed SMA bolts from the material supplier, the next action we need to take is training. The training process could effectively stabilise the hysteretic behaviour of the SMA bolts (McCormick and DesRoches 2006), and more importantly, helps determine the nominal key properties of the bolts during expected earthquakes. While there is currently no particular loading protocols available for determining the nominal SMA material properties for design use, the following equation, which is based on the SAC standard loading protocol for connections (SAC 1997), can be provisionally used:

$$f(\varepsilon) = \varepsilon_{pre} + \frac{h_i \cdot f(\theta)}{L_p} \tag{3.8}$$

where ε_{pre} is the bolt prestrain to be applied, which is unknown at this stage but can be preliminarily taken as the transformation start strain ε_{Ms} for determining the loading protocol; h_i = rotational lever arm, i.e. the distance between the SMA bolt row (or steel angle) and the rotation centre of the connection; L_p is the parallel/working length of the dog-bone tensile coupon, $f(\theta)$ is the SAC standard loading protocol in radians (SAC 1997), and $f(\varepsilon)$ is the proposed training protocol as a function of strain. The required properties such as residual strain ε_{res} and different characteristic stresses could be extracted from the training cycle at the target inter-storey drift. For the same batch of products, tests on at least three SMA tensile coupon specimens with the recommended training protocol are adequate, where the average values can be used for subsequent design.

- *Step 3: Determination of bolt prestrain.* The stiffness, self-centring capability, and possible degradation effect of the connections are all influenced by the prestrain level of the SMA bolts. The following aspects should be kept in mind when we determine the bolt prestrain: (1) the connection needs to satisfy the stiffness/strength requirements, which rely on the prestrain of the SMA bolts; (2) the prestrain should evidently exceed the possible accumulated residual strain ε_{res} at the target inter-storey drift; (3) the maximum strain, with the prestrain inclusive, at the target inter-storey drift should be controlled within a certain threshold to avoid excessive degradation and even fracture of the bolt. In practice, a simple rule of $\varepsilon_{pre} = \varepsilon_{Ms} > \varepsilon_{res}$ may be considered, aided by the 'turn-of-the-nut' procedure when monitoring the prestrain.
- *Step 4: Determination of bolt dimensions.* If the locations of the SMA bolt rows are fixed, the necessary bolt working length L_p is governed by the permitted peak strain ε_{peak} at the target drift of the connection. In addition, the necessary diameter of the SMA bolts should be determined according to the required connection moment resistance. In particular, the connection 'yield' resistance should be reasonably less than the yield moment of the adjacent steel beams. Note that both the SMA bolts and steel angles contribute to the connection moment resistance, i.e. $M = M_{SMA} + M_{ang}$, and therefore the final determination of the SMA bolt and steel angle design may involve an iteration process. An appropriate overstrength factor needs to be considered when calculating the maximum connection moment resistance, such that the hardening effects of the SMA bolts and steel angles are taken into account and thus unexpected early beam yielding is avoided.
- *Step 5: Determination of angle dimensions.* The determination of angle dimensions and layouts depends on the design intention. If no static residual rotation is allowed, the main principle of the design is to make sure that the restoring moment M_{rst} provided by the SMA bolt group is no less than the reverse moment M_{rev} caused by the steel angles. The two moments are depicted in Fig. 3.29. The restoring moment $M_{rst} = \Sigma P_{rst} h_i$ is calculated from the restoring force of each SMA bolt $P_{rst} = A_b \times \sigma_{rst} = A_b \times (\sigma_{Af} + (e_{pre} - e_{Af}) k_{r\text{-}SMA})$, where A_b is the cross section area of the bolt shank, $k_{r\text{-}SMA}$ (unit = MPa) is the reverse transformation stiffness of SMA material according to the material test (Fig. 3.30) and ε_{Af} is the strain at σ_{Af}. Similarly, the reverse moment M_{rev} can be calculated via the reverse force (P_{rev}) of the deformed steel angles (Fig. 3.30). The tensile behaviour of steel angles has been widely studied (e.g. Shen and Astaneh-Asl 2000), where various monotonic and hysteresis models have been developed. As shown in Fig. 3.30, the typical load–deformation (P-δ) response of a steel angle can be idealised by a tri-linear curve including three stages: elastic, transition, and post-yielding stages.

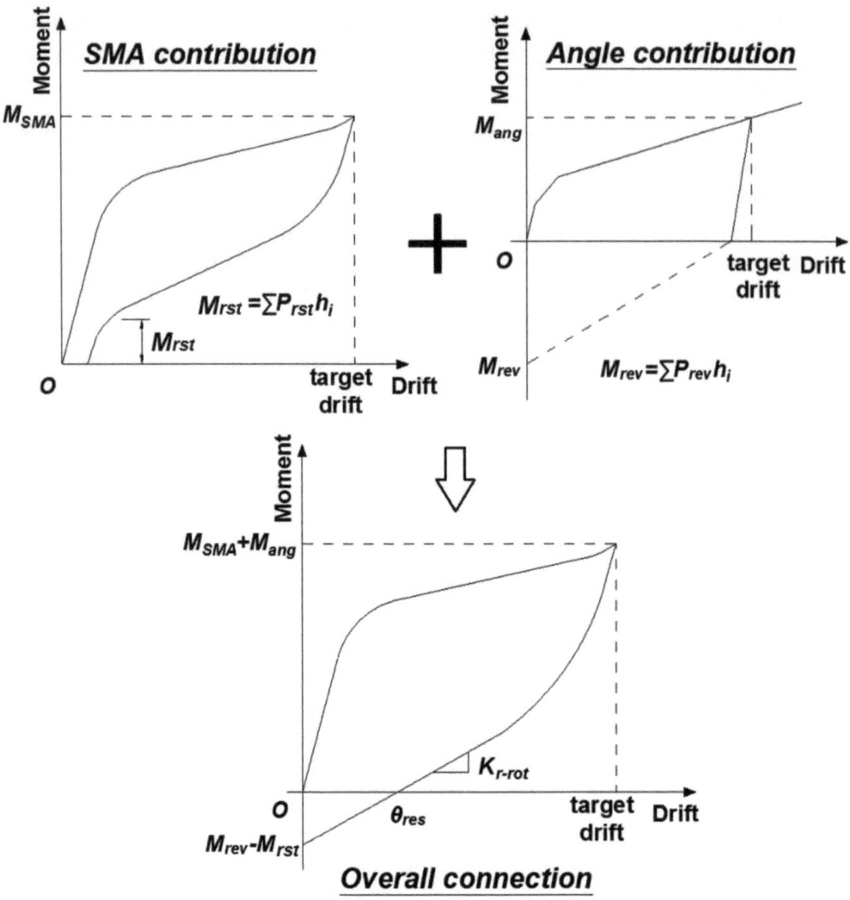

Fig. 3.29 Design strategy for hybrid connections: overall moment–rotation behaviour

The stiffness of the three stages, K_o, K_t and K_g, are expressed by (Shen and Astaneh-Asl 2000)

$$K_o = \frac{12EI_a}{g_1^3}\left[1 - \frac{3g_2}{4(g_1 + g_2)}\right] \tag{3.9}$$

$$K_t = \frac{3EI_a}{g_1^3}\left[1 - \frac{3g_2}{8g_1 + 6g_2}\right] \tag{3.10}$$

$$K_g = \frac{P_u - P_s}{\delta_u - \delta_s} \tag{3.11}$$

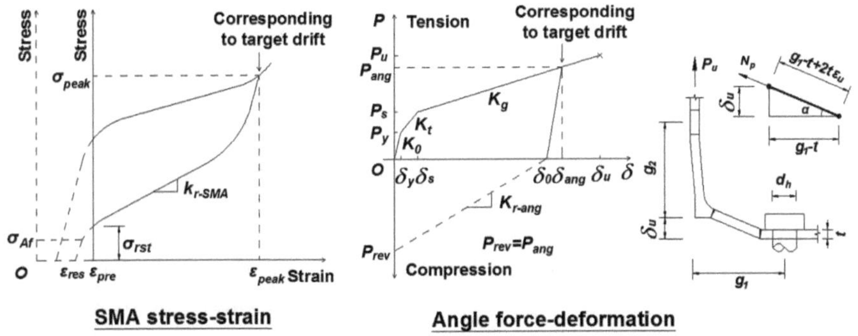

Fig. 3.30 Design strategy for hybrid connections: behaviour of SMA material and individual steel angle

where E = Young's modulus of steel angle, I_a = second moment of area of steel angle section, g_1 and g_2 are defined in Fig. 3.30. The corresponding loads, P_y, P_s and P_u, are given by

$$P_y = \frac{4g_1 + g_2}{g_1(2g_1 + g_2)} M_{ya} \tag{3.12}$$

$$P_s = 2M_{pa} \Big/ \left(g_1 - t - \frac{d_h}{2}\right) \tag{3.13}$$

$$P_u = N_{ya} \sin\alpha \tag{3.14}$$

and the ultimate deformation at failure is

$$\delta_u = 2(g_1 - t)\varepsilon_u \sqrt{\frac{t}{(g_1 - t)\varepsilon_u}} \tag{3.15}$$

In the above equations, M_{ya} and M_{pa} are the yield and plastic moment capacity of steel angle section, respectively, N_{ya} is the yield capacity of steel angle section, ε_u is the tensile strain of angle steel at ultimate stress, t is the angle thickness, d_h is the bolt hole diameter and α is shown in Fig. 3.30. The unloading and reverse paths of the steel angle are also predictable. We may assume that the reverse force P_{rev} is equal to the maximum force P_{ang} at the peak angle deformation according to the design target. Although P_{rev} is normally smaller than P_{ang}, taking $P_{rev} = P_{ang}$ could be on the conservative side in terms of the prediction of the self-centring capability. By satisfying $M_{rst} = M_{rev}$, P_{rev} is obtained employing the relationship of $M_{rev} = \Sigma P_{rev} h_i$, and with the obtained P_{rev}, the angles can be designed accordingly.

- *Step 6: Prediction of residual rotation.* This is an optional step and is only applicable to the design intention that permits static residual rotation. In this case, the above design rules are still applicable, and additionally, the residual rotation can be predicted by

$$\theta_{res} = \frac{M_{rev} - M_{rst}}{K_{r-rot}} = \frac{M_{rev} - M_{rst}}{\sum K_{r-SMA}h_i^2 + \sum K_{r-ang}h_i^2} \tag{3.16}$$

where $K_{r\text{-}rot}$, as illustrated in Fig. 3.29, is the reverse rotational stiffness of the connection and can be calculated based on the reverse stiffness of each component, i.e. each individual SMA bolt and steel angle.

The above design procedure can be explained in a more demonstrative manner. Specimen Out-P1.25-L6 (Wang et al. 2017b), with a moderate bolt prestrain level and a representative angle layout, is selected here as the calculation example. Note that some of the above design steps, e.g. determination of the SMA bolt and angle dimensions, are omitted in this example as the relevant information is already known. The final aim of this example is to predict the residual rotation of the connection.

The measured SMA material stress–strain relationships, as shown in Fig. 3.31, are used for the design. Assuming a 4% target drift of the connection, the maximum strain of the SMA bolts is predicted as $\varepsilon_{peak} = 0.04h_i/L_p + \varepsilon_{pre} = 0.04 \times 375/265 + 0.0125 = 0.07$, and the idealised stress–strain curve for obtaining the key material parameters are shown in Fig. 3.31 (dash line). The contribution of the SMA bolt groups to the moment resistance at the 4% target drift can be derived by $M_{SMA} = \sum \sigma_{peak} A_b h_i = 490 \times 153.9 \times 375 \times 2 = 56.5 \times 10^6$ Nmm = 56.5 kNm. In order to obtain the restoring moment, the restoring force of each SMA bolt is: $P_{rst} = \sigma_{rst} A_b = (\sigma_{Af} + (e_{pre} - e_{Af})k_{r-SMA})A_b = (65 + (0.0125\text{–}0.0089) \times 3829.5) \times 153.9 = 12125\text{N} = 12.1$ kN, where $k_{r\text{-}SMA}$, i.e. the stiffness of the reverse transformation plateau, is directly obtained from the idealised flag-shaped curve as shown in Fig. 3.31. The corresponding restoring moment contributed by the SMA bolts, based on the spring assembly concept, is $M_{rst} = \sum P_{rst} h_i = 2 \times 12.1 \times 0.375 = \mathbf{9.1\,kNm}$.

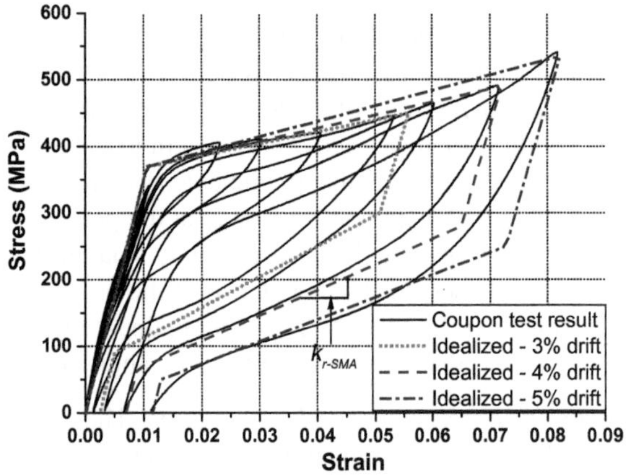

Fig. 3.31 Illustrative example: idealisation of key SMA material properties

The key properties of the steel angles can be derived from Eqs. (3.9) through (3.15), using the measured material properties, i.e. $f_y = 294$ MPa, $f_u = 436$ MPa, $\varepsilon_u = 25\%$. The necessary design parameters obtained from the equations include: $P_y = 2.94$ kN, $P_s = 7.02$ kN, $P_u = 33.47$ kN, $\delta_y = 0.38$ mm, $\delta_s = 2.06$ mm, $\delta_u = 18.00$ mm, $K_0 = 7.23$ kN/mm, $K_t = 2.43$ kN/mm and $K_g = 1.66$ kN/mm. Based on the spring assembly concept, the elongations of the upper and lower rows of the steel angles at 4% drift are 12.08 mm and 1.92 mm, respectively, and the associated angle force P_{ang} are 23.6 kN and 6.7 kN, respectively. Assuming $P_{rev} = P_{ang}$, as recommended by Shen and Astaneh-Asl (2000), the reverse moment can be obtained: $M_{rev} = \Sigma P_{rev} h_i = (23.6 \times 0.302 + 6.7 \times 0.048) \times 2 =$ **14.9 kNm**. As M_{rev} is larger than M_{rst}, a residual rotation is induced and can be predicted.

The difference between the reverse and restoring moments is 5.8 kNm, and this value can be used as a starting point to deduce the residual rotation. Utilising the idealised SMA and angle curves, the reverse stiffness of a single SMA bolt ($K_{r\text{-}SMA}$) is 2224 N/mm, and the values of the individual upper and lower angle ($K_{r\text{-}ang}$) are 2622 N/mm and 6326 N/mm, respectively. According to Eq. (3.16), the residual rotation can be calculated as follows:

$$\theta_{res} = \frac{M_{rev} - M_{inv}}{K_{r-rot}} = \frac{M_{rev} - M_{inv}}{\sum K_{r-SMA} h_i^2 + \sum K_{r-ang} h_i^2}$$
$$= \frac{(14.9 - 9.1) \times 10^6}{\left(2224 \times 375^2 + 2622 \times 302^2 + 6326 \times 48^2\right) \times 2} = \underline{0.51\%} \tag{3.17}$$

Employing the same design procedure and the idealised SMA material properties summarised in Table 3.5, the comparisons between the test and predicted residual rotation results are shown in Fig. 3.32.

Table 3.5 Idealised key properties of SMA for design purpose

Idealised item	3% target drift	4% target drift	5% target drift
σ_{Ms} (MPa)	370	370	370
σ_{peak} (MPa)	450	490	535
σ_{Af} (MPa)	90	65	50
ε_{res}	0.0025	0.007	0.0115

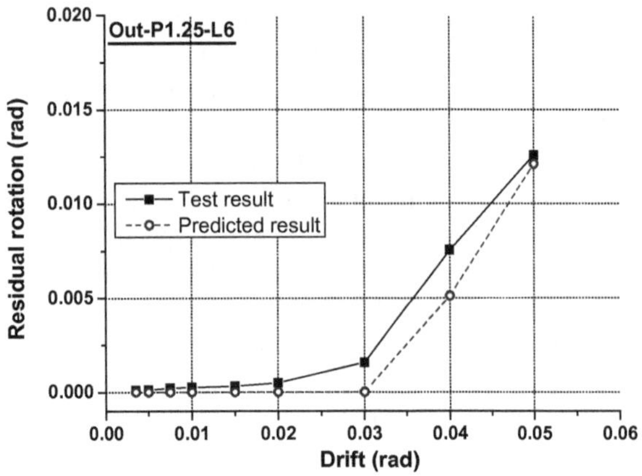

Fig. 3.32 Comparison between test results and design prediction on residual rotation

3.4 Connections with SMA Ring Springs

3.4.1 End-Plate Connections with SMA Ring Springs

An alternative category of self-centring connections is presented in this section. As illustrated in Fig. 3.33, SMA ring springs, which consists of SMA outer rings and HSS inner rings stacked in alternation with mating tapered faces, can serve as the key elements for the self-centring connection. Considering an end-plate type, these ring spring sets are used in conjunction with HS bolts, where the latter are responsible for shear resistance. When the connection rotates, the deformation leads to elongation of the HS bolt rows and hence expansion of the SMA outer rings. The inner rings are strong enough to stay elastic with very minor contraction. The SE promotes a strong self-centring tendency, resulting in deformation recovery of the entire connection. The energy dissipation is provided through inherent damping of the SMA material as well as the friction between the outer and inner rings. The required size and number of the SMA outer rings depend on the required moment resistance and target rotation of the connection, respectively. When the available deformation of the ring springs is fully consumed, further structural deformation occurs in the adjacent steel members. This is called 'locking' mechanism, which protects the SMA outer rings when the connection deformation demand is unexpectedly large.

The design formulae for predicting the stiffness and strength of the connections with the SMA ring springs are similar to those for the other types of SMA connections discussed in the previous sections. It is reasonable to assume that the expected connection deformation is only provided by the SMA ring springs, and the 'thick' end-plate deforms with a rigid V-shape gap opening mode. The design yield moment resistance $M_{j,Rd}$ can be simply calculated by

$$M_{j,Rd} = \sum_{i} F_{r,y} h_i \tag{3.18}$$

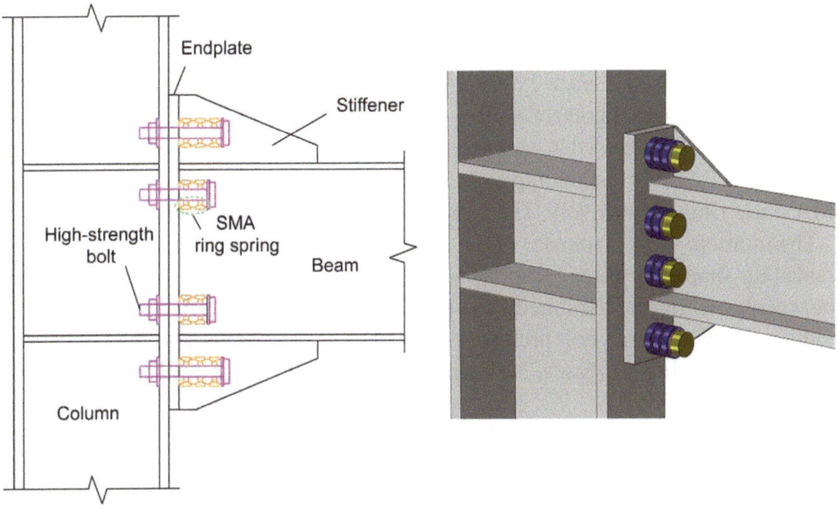

Fig. 3.33 Illustration of end-plate self-centring connections with SMA ring springs

where $F_{r,y}$ is the 'yield' resistance of each SMA ring spring set, where the value can be obtained from Eqs. (2.6) through (2.13); h_i is the distances between the SMA ring springs and the rotation centre of the connection. The design initial stiffness $K_{c,ini}$ of the connections can be calculated using the following equation:

$$K_{c,ini} = \sum_i \frac{F_{r,y}k_r}{n(F_{r,y} - F_{r,pre})} h_i^2 \tag{3.19}$$

where k_r is the initial stiffness of each single SMA ring spring, as can be obtained from Eqs. (2.6) through (2.13); n is the number of SMA outer rings in a ring spring set; $F_{r,pre}$ is the preload applied onto the SMA ring springs.

Three proof-of-concept tests on SMA ring spring-based end-plate connections were conducted by Fang et al. (2018b). Table 3.6 gives the details of the specimens with varying SMA outer ring thicknesses and bolt preloads. Two ring thicknesses,

Table 3.6 Details of connection specimens with SMA ring springs (Fang et al. 2018b)

Specimen code	Δ_p per single SMA ring (mm)	No. of SMA rings per external bolt	No. of SMA rings per internal bolt	Thickness of SMA ring (mm)
P1.5T3	1.5	3	2	3.0
P2.0T3	2.0	3	2	3.0
P1.5T5	1.5	3	2	5.0

Note Δ_p = pre-compression deformation via bolt preload i.e. $t = 3$ and 5 mm, were considered. The other dimensions of the SMA outer rings are identical to those described in Chap. 2. As shown in Figs. 3.34–3.35, three SMA outer rings were used in each external HS bolt, providing a deformation capacity of 15 mm. As the internal bolt rows were expected to experience less deformation, two SMA outer rings were used in each internal HS bolt. Two bolt preloads were adopted, i.e. P1.5 and P2.0, corresponding to precompressions of 1.5 mm and 2.0 mm for each SMA outer ring, respectively. The target drift of the connection specimens was 4%.

The moment–drift responses of the connection specimens are shown in Fig. 3.36. Stable flag-shaped hysteretic responses with negligible residual deformation and pronounced hardening are exhibited. An increase in bolt preload leads to increased yield moment. With the same bolt preload, the yield moment increases as the thickness of the SMA outer ring increases. No damage occurs to the main steel members of the specimens.

The connection specimens well satisfy the DCH and SMF criteria, as stipulated in Eurocode 8 (2004) and AISC (2010), respectively. The self-centring tendency is evaluated in Fig. 3.37 via restoring moment M_{rst}. It is seen that an increase in the precompression of the ring springs leads to increased M_{rst} at small amplitudes, but the influence of precompression becomes less significant at larger drifts due to more rapid TIF degradation under a high precompression level. The EVD first increases with drift, and then stabilised between 12 and 16%, as shown in Fig. 3.38. It is of interest to find that even at 0.375% drift, 6–10% EVDs can be achieved, indicating very early participation in energy dissipation. This behaviour is quite different from that of the connections equipped with other SMA elements such as SMA bolts. The improved EVD is attributed to the friction effect over the tapered face between the SMA outer rings and HSS inner rings.

Fig. 3.34 SMA ring spring sets together with HS bolts

Fig. 3.35 Typical deformation mode of end-plate self-centring connections with SMA ring spring system

Fig. 3.36 Moment–drift responses with SMA ring spring system: **a** P1.5T3, **b** P2.0T3, **c** P1.5T5

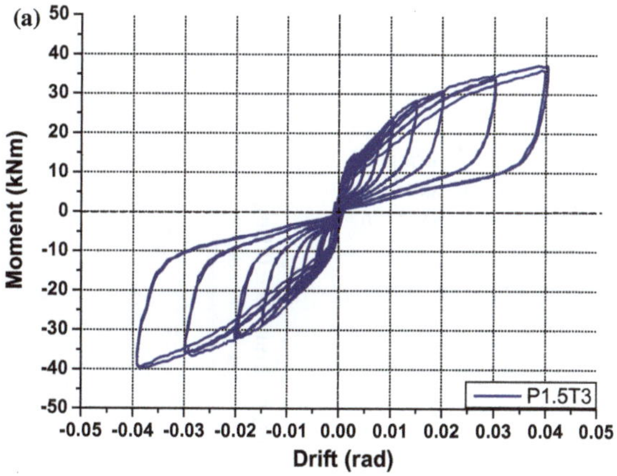

Fig. 3.36 (continued)

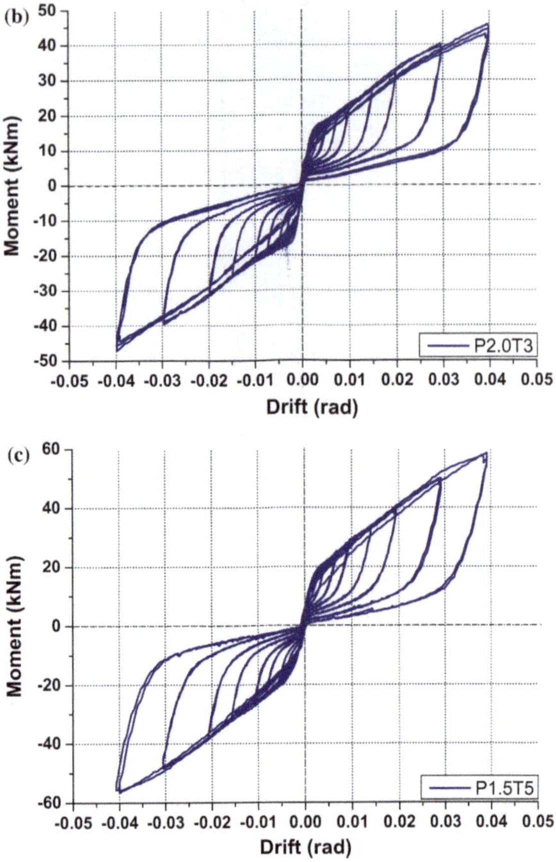

Fig. 3.37 Self-centring capability of self-centring connections with SMA ring springs

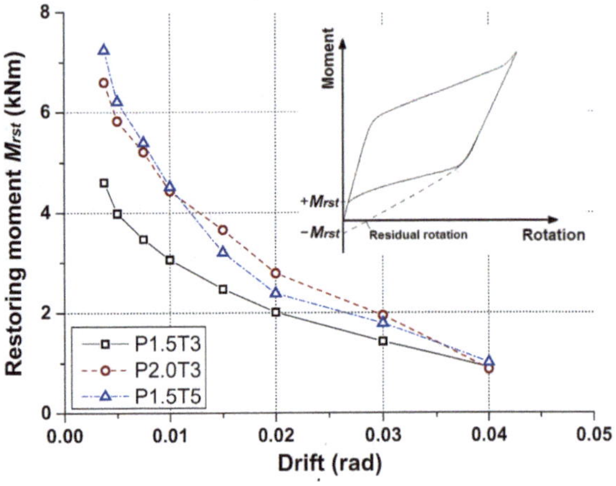

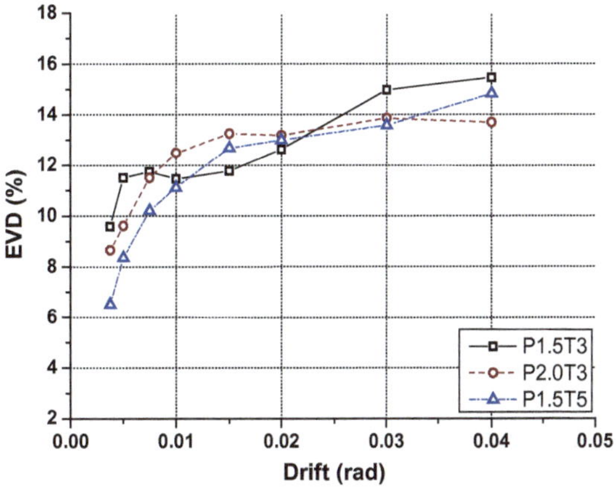

Fig. 3.38 Energy dissipation of self-centring connections with SMA ring springs

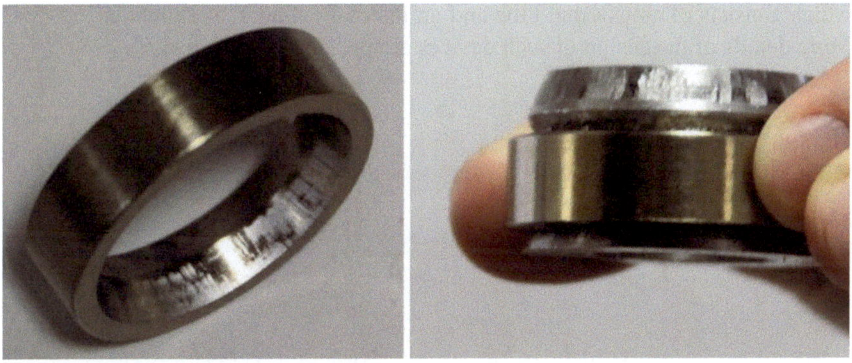

Fig. 3.39 Scratches over contacting surfaces between inner and outer rings

Incidentally, scratches are found over the greased contacting surfaces between the inner and outer rings, as shown in Fig. 3.39, although the overall connection behaviour is minimally affected. The scratches may be caused by the inclination and uneven deformation of the SMA outer rings during the rotation of the connections.

3.4.2 Improved Connection Type Free from Frame Expansion

Inspired by the pioneering work done by Khoo et al. (2012, 2013), the end-plate connections with SMA ring springs can be modified such that the connections only rotate about the top flange of the beam, and as a result, frame expansion is eliminated (Fang et al. 2019; Feng et al. 2019). As shown in Fig. 3.40, a top flange cover plate can be designed to connect the column flange surface to the top flange of the steel beam. A shear tab should be used to resist the shear force applied to the connection. Slotted bolt holes are necessary to allow certain rotational flexibility. Specially designed SMA ring spring devices are the kernel components that provide strength, stiffness, energy dissipation and self-centring capability for the connections. These devices are placed underneath the beam bottom flange, with one end connected to the column flange and the other end connected to the bracket that can be welded to the beam bottom flange.

The SMA ring spring system within the device should always be compressed regardless of the loading direction. To achieve this, a piston-type device is viable which consists of an external tube and an internal cylinder, as shown in Fig. 3.41. More details of the design of such devices are presented in Chap. 4.

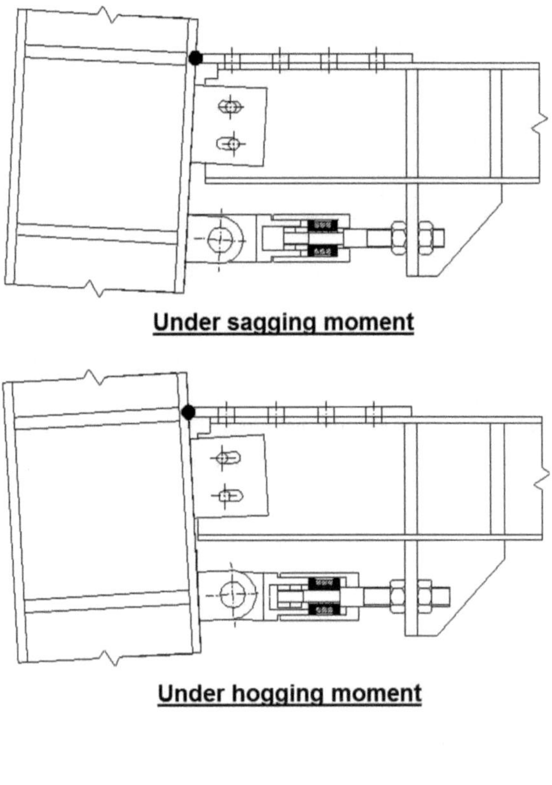

Fig. 3.40 Concepts of frame expansion free connections with SMA ring spring devices

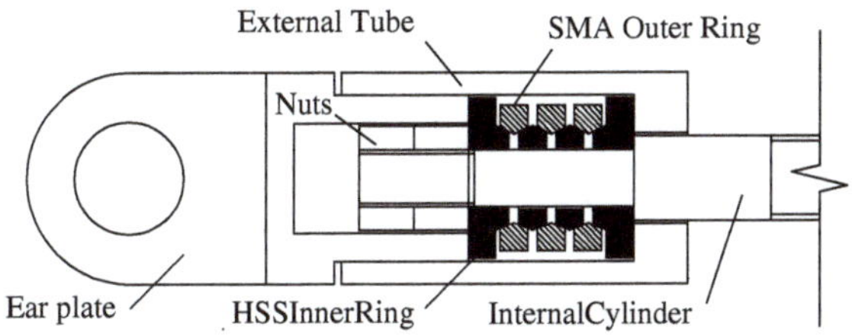

Fig. 3.41 Details of SMA ring spring device placed underneath beam

Fig. 3.42 Proof-of-concept frame expansion-free bare steel connection specimen

Figures 3.42 and 3.43 show the deformation mode of proof-of-concept bare steel and composite connection specimens, respectively, both employing the new frame expansion-free concept (Fang et al. 2019; Feng et al. 2019). The test setup is similar to that used in the other connections discussed above, and is therefore not elaborated here. We can clearly see that compared with the SMA-bolted end-plate connections where significant slab damage occurs, as previously shown in Fig. 3.11, only minor cracks are identified in the slab of the improved SMA ring spring connections. The devices undergo anticipated elongation and compression under sagging and hogging moments, respectively, and the top flange cover plate rotates quite flexibly and reliably.

The moment–drift responses of the two connection specimens are shown in Figs. 3.44 and 3.45, respectively. An excellent self-centring capability is observed. The presence of the slab affects the shape of the hysteretic curves, which is not sur-

Fig. 3.43 Proof-of-concept frame expansion-free composite connection specimen

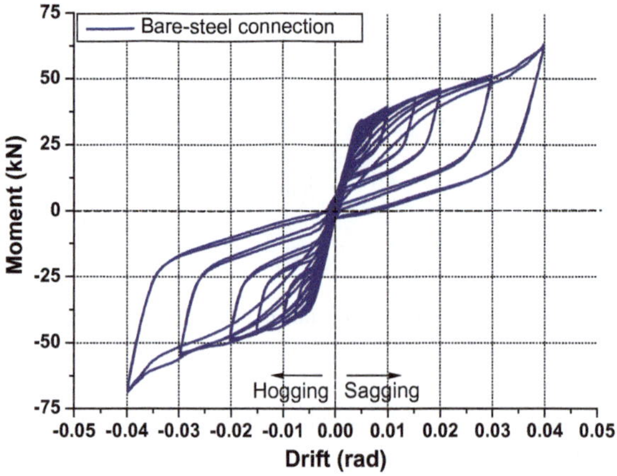

Fig. 3.44 Moment–drift response of frame expansion-free bare steel connection specimen

prising, as the slab contributes to the moment resistance and it also shifts the neutral axis of the connection. The ductility and energy dissipation characteristics are similar to those of the corresponding end-plate connection types, and hence, we skip over these details.

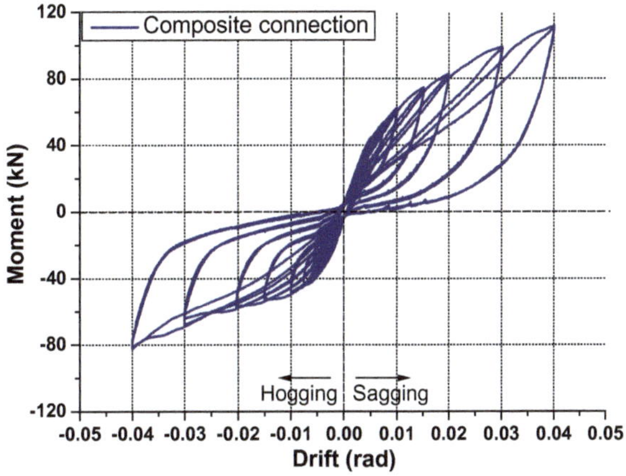

Fig. 3.45 Moment–drift response of frame expansion-free composite connection specimen

References

Abolmaali A, Treadway J, Aswath P, Lu FK, Emily McCarthy (2006) Hysteresis behavior of t-stub connections with superelastic shape memory fasteners. J Constr Steel Res 62(8):831–838

AISC (2010) Seismic provisions for structural steel buildings, (ANSI/AISC 341-10). American Institute of Steel Construction, Chicago, IL, USA

Chou CC, Chen JH (2011a) Seismic design and shake table tests of a steel post-tensioned self-centering moment frame with a slab accommodating frame expansion. Earthq Eng Struct D 40(11):1241–1261

Chou CC, Chen JH (2011b) Development of floor slab for steel post-tensioned self-centering moment frames. J Constr Steel Res 67(10):1621–1635

Christopoulos C, Filiatrault A, Uang CM, Folz B (2002) Posttensioned energy dissipating connections for moment-resisting steel frames. J Struct Eng-ASCE 128(9):1111–1120

Dowden DM, Clayton PM, Li CH, Berman JW, Bruneau M, Lowes LN, Tsai KC (2016) Full-scale pseudodynamic testing of self-centering steel plate shear walls. J Struct Eng-ASCE 142(1):04015100

Dowden DM, Bruneau M (2016) Dynamic shake-table testing and analytical investigation of self-centering steel plate shear walls. J Struct Eng-ASCE 142(10):04016082

Eatherton MR, Hajjar JF (2011) Residual drifts of self-centering systems including effects of ambient building resistance. Earthq Spectra 27(3):719–744

European Committee for Standardization (CEN) (2004) EN 1998-1, Eurocode 8: design of structures for earthquake resistance—part 1: general rules, seismic actions and rules for buildings. Belgium, Brussels

European Committee for Standardization (CEN) (2005) EN 1993-1-8, Eurocode 3: design of steel structures—part 1-8: design of joints. European Committee for Standardization, Brussels, Belgium

Fang C, Yam MCH, Lam ACC, Xie LK (2014) Cyclic performance of extended end-plate connections equipped with shape memory alloy bolts. J Constr Steel Res 94:122–136

Fang C, Wang W, He C, Chen YY (2017) Self-centring behaviour of steel and steel-concrete composite connections equipped with NiTi SMA bolts. Eng Struct 150:390–408

Fang C, Yam MCH, Chan TM, Wang W, Yang X, Lin XM (2018a) A study of hybrid self-centring connections equipped with shape memory alloy washers and bolts. Eng Struct 164:155–168

Fang C, Wang W, Ricles J, Yang X, Zhong QM, Sause R, Chen YY (2018b) Application of an innovative SMA ring spring system for self-centering steel frames subject to seismic conditions. J Struct Eng-ASCE 144(8):04018114
Fang C, Wang W, Feng W (2019) Experimental and numerical studies on self-centring beam-to-column connections free from frame expansion. Eng Struct Under Rev
Feng W, Fang C, Wang W (2019) Behavior and design of top flange-rotated self-centering steel connections equipped with SMA ring spring dampers. J Constr Steel Res Under Rev
Federal Emergency Management Agency (FEMA) (2000a) State of the art report on connection performance. FEMA 355D, prepared by the SAC Joint Venture for FEMA, Washington, DC
Federal Emergency Management Agency (FEMA) (2000b) Recommended seismic design criteria for new moment-resisting steel frame structures. FEMA 350, prepared by the SAC Joint Venture for FEMA, Washington, DC
Fromm E, Kleiner W (2003) Handbook for disc springs. Schnorr Corporation, Hela Werbung, Heilbronn
Garlock MEM, Li J (2008) Steel self-centering moment frames with collector beam floor diaphragms. J Constr Steel Res 64(5):526–538
Garlock MEM, Ricles JM, Sause R (2008) Influence of design parameters on seismic response of post-tensioned steel MRF systems. Eng Struct 30(4):1037–1047
Garlock MEM, Ricles JM, Sause R (2005) Experimental studies of full-scale posttensioned steel connections. J Struct Eng-ASCE 131(3):438–448
Huang XG, Zhou Z, Xie Q, Xue RL, Zhu DP (2017) Force distribution analysis of self-centering coupled-beams for moment-resisting-frames without floor elongation. Eng Struct 147:328–344
Khoo HH, Clifton GC, Butterworth J, MacRae GA, Gledhill S, Sidwell G (2012) Development of the self-centering sliding hinge joint with friction ring springs. J Constr Steel Res 78:201–211
Khoo HH, Clifton GC, Butterworth J, MacRae GA (2013) Experimental study of full-scale self-centering sliding hinge joint connections with friction ring springs. J Earthq Eng 17(7):972–997
Lagoudas DC (2008) Shape memory alloys: modeling and engineering applications. Springer, TX, USA
Lin YC, Sause R, Ricles JM (2013a) Seismic performance of a steel self-centering, moment-resisting frame: hybrid simulations under design basis earthquake. J Struct Eng-ASCE 139(11):1823–1832
Lin YC, Sause R, Ricles JM (2013b) Seismic performance of a large-scale steel self-centering moment-resisting frame: MCE hybrid simulations and quasi-static pushover tests. J Struct Eng-ASCE 139(7):1227–1236
Ma HW, Wilkinson T, Cho C (2007) Feasibility study on a self-centering beam-to-column connection by using the superelastic behavior of SMAs. Smart Mater Struct 16(5):1555–1563
MacRae GA (2010) Some steel seismic research issues. In: Proceedings of the steel structures workshop 2010, University of Canterbury, April 2010
Maurya A, Eatherton MR (2016) Experimental study of the restoring force mechanism in the self-centering beam (SCB). Front Struct Civ Eng 10(3):272–282
McCormick J, Aburano H, Ikenaga M, Nakashima M (2008) Permissible residual deformation levels for building structures considering both safety and human elements. In: Proceedings of 14th world conference on earthquake engineering, Seismological Press of China, Beijing
McCormick J, DesRoches R (2006) The effect of training, pre-straining, and loading history on the properties of NiTi shape memory alloys for protective systems in civil structures. In: Proceedings of structures congress 2006: structural engineering and public safety, American Society of Civil Engineers, St. Louis
Murray TM, Sumner EA (2003) Extended end-plate moment connections seismic and wind applications. AISC Design Guide 4, 2nd edn
National Academy of Sciences (NAS) (2011) National earthquake resilience—research, implementation, and outreach. The National Academies Press, Washington
Ocel J, DesRoches R, Leon RT, Hess WG, Krumme R, Hayes JR, Sweeney S (2004) Steel beam-column connections using shape memory alloys. J Struct Eng-ASCE 130(5):732–740

Penar BW (2005) Recentering beam-column connections using shape memory alloys. Master thesis, Georgia Institute of Technology

Ricles JM, Sause R, Garlock MEM, Zhao C (2001) Post-tensioned seismic-resistant connections for steel frames. J Struct Eng-ASCE 127(2):113–121

SAC Joint Venture (SAC) (1997) Protocol for fabrication, inspection, testing and documentation of beam-column connection tests and other experimental specimens. SAC/BD-97/02, California

Shen J, Astaneh-Asl A (2000) Hysteresis model of bolted-angle connections. J Constr Steel Res 54(3):317–343

Speicher MS, DesRoches R, Leon RT (2011) Experimental results of a NiTi shape memory alloy (SMA)-based recentering beam-column connection. Eng Struct 33(9):2448–2457

Stojadinović B, Goel SC, Lee KH, Margarian AG, Choi JH (2000) Parametric tests on unreinforced steel moment connections. J Struct Eng-ASCE 126(1):40–49

Wang W, Chan TM, Shao HL, Chen YY (2015a) Cyclic behavior of connections equipped with NiTi shape memory alloy and steel tendons between H-shaped beam to CHS column. Eng Struct 88:37–50

Wang W, Chan TM, Shao HL (2015b) Seismic performance of beam–column joints with SMA tendons strengthened by steel angles. J Constr Steel Res 109:61–71

Wang W, Du XL, Zhang YF, Chen YY (2017a) Experimental investigation of beam-through steel frames with self-centering modular panels. J Struct Eng-ASCE 143(5):04017006

Wang W, Fang C, Liu J (2017b) Self-centering beam-to-column connections with combined superelastic SMA bolts and steel angles. J Struct Eng-ASCE 143(2):04016175

Wang W, Kong J, Zhang Y, Chu G, Chen Y (2018) Seismic behavior of self-centering modular panel with slit steel plate shear walls: experimental testing. J Struct Eng-ASCE 144(1):04017179

Yam MCH, Fang C, Lam ACC, Zhang YY (2015) Numerical study and practical design of beam-to-column connections with shape memory alloys. J Constr Steel Res 104:177–192

Youssef NFG, Bonowitz D, Gross JL (1995) A survey of steel moment-resisting frame buildings affected by the 1994 Northridge earthquake. Report NISTIR 5625, Gaithersburg, MD

Zhang AL, Zhang YX, Li R, Wang ZY (2016) Cyclic behavior of a prefabricated self-centering beam-column connection with a bolted web friction device. Eng Struct 111:185–198

Chapter 4
Self-centring Braces with SMA Elements

Abstract This chapter proceeds with discussions of the application of SMA elements in self-centring bracing members in framed structures. First, the existing solutions for self-centring braces are briefly introduced, and the potential limitations are also outlined. A series of newly proposed braces, employing SMA wires, tendons or ring springs, are subsequently discussed in detail. The main focus of this chapter is on the design principle, working mechanism, and fundamental mechanical behaviour of the kernel devices for the braces. Some technical issues such as the manufacturing process and annealing scheme are particularly addressed for the devices equipped with SMA ring spring systems.

4.1 Introduction

Concentrically braced frames are one of the most popular lateral resisting systems employed for multi-storey and high-rise building structures (Fig. 4.1). However, past earthquakes revealed that the bracing members displayed unsatisfactory cyclic behaviour due to their limited buckling resistance in compression, leading to poor ductility and energy dissipation performances (Sabelli et al. 2003). Alternatively, eccentrically braced frames permit the linking beams to experience plastic deformation, although fracture failure could potentially occur in these links (Hjelmstad and Popov 1984).

Buckling-restrained braced frames (BRB frames) have received great attention in the past decade. By allowing yielding in both tension and compression, BRBs provide stable and full hysteresis with excellent energy dissipation capability and sufficient ductility (Tremblay et al. 2006; Fahnestock et al. 2007; Chou and Chen 2010; Takeuchi et al. 2010). Although BRB frames are generally superior to conventional concentrically braced frames in terms of energy dissipation, one potential issue of the former is the noticeable permanent inter-storey drift. Past studies revealed that, depending on the number of stories, the mean residual drifts of typical BRB frames subjected to strong earthquakes could range between 2 and 6% (Erochko et al. 2011), recalling that a residual inter-storey drift exceeding 0.5% implies uneconomical repair work, as suggested by McCormick et al. (2008). A similar threshold is

C. Fang and W. Wang, *Shape Memory Alloys for Seismic Resilience*, https://doi.org/10.1007/978-981-13-7040-3_4

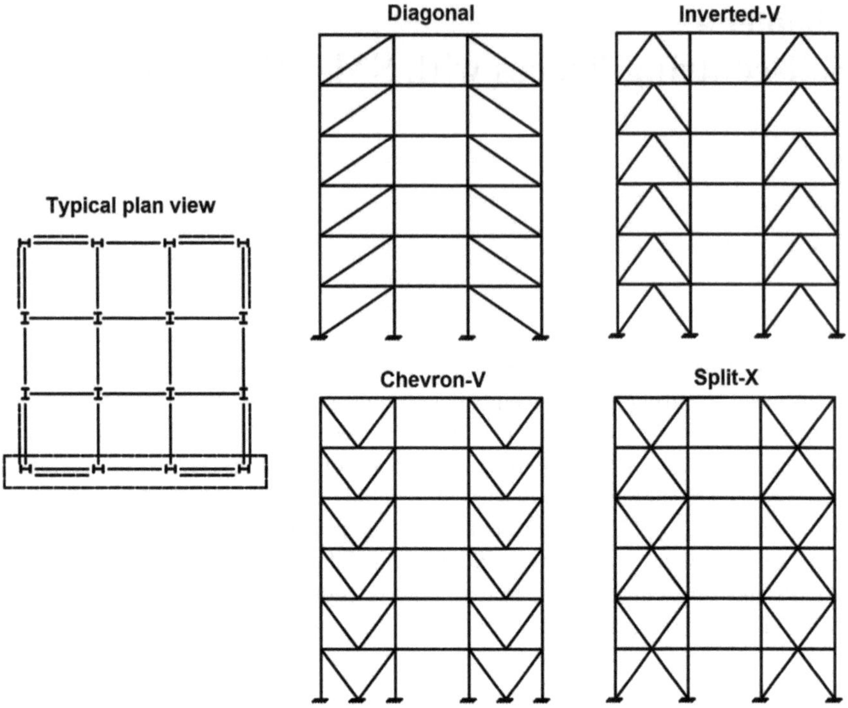

Fig. 4.1 Illustration of typical concentrically braced frames

recommended by FEMA P-58 (FEMA 2012), which stipulates that 'the required realignment and repair of the structure may not be economically and practically feasible' if the residual drift exceeds 0.5%.

There is an urgent need to develop new earthquake resilient braced frames with reduced residual inter-storey drifts. To this end, a novel category of braced frames, namely self-centring braced frames (SCBFs), have been proposed by seismic engineers. Post-tensioning (PT) technology has been adopted as one of the possible ways for developing self-centring braces (Chou et al. 2014, 2016; Christopoulos et al. 2008; Zhou et al. 2015; Erochko et al. 2015a, b). PT-based self-centring braced frames may have certain advantages over the structural systems with PT beam-to-column connections. For example, there are no additional challenges associated with the complex construction and the incompatibility between the gravity system and the lateral load resisting frame due to connection gap opening, for self-centring braced frames. Nevertheless, yielding or damage to the adjacent members such as steel beams and connections could still happen in self-centring braced frames if the peak deformation is large.

A PT-based bracing member typically consists of tensioning elements, energy dissipation devices, and a couple of accommodating steel tubes which ensure that the PT tendons are always in tension (e.g. Fig. 4.2). Both steel and composite materials

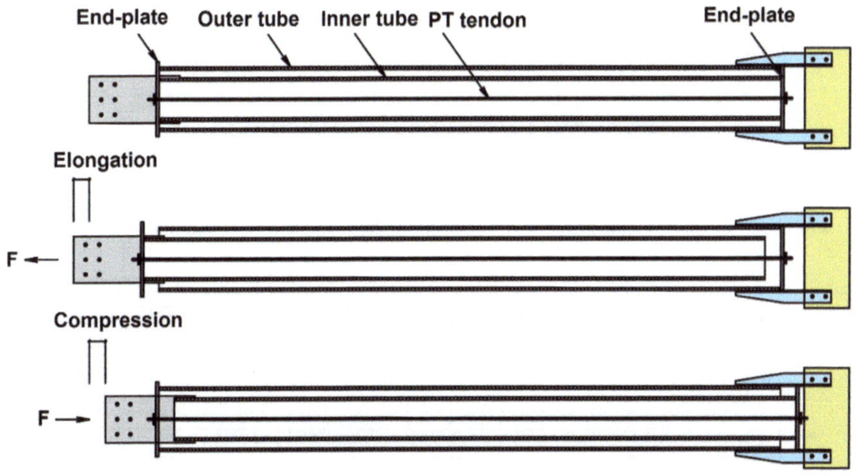

Fig. 4.2 Working mechanism of PT-based self-centring bracings (Erochko et al. 2015a)

(e.g. fibre-reinforced polymer) can be used for the PT tendons; however, the overall deformability of a PT-based self-centring brace is often limited by the small elastic strain range of the tendon material, not to mention that a large proportion of the elastic strain needs to be 'pre-consumed' during the initial post-tensioning procedure. Once the PT tendons experience yielding, the self-centring capability of the brace is significantly deteriorated. To address this issue, a parallel arrangement of the PT tendons with dual or multiple cores has been proposed (Fig. 4.3). This solution increases the deformability of the brace but makes the detailing of the brace more complicated.

Figure 4.4 shows an idealised flag-shaped hysteretic behaviour of PT-based self-centring braces. The load–deformation behaviour first follows a linear stage and then reaches to a 'yield' point before advancing into the 'post-yield' stage. In the unloading stage, the load first decreases linearly, and then enters into the unloading plateau and finally decreases to zero. The force–deformation relationship can be adequately characterised by four key parameters, namely, initial stiffness $K_{b,ini}$, yield strength F_y, post-yield stiffness ratio α, and energy dissipation factor β. The energy dissipation factor (β) describes the shape of the hysteretic loop, where $\beta = 0.0$ indicates an elastic bilinear system with no energy dissipation, and $\beta = 2.0$ makes an idealised elastoplastic behaviour with full hysteresis. For self-centring braces designed with no static residual deformation, the value of β is expected to range between 0.0 and 1.0.

Another promising candidate which can be used in self-centring braces is superelastic SMA. The flag-shaped hysteretic response of SMA elements promotes self-centring and energy dissipation capabilities for the braces. The SMA elements can be used either individually or in combination with other energy dissipative elements such as metallic angles and friction pads. Up to now, several conceptual SMA-based

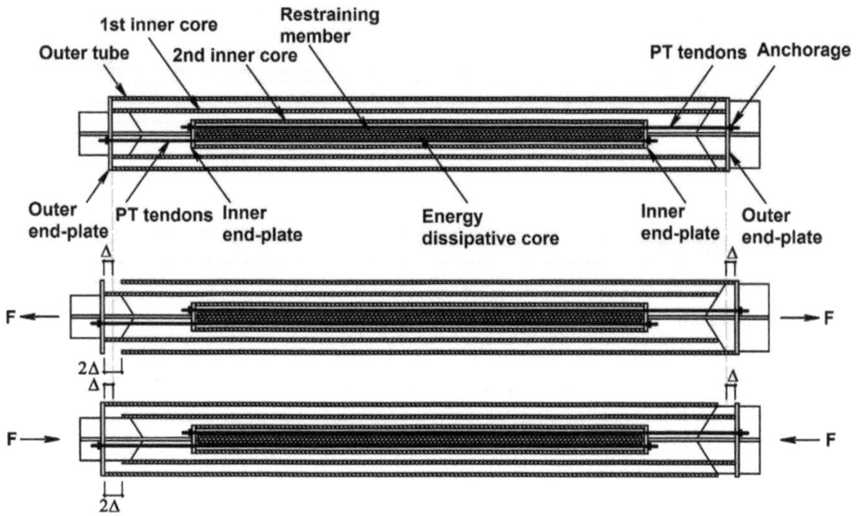

Fig. 4.3 Working mechanism of self-centring bracings with parallel PT tendons (Chou et al. 2016)

Fig. 4.4 Idealised axial load–deformation response of self-centring bracings

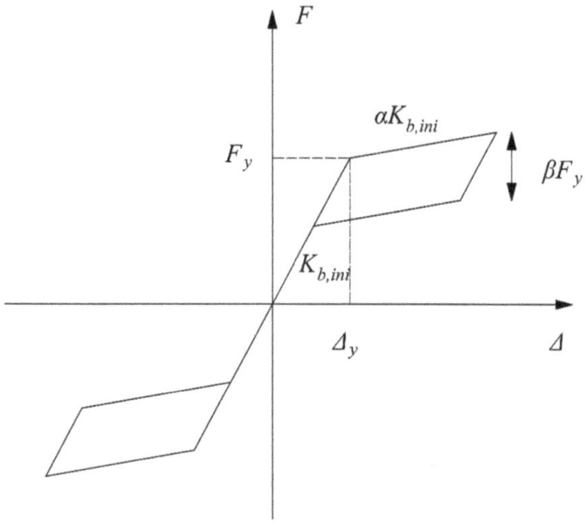

self-centring braces have been successfully developed and tested (Miller et al. 2012; Xu et al. 2016; Qiu and Zhu 2017), where satisfactory hysteretic performances were generally obtained. The repeatable and stable flag-shaped hysteretic behaviour makes these braces reusable even after strong earthquakes, with no replacement or repair work required.

In contrast to the limited deformability provided by the PT braces, SMA-based braces can have more flexible axial deformations thanks to the large recoverable deformation of the SMA. A system-level analysis conducted by Moradi et al. (2014)

also showed that 'weak floor' is less likely to occur in structures with SMA braces due to the unique martensitic hardening response. With the aim of increasing the energy dissipation whilst controlling the maximum residual deformation of the system, a combined use of SMA-based braces and BRBs have been attempted by Kari et al. (2011). Under certain circumstances, however, either the SMA- or PT-based self-centring braces could lead to an increase in the absolute floor acceleration. Floor acceleration is a key factor that governs the non-structural building damage and the condition of the building contents. Increasing the energy dissipation capability (i.e. increasing β) helps suppress the peak floor acceleration response.

The findings from the existing studies are helpful for the future design of self-centring braces and structural systems. In the following sections, the design concept, working principle, and mechanical behaviour of typical SMA-based self-centring braces are presented. The discussion of the system-level response of self-centring braced frames is reserved for Chap. 6.

4.2 Self-centring Braces with SMA Wires

As wire has been the most readily available form of SMA products in the market, most of the early small-scale prototype self-centring brace specimens were developed employing SMA wire groups. This solution can benefit from the high ductility of the SMA wires in comparison with larger scale SMA bolts. As SMA wires have good ductility, it is often unnecessary to use full-length SMA wires in the bracing members. A common approach is to develop a shorter kernel part, which is essentially an SMA-based damper, to be spliced to adjacent members for the remaining necessary length of the brace. The maximum load resistance of the kernel part should be less than the overall buckling resistance of the entire bracing member. The braces should behave symmetrically in tension and compression, and for both cases, the SMA wires are preferably in tension only. The SMA wires can be prestrained in order to obtain a higher initial stiffness of the brace.

Figure 4.5 shows the configuration and constituent components of a small-scale proof-of-concept device designed as the kernel part for self-centring braces (Qiu 2016; Qiu and Zhu 2017). The device consists of two sliding steel blocks, two steel rods and a pair of SMA wire groups. As illustrated in the figure, the two steel rods that travel through the slots transfer the resisting force between the SMA wires and the steel blocks by elongating the SMA wires. More specifically, when the device is in tension, the two sliding blocks tend to move away from each other, causing elongation of the SMA wires when one steel rod moves with the outer block while the other rod is constrained to the inner block. Similarly, when the damper is in compression, the two sliding blocks tend to move towards each other, where the SMA wires are again elongated when one steel rod moves away from the other rod which is constrained. In other words, the SMA wire groups are always in tension when the device is subjected to either tension or compression. Such a mechanism enables efficient use of the SMA wires.

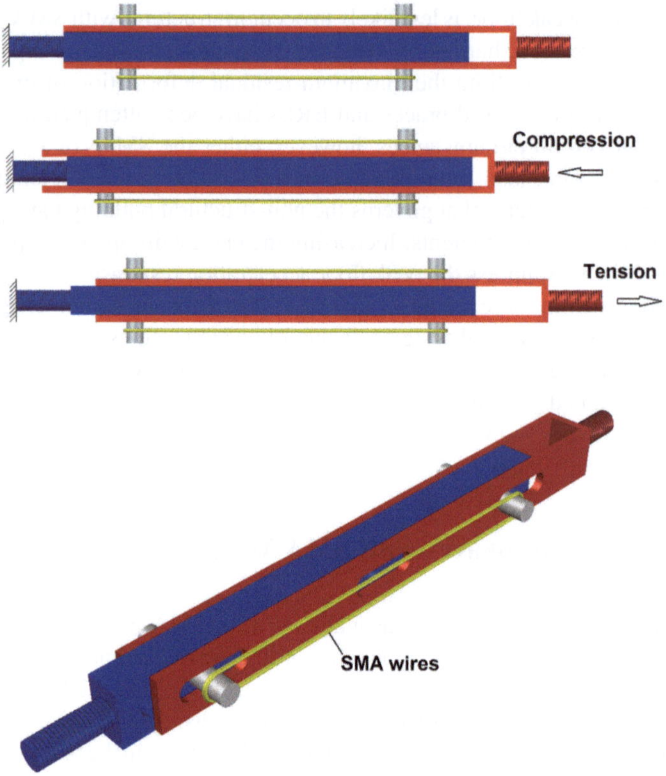

Fig. 4.5 Configuration and working mechanism of a proof-of-concept device used for self-centring bracings (Qiu and Zhu 2017)

To ensure that the deformation of the device is effectively provided by the SMA wires, the steel rods have to be sufficiently stiff. The wire length, which is approximately equal to the distance between the two rods, should be carefully determined such that the required deformation capacity of the brace is achieved. The wire ends can be connected using U-connectors in each loop of the SMA wires (Fig. 4.6). Caution should be exercised that these connectors are not very effective for smooth wires with diameters larger than 1.0 mm. Therefore, small diameter SMA wires are preferred in this case to ensure an effective end gripping. Alternatively, larger diameter bundled wires may be used instead. To make the hysteretic behaviour stable, the SMA wires may be trained for several cycles before their long-term service.

Figure 4.7 further shows the configuration and dimension of the entire self-centring brace specimen incorporating the kernel device together with two extended parts made of steel square tubes. This small-scale specimen was designed and tested by Qiu and Zhu (2017). The kernel device was welded to the two steel square tubes to extend to the required length of the brace. Splice plates are made at the two ends of the brace which is supposed to connect to the frame via gusset plates. The steel

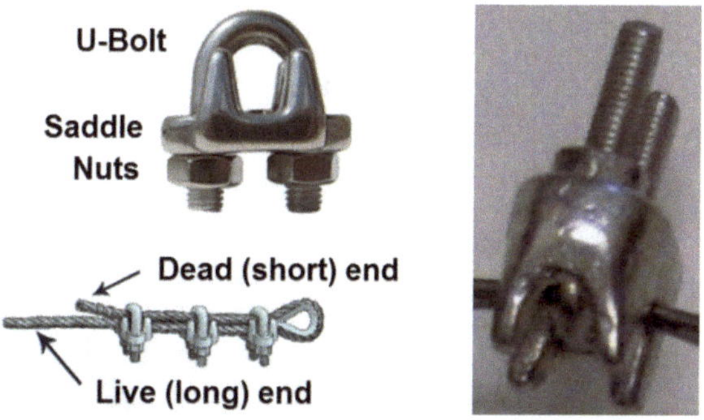

Fig. 4.6 U-connector used for clamping the SMA wire ends (Qiu and Zhu 2017)

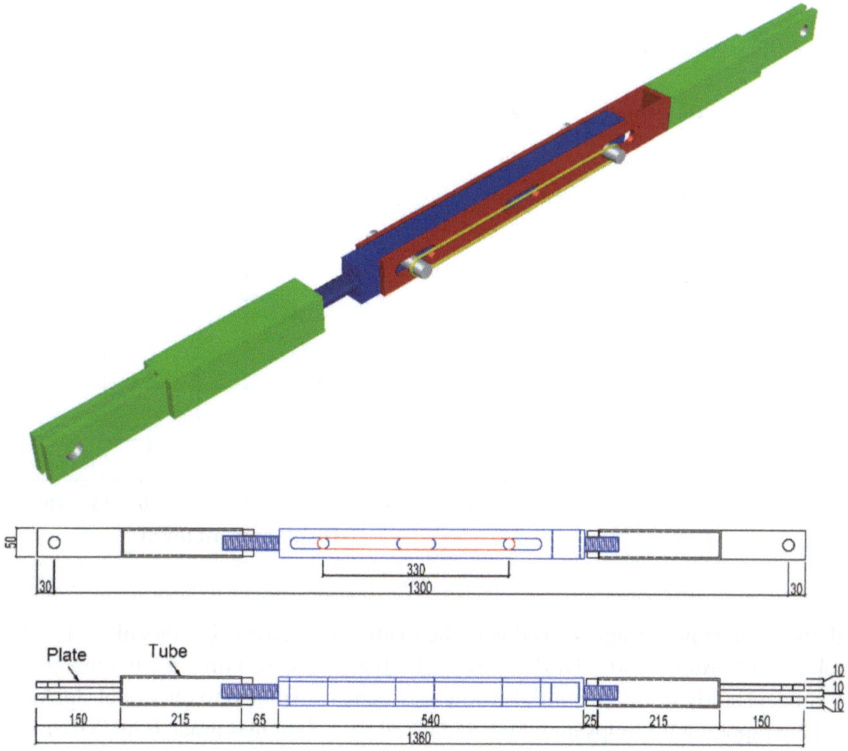

Fig. 4.7 Self-centring brace incorporating kernel device with SMA wires (Qiu and Zhu 2017)

square tubes were designed to remain elastic so that the inelastic deformation is concentrated in the middle kernel device segment. Figure 4.8 shows alternative layouts of the kernel devices within a steel frame, depending on the designers' preference.

The typical cyclic behaviour of the self-centring brace specimen conducted by Qiu and Zhu (2017) under a loading frequency of 2 Hz is shown in Fig. 4.9. The

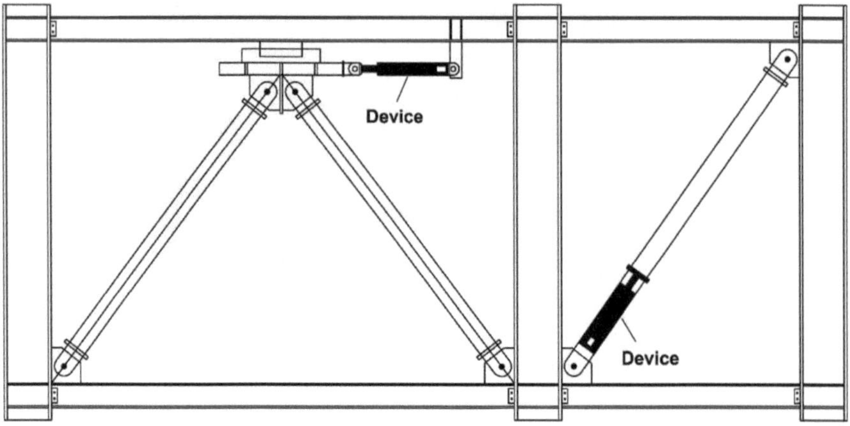

Fig. 4.8 Arrangements of kernel device for self-centring braces

Fig. 4.9 Cyclic behaviour of self-centring brace incorporating kernel device with SMA wires (Qiu and Zhu 2017)

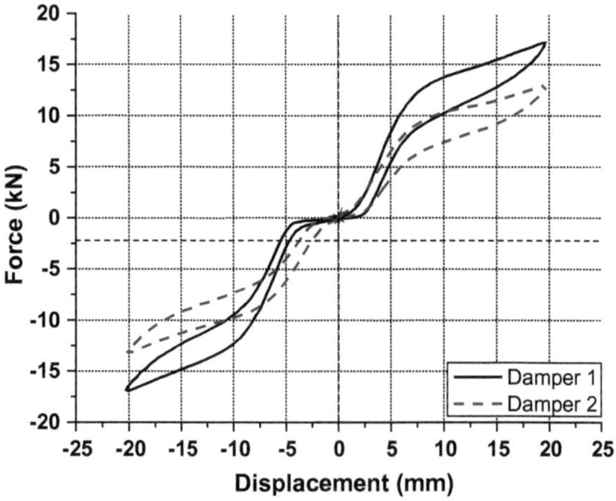

relatively 'narrow' hysteresis is due to the loading rate effect (Piedboeuf et al. 1998; Dolce and Cardone 2001; DesRoches et al. 2004). A lesson drawn from this proof-of-concept study is the low initial stiffness which is attributed to the initial slackness of the wrapped wire loops and the backlash of the fabricated brace (Qiu 2016). The initial slackness is approximately 2 mm, which is less than 1% of the wire length. Such a low initial stiffness is certainly undesirable and should be avoided in practice by pre-tensioning the SMA wires and precise machining.

As mentioned, the force–deformation relationship of a self-centring brace can be characterised by the initial stiffness $K_{b,ini}$, yield strength F_y, post-yield stiffness ratio α and energy dissipation factor β. According to the test result reported by Qiu and Zhu (2017), the four parameters for the considered brace at a maximum displacement of 20 mm under a loading frequency of 2 Hz are approximately $K_{b,ini} = 2530$ kN/m (excluding the initial slackness), $F_y = 11.9$ kN, $\alpha = 0.15$ and $\beta = 0.3$. In addition, the EVD at this displacement level is approximately 5%.

4.3 Self-centring Braces with SMA Bolts

The relatively low load-carrying capacity of SMA wires makes them less suited to full-scale bracing members. An alternative solution is to use larger size SMA bolts/tendons. The SMA bolts are prone to buckling in compression and should be subjected to tension only. An ideal device/brace is the one that allows the elongation of the SMA bolts when the brace is loaded either in tension or compression. The existing PT-based self-centring brace solutions, such as those illustrated in Fig. 4.2, can be easily modified to incorporate the SMA bolts.

Perhaps a more feasible solution is to use SMA elements in combination with other energy dissipative sources such as BRB or friction systems, leading to a hybrid bracing system. This concept stems from mounting concern that SMA elements alone may provide insufficient energy dissipation, in which case larger peak inter-storey drifts are expected compared with more conventional structural systems such as MRFs and BRB frames (Ozbulut and Hurlebaus 2012). People also realise the fact that even if a self-centring brace exhibits certain residual deformation under a 'static' unloading process, the final system-level residual deformation would still be suppressed thanks to the probabilistic self-centring phenomenon during dynamic shakedown (Eatherton and Hajjar 2011; Eatherton et al. 2014). This forms the basis of the hybrid self-centring brace design concept which intentionally encourages an increased energy dissipation, even at the cost of compromised self-centring capability.

Figure 4.10 shows the typical assembly of the hybrid self-centring brace proposed by Miller et al. (2012). The key components include an inner BRB, a pair of movement guiding-and-constraining tubes, a pair of end-plates with slotted holes, and a series of SMA tendons anchored between the two end-plates. The SMA tendons can be connected to high-strength steel tendons via couplers to extend to the required brace length. Like any other BRB systems, the inner BRB consists of a steel core, a nominally unstressed inner constraining tube, and concrete infill which is unbonded to the steel core. The steel core yields without buckling when loaded in compression. The cruciform end connections of the inner BRB pass through the slotted end-plates which are free to slide longitudinally. The SMA tendons are anchored and prestrained between the two end-plates, where the distance between the end-plates is maintained (constrained) by the middle and outer tubes during the prestraining process. Another key detail is that the middle tube must be welded only to the left cruciform end connection of the inner BRB (see the orientation of the brace as shown in Fig. 4.10), while the outer tube should be welded only to the right cruciform end connection of the inner BRB.

Figure 4.11 provides a more detailed demonstration of the deformation mode of the brace. When in compression, the inner BRB is shortened. Assuming that the left

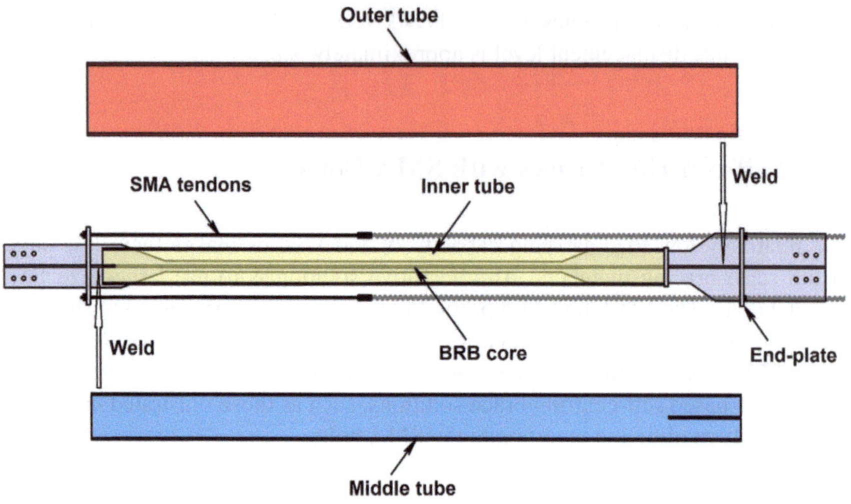

Fig. 4.10 Typical assembly of hybrid self-centring brace (Miller et al. 2012)

end of the brace is fixed, the outer tube has to move leftwards with the right cruciform end connection of the inner BRB, recalling that this tube is welded to this right end. The leftward movement of the outer tube pushes the left slotted end-plate to move, but the right slotted end-plate is constrained by the middle tube (which is welded to the left side) and therefore cannot be moved. Such a separation of the two end-plates elongates the SMA tendons.

When the brace is in tension, the inner BRB is stretched. Again, assuming that the left end of the brace is fixed, as shown in Fig. 4.11, the outer tube has to move rightwards with the right cruciform end connection of the inner BRB, and this action pushes the right end-plate to move rightwards. However, the left end-plate is constrained by the middle tube which is welded to the left cruciform end connection of the inner BRB. Such a separation of the two end-plates also elongates the SMA tendons. With the above working principle, the SMA tendons act only in tension no matter how the brace is deformed.

As the load resistance mechanism of the hybrid self-centring brace is quite straightforward, a simple design rule can be followed. The inner BRB, which offers energy dissipation, and the SMA tendons, which produce restoring forces and extra energy dissipation, essentially work in parallel with each other. The design yield resistance of the brace ($F_{Rd,y}$) can be defined as the summation of the yield resistance of the BRB steel core ($F_{BRB,y}$) and the yield (forward transformation) force of the SMA bolts/tendons ($F_{b,y}$), as expressed by

$$F_{Rd,y} = F_{BRB,y} + F_{b,y} = f_y A_{BRB} + \sigma_{Ms} A_b \tag{4.1}$$

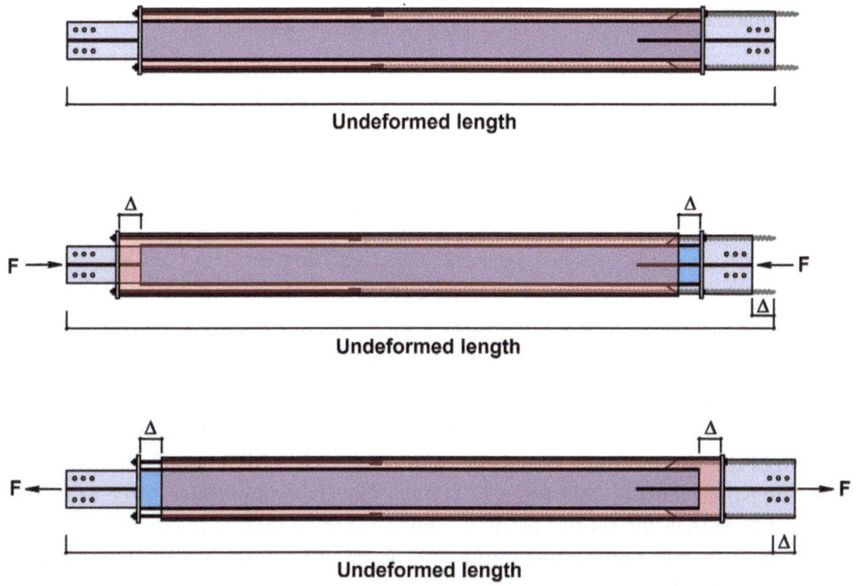

Fig. 4.11 Illustration of deformation mode of hybrid self-centring brace (Miller et al. 2012)

where f_y is the steel yield strength, A_{BRB} is the cross-sectional area of the BRB steel core, A_b is the total area of the SMA bolt/tendon. Taking account of the prestrain of the SMA bolts/tendons, the design initial stiffness $K_{b,ini}$ of the brace can be approximated by the following expression:

$$K_{b,ini} = \frac{EA_{BRB}}{L_{e,BRB}} + \frac{\sigma_{Ms}E_A A_b}{(\sigma_{Ms} - \sigma_{pre})L_p} \tag{4.2}$$

where E and E_A are the Young's moduli of steel and SMA, respectively, $L_{e,BRB}$ is the effective working length (yielding region) of the inner BRB, L_p is the working length of the SMA tendons, σ_{pre} is the prestress applied to the SMA tendons.

To ensure a reasonable balance between self-centring capability and energy dissipation of the brace, the contributions from each functional component to the overall brace behaviour should be considered in the design. The necessary initial preload applied to the SMA tendons can be determined using the following equation (Miller et al. 2012), which is based on the expected maximum BRB force under design-level earthquakes according to the AISC Provisions (AISC 2010):

$$\sigma_{pre}A_b = \alpha_{sc}\beta_c\omega f_y A_{BRB} \tag{4.3}$$

where α_{sc} is a parameter that reflects the separation between the preload of the SMA tendons and the expected BRB steel core force taking account of the strain hardening effect as well as the difference between the tensile and compressive behaviour. $\alpha_{sc} \geq$

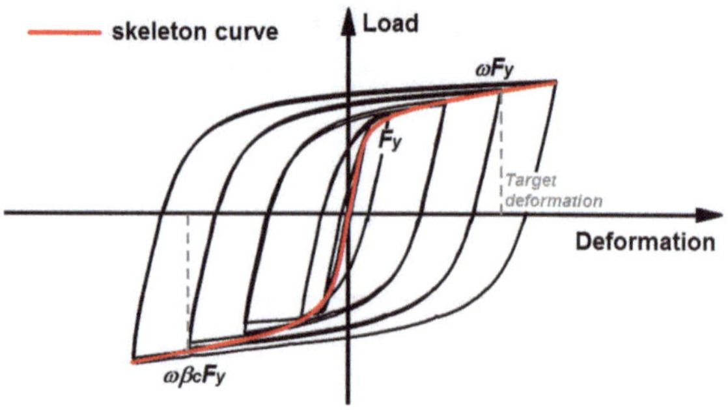

Fig. 4.12 Illustration of typical BRB force–deformation response

1.0 is recommended such that adequate self-centring behaviour is achieved (Miller et al. 2012). As illustrated in Fig. 4.12, the strain hardening adjustment factor, ω, is the ratio of the maximum tension force of the considered BRB at target deformation to the measured tensile yield force F_y. Given that the compressive force of a BRB is often larger than the tensile force, a compression strength adjustment factor, β_c, is defined as the ratio of the maximum compression force to the maximum tension force. For a BRB frame, the product $\omega\beta_c$ is used as part of the capacity design process for proportioning the beams, columns, and connections so that they remain essentially elastic. According to the industrial survey, ω typically ranges from 1.3 to 1.5, and β_c ranges from 1.05 to 1.15 (Kersting et al. 2015).

Based on an experimental study conducted by Miller et al. (2012), the typical load–deformation hysteretic curve of the hybrid brace is shown in Fig. 4.13. The brace exhibits a very stable hysteretic behaviour with acceptable residual deformation. Compared with BRBs, the hybrid brace exhibits much smaller residual deformation due to the recovery action provided by the SMA tendons. At 2% brace elongation (or compression), the residual deformations (zero force crossings) are approximately 0.93% and 0.6% in tension and compression, respectively, corresponding to an average recovery rate on the order of 60%. It is also noted that the unloading stiffness at the zero force crossings is quite low, a case which is expected to have limited detrimental influence on the self-centring behaviour of the overall system during dynamic shakedown. In fact, past studies (Eatherton and Hajjar 2011; Eatherton et al. 2014) confirmed that braces exhibiting such hysteretic behaviour can practically eliminate the residual inter-storey drifts. Figure 4.14 shows the typical EVD values of the brace specimens examined by Miller et al. (2012) as a function of axial displacements. It is seen that the maximum EVD can be close to 25%, which is larger than the typical value of 8–15% for pure SMA-based devices. This is clearly attributed to the effective supplementary energy dissipation offered by the inner BRBs.

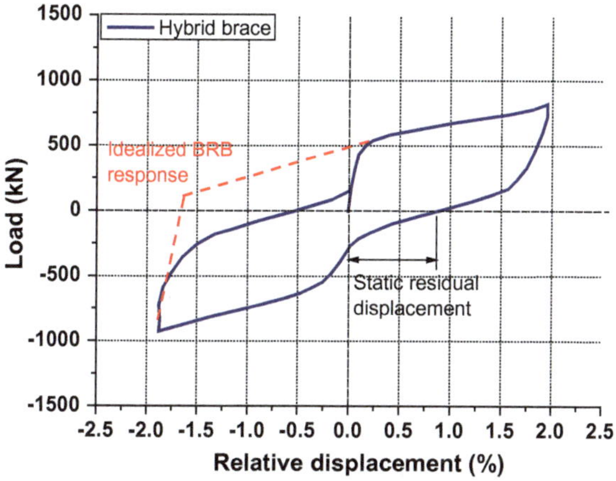

Fig. 4.13 Typical hysteretic response of hybrid self-centring brace with combined BRB system and SMA tendons (Miller et al. 2012)

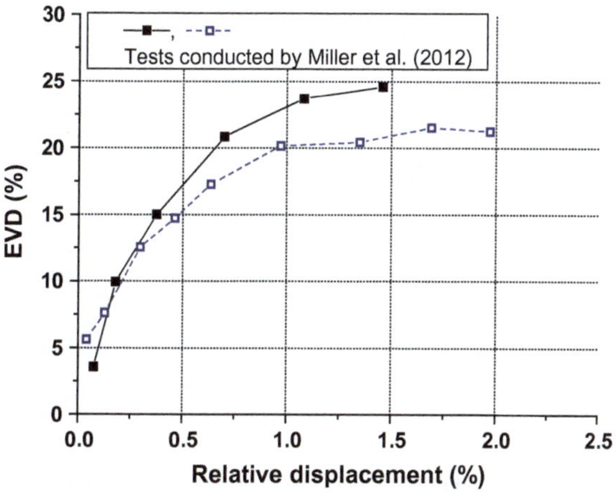

Fig. 4.14 Energy dissipation characteristics of hybrid self-centring brace with combined BRB system and SMA tendons (Miller et al. 2012)

4.4 Self-centring Braces with SMA Ring Springs

4.4.1 Working Principle and Fundamental Behaviour

This section introduces another type of self-centring braces equipped with superelastic SMA ring spring-based devices. The working mechanism and fundamental property of individual SMA ring springs have been discussed in Chap. 2, and now, we will further explain how these springs are used for bracing members. In contrary to the SMA wires and bolts which are effective in tension, the ring spring system is only effective in compression. Therefore, the kernel devices should be appropriately developed such that the ring springs are always under compression regardless of the loading direction of the brace.

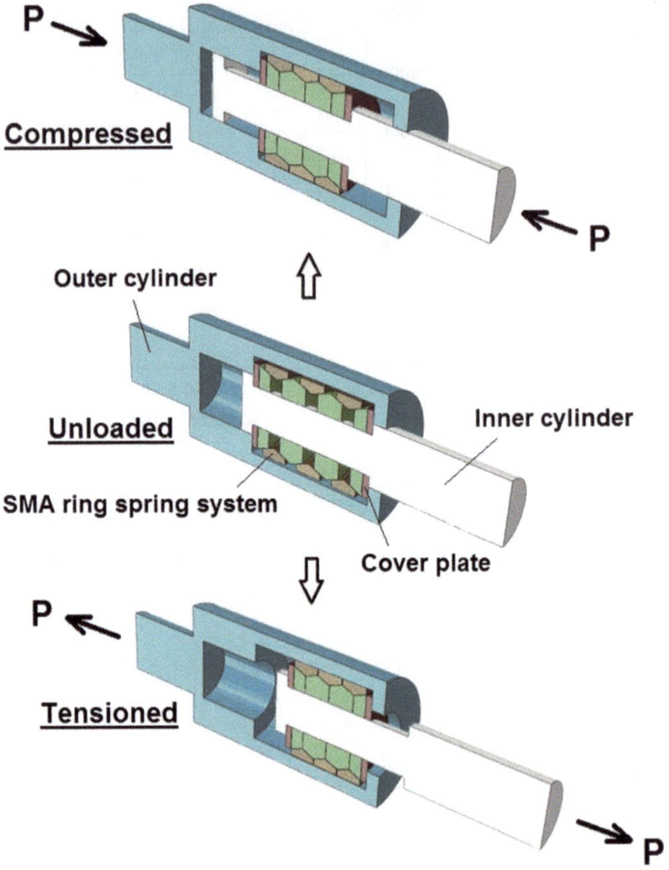

Fig. 4.15 Fundamental working principle of kernel device incorporating SMA ring spring system

Figure 4.15 illustrates the possible working principle of the device acting as a kernel part of a self-centring brace. The device is specially designed for accommodating the SMA ring springs, and it is essentially a bidirectional damper where the key components include an outer cylinder, an inner cylinder, an SMA ring spring system and two cover plates. The inner cylinder passes through the HSS inner rings of the ring spring system, and the outer SMA rings are constrained via the two cover plates at the two ends of the ring spring system. The external diameters of the inner cylinder, as well as the internal diameters of the outer cylinder, are appropriately stepped to guide the movement of the ring spring system.

Assuming that the right end of the outer cylinder is fixed (see the orientation of the device as shown in Fig. 4.15), the inner cylinder moves leftwards when the device is in compression. The right cover plate is pushed to move leftwards, but the left cover plate is constrained by the step of the outer cylinder. This action leads to compression of the ring spring system. When the device is in tension, the movement

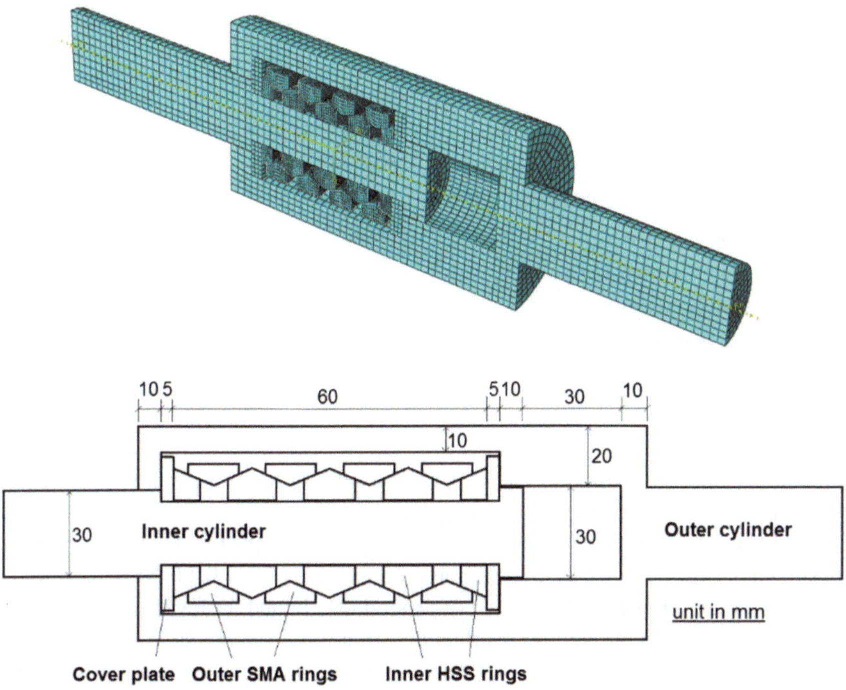

Fig. 4.16 Numerical model of a kernel device with SMA ring spring system

of the inner cylinder pulls the left cover plate to move rightwards. At the meantime, the right cover plate is constrained by the step of the outer cylinder, and as a result, the ring spring system is also in compression.

One advantage of the SMA ring spring-based self-centring braces is the adjustable deformability which is determined by the number of rings. Taking a 5-metre brace with a target relative elongation/compression of 3% under the maximum considered earthquake (MCE) for instance, the required axial displacement of the brace is 150 mm. If SMA tendons are used in this brace, and considering that a maximum 6% strain is permitted for the tendons, the required working length of the SMA tendons is 0.15 m/0.06 = 2.5 m which takes up half of the entire brace length. If SMA ring springs are used, where each SMA outer ring can provide a displacement of, say, 10 mm; then only 15 rings are required. Depending on the detailed design configuration of the rings, the total required length of the ring spring columnar assembly is typically less than 1.0 m.

The fundamental behaviour of the device can be first understood via a numerical study (Fang et al. 2015). The model of the device is shown in Fig. 4.16, and the representative material properties for the SMA outer rings are given in Table 4.1. Two values of coefficient of friction (C_F) between the outer and inner rings are

Table 4.1 Material properties of SMA used in ABAQUS Fang et al. (2015)

Material properties	Value
Forward transformation (start) stress σ_{Ms}	400 MPa
Forward transformation (finish) stress σ_{Mf}	500 MPa
Reverse transformation (start) stress σ_{As}	250 MPa
Reverse transformation (finish) stress σ_{Af}	150 MPa
Young's modulus (austenite) E_A	70 GPa
Young's modulus (martensite) E_M	40 GPa
Maximum transformation strain ε_L	5%
Poisson's ratio (austenite) v_A	0.33
Poisson's ratio (martensite) v_M	0.33

considered, i.e. $C_F = 0.1$ and $C_F = 0.4$, representing the typical cases of greased and non-greased surface treatments, respectively.

The hysteretic responses of the FE models considering the two friction conditions are shown in Fig. 4.17. When low friction is considered ($C_F = 0.1$), the device model provides a maximum load resistance of around 40 kN and an EVD of 19.8% at its maximum displacement, and no residual deformation is observed. When the coefficient of friction increases to $C_F = 0.4$, a completely different hysteretic response is exhibited. Due to the more significant friction action, the maximum load resistance attains 70 kN during the first cycle. However, the recoverability of the SMA outer rings is not enough to overcome the friction force upon unloading, and as a result, a remarkable residual displacement is induced in the device, causing a gap between the outer and inner cylinders. This gap would be permanent because the device itself is unable to pull the SMA spring system back to its original shape, unless the device is dismantled and the SMA ring springs are repositioned and greased. The unrecoverable gap leads to complete losses of the initial stiffness, self-centring capability and energy dissipation of the device in the subsequent cycles.

The numerical study emphasises the importance of maintaining a reasonable balance between self-centring capability and energy dissipation of the device, and the possible benefits from friction for energy dissipation should be carefully utilised. For the SMA ring spring systems designed with practical geometric dimensions, relatively low friction between the outer and inner rings, i.e. a greased contact face, is recommended. The presence of the grease not only encourages a larger restoring force, but also helps reduce the scratch over the taper surfaces.

4.4.2 Experimental Investigation of Kernel Device

Figure 4.18 shows a prototype kernel device equipped with the SMA ring spring system (Fang et al. 2019; Wang et al. 2019). The main components and the working

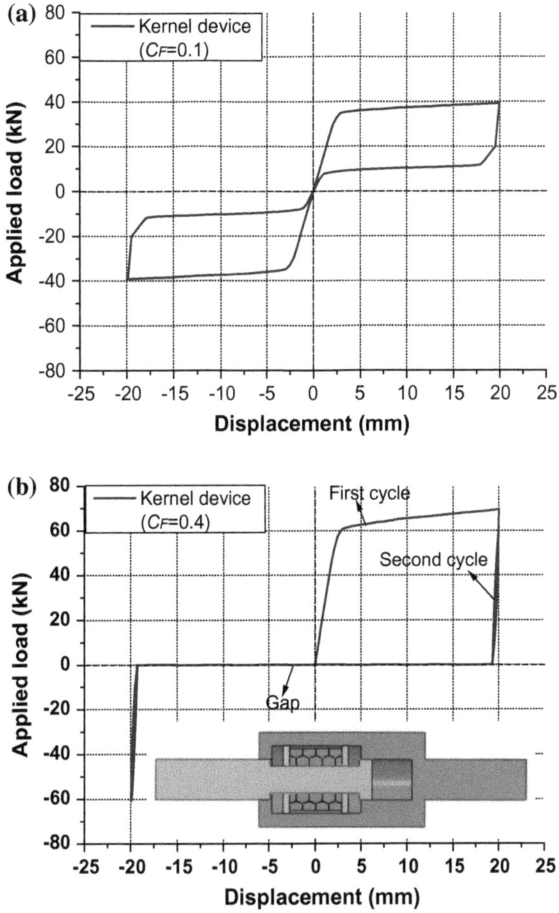

Fig. 4.17 Numerical prediction of hysteretic responses of kernel device with SMA ring spring system: **a** $C_F = 0.1$, **b** $C_F = 0.4$

principle of the device are illustrated in more detail in Fig. 4.19, in comparison to the simplified illustration previously given in Fig. 4.15. The main technical challenges associated with the fabrication of such devices are the determination of the appropriate manufacturing processes (including annealing) of the large-scale SMA outer rings, as well as the steps to assemble the entire device.

As discussed in Chap. 2, the optimum annealing scheme is related to the size of the SMA elements. The required annealing temperature and annealing duration for relatively large size SMA elements are often higher or longer than those for the ones with smaller sizes. For the prototype device, the SMA outer rings are made of commercial Ti-50.8at.%Ni superelastic SMA material. The size of the SMA outer rings is larger than that presented in Chap. 2, where the external diameter (D_e) of the outer ring is 80 mm, and the ring thickness (t) is 6 mm. The taper angle (α) is 21.8°, and the height (H) of each ring is 20 mm. The alloy was processed by hot rolling at temperatures between 800 and 900 °C, and was then forged and straightened to

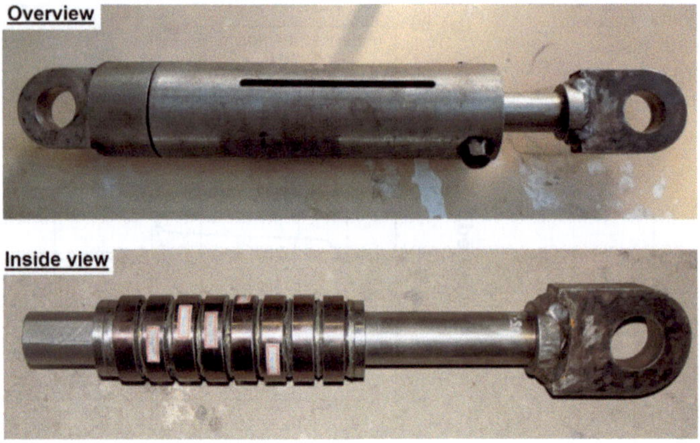

Fig. 4.18 Prototype kernel device with SMA ring springs for self-centring braces

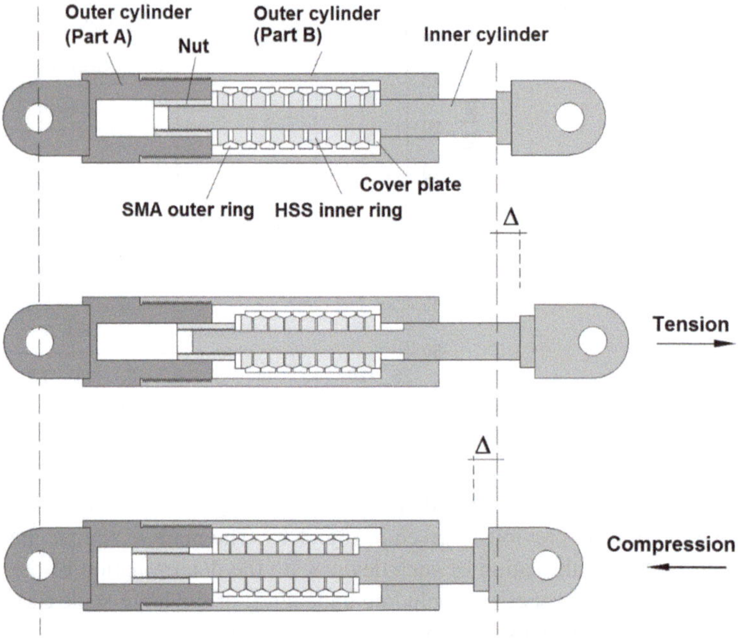

Fig. 4.19 Basic components and working principle for prototype kernel device with SMA ring springs

a final diameter of 80 mm. The nominal austenite finish temperature of the material is around 0 °C, ensuring superelasticity at room temperature. The raw bars were subsequently cut into a series of 20 mm-thick pieces which were then hollowed and machined into the desired ring shape. The inner rings for the prototype device were produced from 38CrMoAl alloy.

Past experience revealed that an annealing temperature of 450 °C could lead to satisfactory hysteretic behaviour of large scale commercial Ti-50.8at.%Ni SMA elements (Wang et al. 2016). Therefore, the individual SMA outer rings for the device were annealed at 450 °C for three different trail durations (i.e. 15, 23 and 30 min) before water quenching. The test results of the three cases are shown in Fig. 4.20. It seems that an annealing duration of 30 min could achieve the best mechanical performance in terms of recoverability.

Figure 4.21 illustrates the recommended installation steps for the kernel device. To facilitate assembly, the outer cylinder can be divided into two connectable parts. The inner cylinder is first placed in position (Step 1), and then part of the outer cylinder is installed (Step 2). At this stage, the outer cylinder is free to move in the axial direction. Then the bottom cover plate, the SMA ring spring system, and the top cover plate are successively placed in position by passing through the inner cylinder (Steps 3–4). Preload is applied subsequently by gradually tightening up the nut, until the required precompression is attained (Step 5). For the prototype device specimen shown in Fig. 4.18, a precompression of 3.5 mm was applied to each SMA outer ring. The last step is to install the remaining part of the outer cylinder, until it firmly touches the top cover plate. The typical test setup for examining the kernel device specimen is shown in Fig. 4.22.

Figure 4.23 shows the load–displacement response of the prototype kernel device. Symmetrical flag-shaped hysteretic responses are clearly observed. The device also demonstrates excellent self-centring capability with no residual displacement. The ‘yield’ resistance is approximately 90~100 kN, which is followed by an obvious hardening effect. The hysteretic response is generally stable, although a slight decrease in the transformation plateaus is observed, which is caused by TIF degradation.

Being in line with the design intention, the SMA ring springs are always compressed when the device is subjected to both compression and tension, as shown in Fig. 4.24. Inelastic deformation only occurs in the SMA ring spring group, while the other components such as the outer and inner cylinders remain elastic. The difference in the gap width between neighbouring SMA outer rings could be caused by inconsistent individual SMA ring spring behaviour, as well as, uncertainties related to friction condition and possible minor inclination of the individual rings at large deformation. The uneven displacement is often unavoidable but has little influence on the overall behaviour of the device, so we do not need to worry too much about it. Provided that the SMA ring spring system is appropriately designed, no damage to the device is expected under a fully compressed condition.

It can be observed from the load–displacement curve that the device specimen exhibits a large initial stiffness and an adequate level of ‘yield’ resistance due to the relatively high decompression force of the preloaded SMA ring spring group. The initial tangent stiffness is approximately 223.2 kN/mm. Figure 4.25a further shows

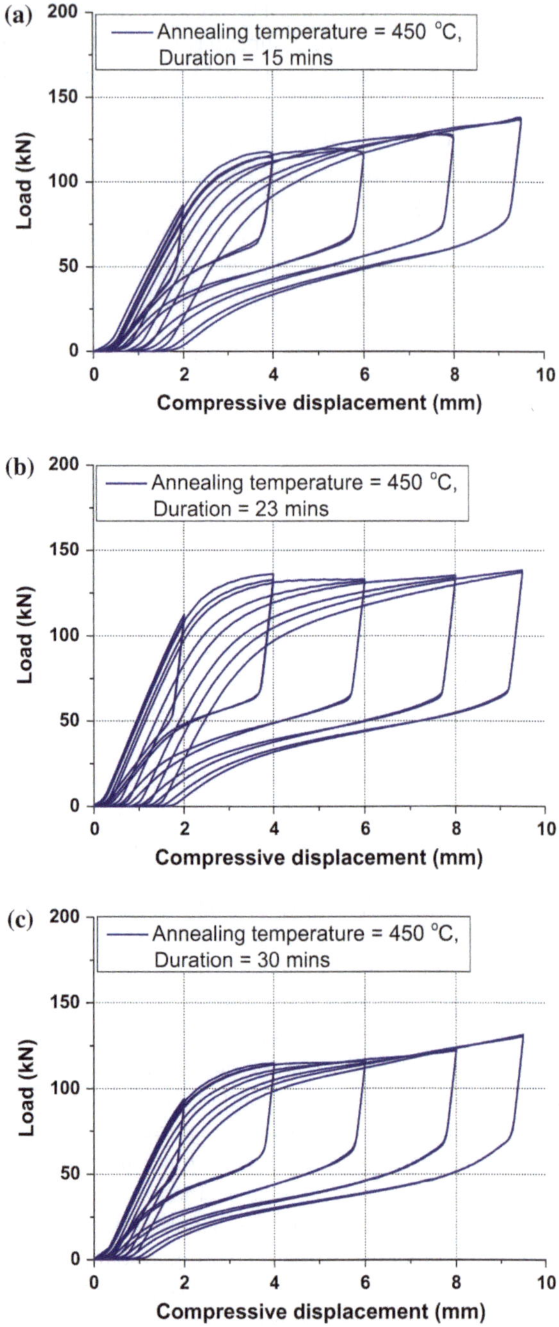

Fig. 4.20 Influence of annealing duration on the mechanical behaviour of SMA outer rings: **a** 15 min, **b** 23 min, **c** 30 min

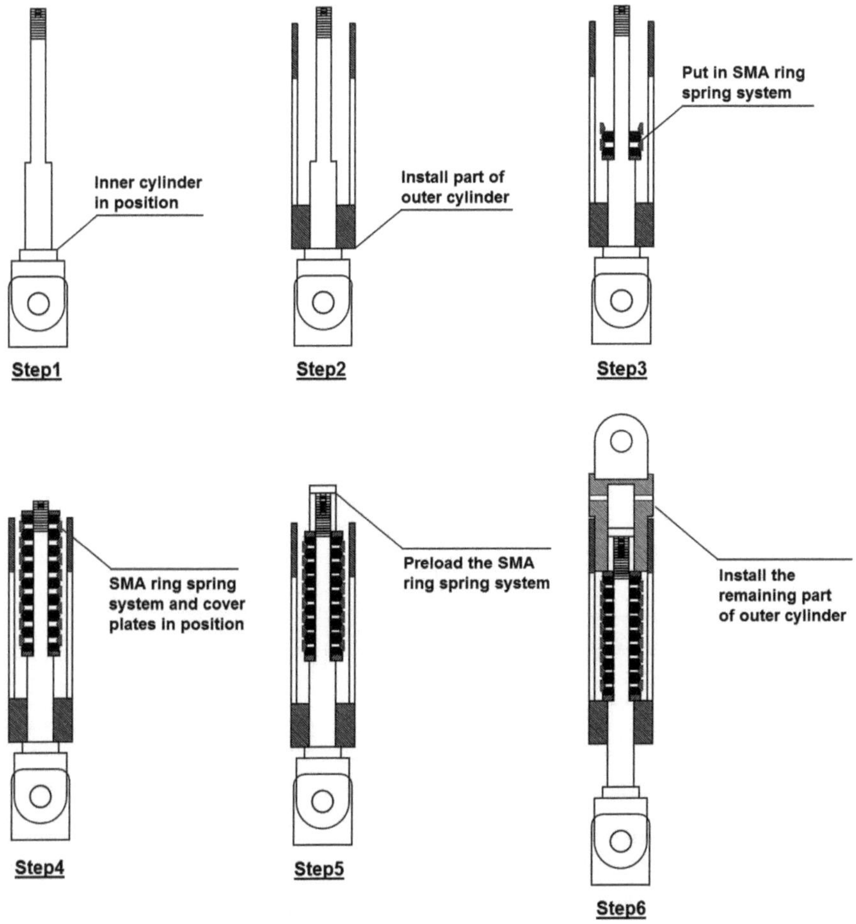

Fig. 4.21 Installation process of kernel device

the variation of the secant stiffness which is defined as the ratio of the peak load to the maximum displacement of the considered cycle. With more pronounced nonlinearity as the amplitude increases, the secant stiffness decreases from around 20 kN/mm in the first cycle to 5 kN/mm in the last cycle.

Figure 4.25b shows that the yield resistance of the device specimen decreases with the increase in amplitude. By the end of the last loading cycle, at least 30% of the yield resistance is lost compared to the initial value at the first cycle. On the other hand, the peak load keeps increasing and exceeds 150 kN at the final loading cycle. This corresponds to a maximum over-strength ratio, i.e. the ratio of peak load to the initial yield load, of more than 1.60. It is recognised that a reasonable over-strength can effectively suppress weak storey response especially when the P-Δ effect is significant at large inter-storey drifts (Eatherton et al. 2014). Nevertheless,

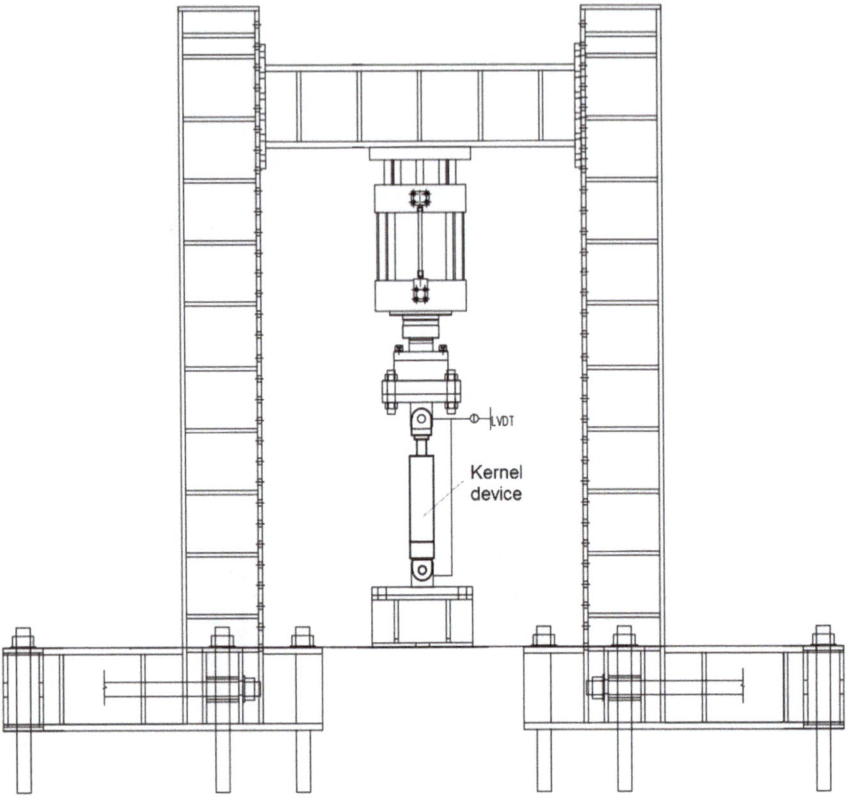

Fig. 4.22 Typical test setup for kernel device

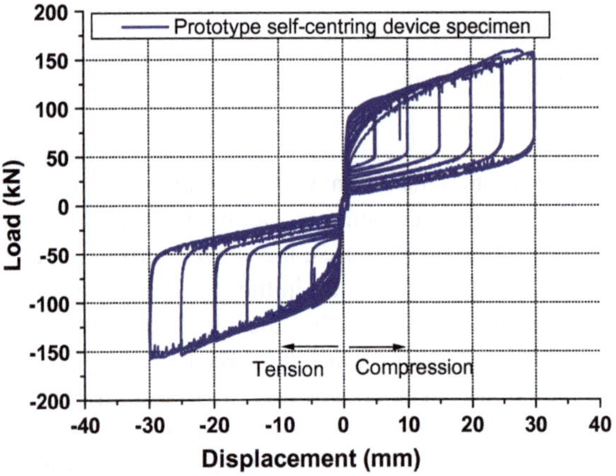

Fig. 4.23 Load–displacement curve of kernel device

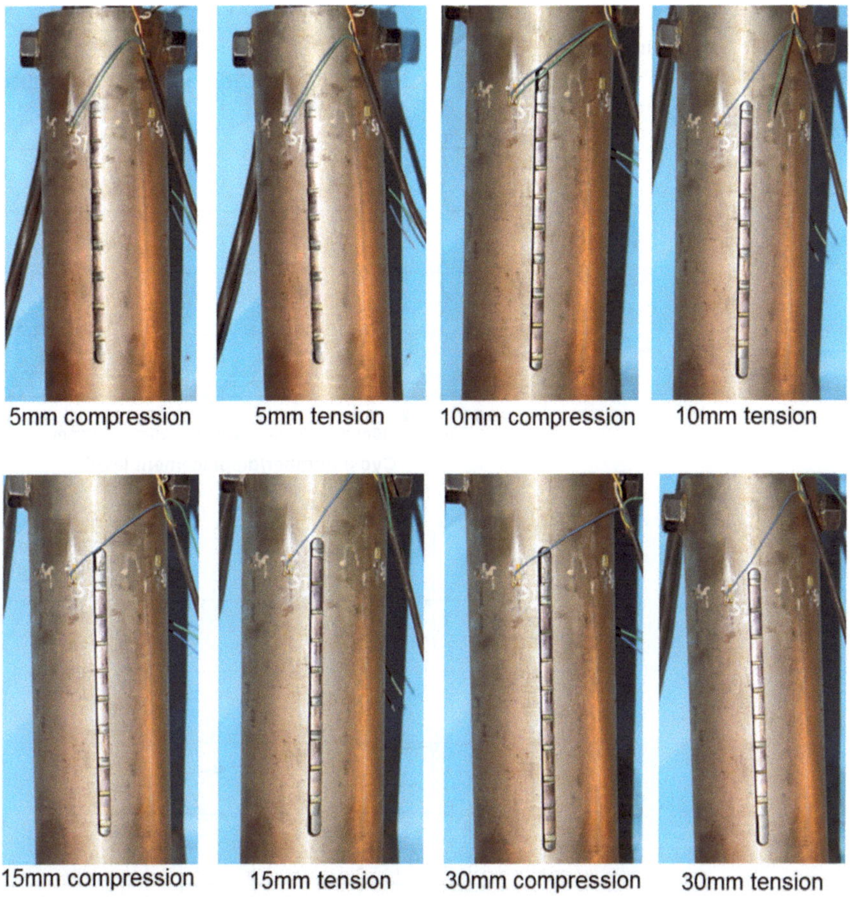

Fig. 4.24 Deformation behaviour of SMA ring springs and kernel device

the effect of over-strength should also be considered when designing the adjacent beams, columns and connections.

As self-centring braces are supposed to effectively control the residual deformation and are preferably reusable against aftershocks or future earthquakes, no residual deformation is wanted during their service life. The self-centring tendency is quantitatively evaluated in Fig. 4.26 through the restoring force F_{rst}, a concept which is similar to the restoring moment M_{rst} discussed previously. An increase in F_{rst} indicates increased self-centring tendency but reduced energy dissipation. For braced frames, system-level analysis suggests that the ratio of the restoring force to the yield resistance should be less than 0.5 to ensure sufficient energy dissipation, and more importantly, to suppress the high-mode effect for high-rise building structures (Qiu and Zhu 2016). As shown in Fig. 4.26, the restoring force of the device ranges

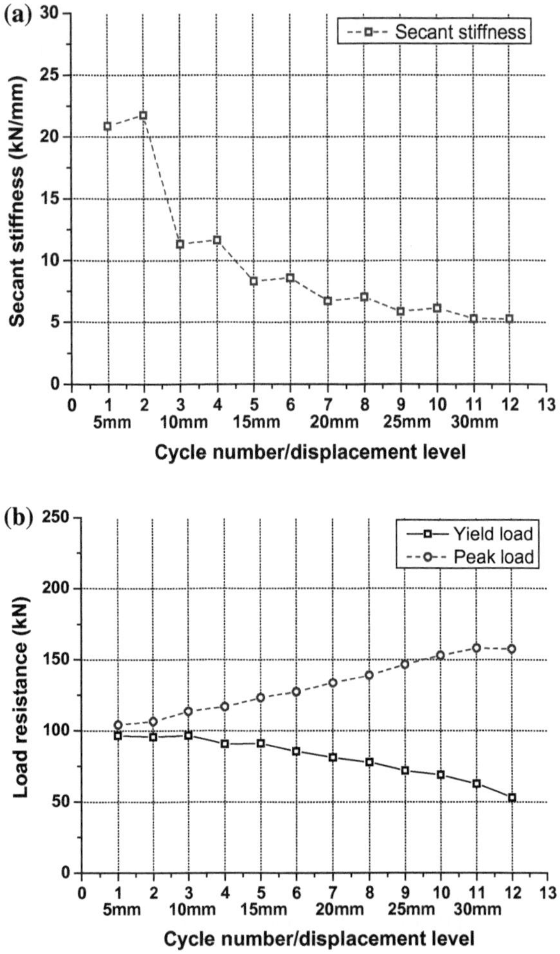

Fig. 4.25 Behaviour of kernel device with loading cycles: **a** secant stiffness, **b** load resistance

from 10 to 35 kN during the entire loading process, corresponding to the restoring force to yield resistance ratios falling between 0.18 and 0.37.

The energy dissipation characteristics of the kernel device are shown in Fig. 4.27. The energy loss per cycle ranges from 0.6 to 5.5 kJ with varying amplitudes, and the EVD is stabilised between 16 and 18.5% during the entire loading procedure. Again, it is good to see that a relatively high EVD is motivated from the beginning of the loading procedure, implying that the device could make a very early contribution to energy dissipation. As mentioned in Chap. 3, the increased EVDs at small displacements largely build on the hybrid friction-superelasticity energy dissipation mechanism of the SMA ring springs. The device can be deemed as an added damping and stiffness (ADAS) damper that is commonly used in braced frames, such as those with chevron type braces, against the lateral load.

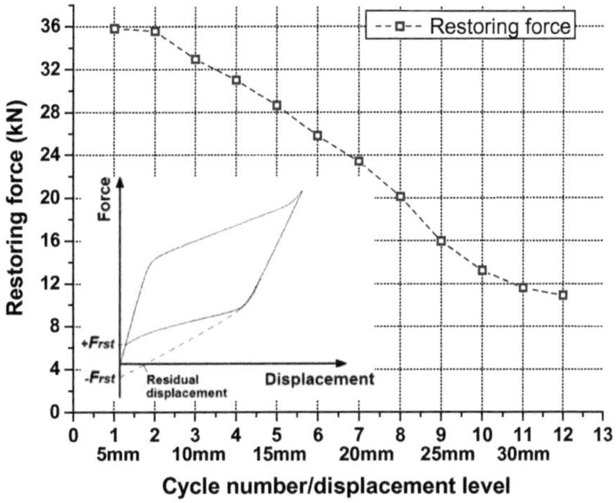

Fig. 4.26 Self-centring capability of kernel device

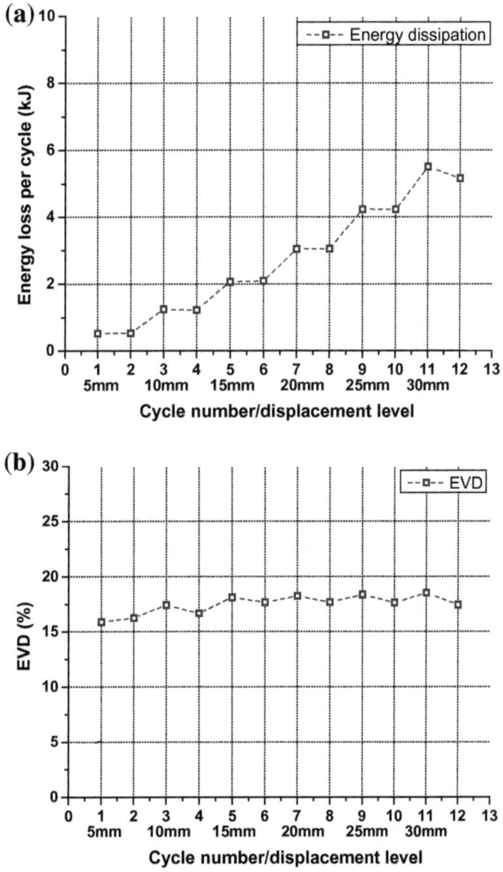

Fig. 4.27 Energy dissipation characteristics of kernel device: **a** energy loss per cycle, **b** EVD

4.4.3 Design of Kernel Device

The performance of the entire kernel device depends significantly on the behaviour of individual SMA ring springs. The analytical expressions presented in Chap. 2 offer a clear relationship between the applied load (P) and the average hoop stress of individual SMA ring springs, as copied in Eqs. (4.4)–(4.5), as well as the relationship between the compressive displacement and the average hoop strain of the outer ring, i.e. Equation (4.6). If the uncertainties related to the material property and the degradation effect of SMA are not particularly considered in practical design, which is a reasonable consideration if the SMA elements are appropriately trained, the hysteretic rule of the kernel device can be determined by these simple equations using the representative material parameters of SMA ascertained by the material supplier. The key design philosophy involves the following major steps and considerations (Fang et al. 2019):

$$P = \Sigma P_V + \Sigma F_V = \pi \sigma A(\tan\alpha + C_F) \quad \text{for loading stage} \tag{4.4}$$

$$P = \Sigma P_V - \Sigma F_V = \pi \sigma A(\tan\alpha - C_F) \quad \text{for unloading stage} \tag{4.5}$$

$$\varepsilon = \frac{\pi \cdot \delta D}{\pi D} = \frac{2\Delta \tan\alpha}{D} \tag{4.6}$$

Step 1: Determination of performance target. The expected performance target of a self-centring brace needs to be determined first, where there are two important aspects, namely minimum yield resistance and deformation demand. The former is essentially the required proportion of the load resistance contributed by the device against normal design actions such as wind, and the latter is determined according to the inter-storey drift demand of the building frame that adopts the self-centring devices. A precompression level, which leads to the associated hoop strain of the SMA outer ring exceeding or being close to the forward transformation strain of the SMA, is recommended (Wang et al. 2017a). A good starting point is to select an appropriate type of SMA ring spring which provides the required yield resistance at the applied precompression. Equations (4.4) and (4.5), which have been elaborated in Chap. 2, can be used for such preliminary calculation, where a provisional friction coefficient of 0.15 may be used (Wang et al. 2017b). Another practical consideration for SMA ring spring selection is that the maximum hoop strain of the SMA outer rings at a fully compressed condition is preferably less than the forward transformation finish strain ε_{Mf} (typically 6–7%). Given this restriction and according to the intended target inter-storey drift, the required number of the SMA outer rings can be determined. It is worth pointing out that the final design of the SMA ring spring systems may involve an iteration and optimization process, as the precompression, yield load and deformability, interrelate with each other. Furthermore, we may simply assume that the remaining components such as the HSS inner ring and the outer and inner cylinders behave rigidly for ease of design.

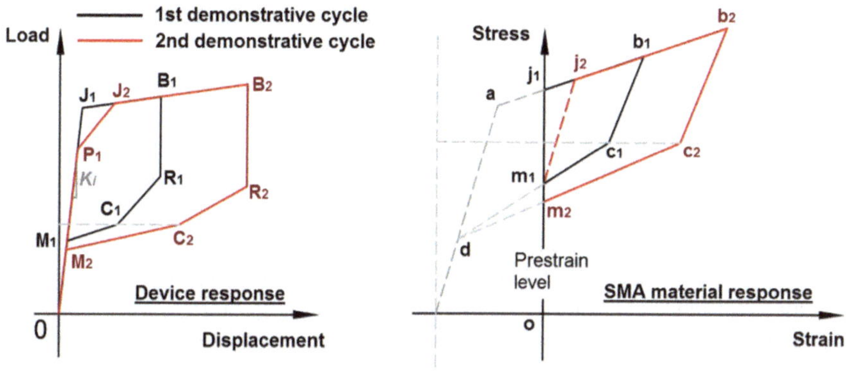

Fig. 4.28 Schematic illustration of design model for kernel device

Step 2: Building load–displacement hysteresis. Once the design of the SMA ring spring systems is ascertained, the representative load–displacement hysteretic curves of the device can be derived. A demonstrative load–displacement hysteretic curve and the corresponding stress–strain responses of the SMA material are schematically illustrated in Fig. 4.28. Let's focus on the 1st demonstrative cycle first. With the obtained stress–strain curve of the SMA material, information which is provided by the material supplier, the construction of the load–displacement hysteresis includes the following featured stages:

(1) Stage OJ_1: this is the initial loading stage of the device, where the yield load (point J_1) corresponds to the prestress of the SMA outer ring (point j_1). The main challenge is the prediction of the initial stiffness which is not only influenced by the elastic stiffness of the various device components but is also affected by possible machining tolerance. This issue also occurs in many other damping and self-centring devices. An empirical value can be used according to the limited test data (Fang et al. 2019):

$$K_{d,ini} = \alpha_{is} F_y \tag{4.7}$$

where $K_{d,ini}$ is the design initial stiffness of the device (unit = kN/mm), F_y is the yield load of the device (point J_1, unit in kN), α_{is} is an empirical factor, which is taken as 2.3 for elastic global analysis (e.g. wind load checking), and 1.0 for inelastic global analysis (e.g. non-linear time-history seismic analysis). Such simplification can be preliminarily used in the absence of detailed test data, although experimental verification is strongly recommended. As the research of self-centring devices is still in its infancy, the most accurate way to determine $K_{d,ini}$, at this stage, is to directly measure the initial stiffness of each device product before its installation and long-term service. This could be deemed as a standard quality conformance inspection procedure. Some other requirements related to machining accuracy (flatness of contact surfaces) may also be stipulated in future design handbooks. It should also be emphasised that Eq. (4.7) is only for estimating the stiffness of the kernel device itself, while the stiffness of the connected (extended) bracing members needs to be calculated additionally in order to obtain the stiffness of the entire brace.

(2) Stage J_1B_1: this corresponds to a 'post-yield' plateau induced by the forward transformation of the SMA material. The load resistance at point B_1 can be derived from Eqs. (4.6) and (4.4) successively.
(3) Stage B_1R_1: this stage corresponds to the immediate reversal of the friction force, where the stress and strain conditions at point b_1 are not changed. The load resistance at point R_1 can be calculated from Eq. (4.5).
(4) Stage R_1C_1: this stage corresponds to the martensitic unloading path, where the stress–strain point of the SMA material moves from b_1 and ends at c_1 which is the beginning of reverse transformation. The load and displacement at point C_1 can be calculated via Eqs. (4.5) and (4.6), respectively.
(5) Stage C_1M_1: this stage corresponds to a reverse transformation path where the stress–strain point of the SMA material moves from c_1 to m_1, starting from the occurrence of reverse transformation and ending at the initial prestrain level. Again, the load at point M_1 is obtained from Eq. (4.5).
(6) Stage M_1O: the device finally returns to its origin with no residual displacement being induced.

Figure 4.29 shows the comparison between the test result and the predicted load–displacement hysteretic curve of the device, noting that the SMA material properties used in the prediction model are the representative ones according to Wang et al. (2016). A good agreement is generally observed, although the load resistance is underestimated by the prediction model at large amplitudes. The uncertainties related to the SMA material property and friction coefficient may form the main sources of discrepancy. In particular, using a consistent friction coefficient (i.e. $C_F = 0.15$) during the entire loading procedure tends to underestimate the actual friction condition between the inner and outer rings when the ring spring system is significantly compressed. The friction effect is even more difficult to predict when the SMA ring spring assembly exhibits a transverse 'global buckling' tendency. This action also causes extra friction between the rings and the surface of the cylinder, noting that the rings are laterally constrained by the cylinders. According to the available test data, an increased friction coefficient of 0.2 could be a more appropriate value for predicting the maximum load resistance of the device at large amplitudes.

Step 3: Consideration of degradation. This is an optional step which allows engineers to include the degradation effect in the design model for the device. As the degradation phenomenon is complex and difficult to capture, an empirical method instead of a fully justified hysteresis rule could be a more feasible solution. Let's have a look at the 2nd demonstrative cycle shown in Fig. 4.28. A reduced yield load is expected for the reloading stage OP_1 after experiencing the 1st demonstrative cycle. In order to predict the load at P_1, an extra variable, namely, a reduction factor χ, can be employed. The reduction factor is related to the peak transient hoop strain (ε_{peak}) achieved during

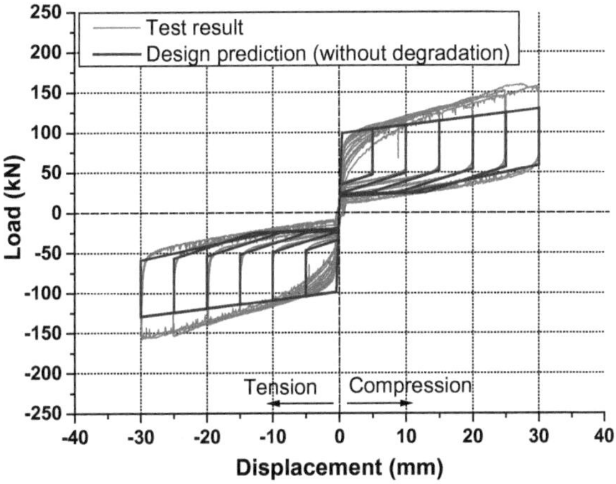

Fig. 4.29 Result from design model without degradation

the previous loading history, and the value can be determined empirically according to the available test data (Fang et al. 2019):

$$\chi = 1 - \alpha_r \left(\varepsilon_{peak} - \varepsilon_{pre}\right) \leq 1.0 \tag{4.8}$$

where ε_{pre} is the hoop strain due to preload and α_r is an empirical parameter with a suggested value ranging between 15 and 25. Employing this simplified approach, the yield load at any cycle can be derived, provided that the peak transient hoop strain during the previous loading cycles is traced. A transitional stage P_1J_2 can be determined based on the idealised ascending stiffness along m_1j_2 of the stress–strain relationship of the SMA. The subsequent stages from J_2 back to M_2 via the points B_2, R_2, and C_2 are similar to those described in the 1st demonstrative cycle. With the above supplementary steps, engineers may choose to take account of such degradation effect in the design model. The 'refined' design prediction is given in Fig. 4.30, where a better agreement with the test result in terms of the yield load can be found.

Step 4: Design for other components. After finalising the design for the SMA ring spring systems, the remaining task is to determine the details of the other components. The extended part of the brace and the connection between the device and the extended bracing member (and the connections between the brace and the frame members) should also be designed. For the device itself, the components such as the outer and inner cylinders should always stay elastic, such that the deformation mainly occurs in the SMA ring springs. Other design considerations include: (1) adequate internal space within the device should be guaranteed to satisfy the maximum stroke of the device, (2) all the threaded parts within the device should be checked to ensure that no failure occurs at these junctions, (3) either an outer layer or an inner layer guide is necessary to prevent the 'global buckling' tendency which may occur in 'long' SMA ring spring columnar assemblies. For the device examined by Fang et al. (2019), the inner cylinder was considered as an inner layer guide. Of course, a certain clearance between the guide and the rings is still necessary to allow free expansion or contraction of the ring springs, (4) the over-strength of the

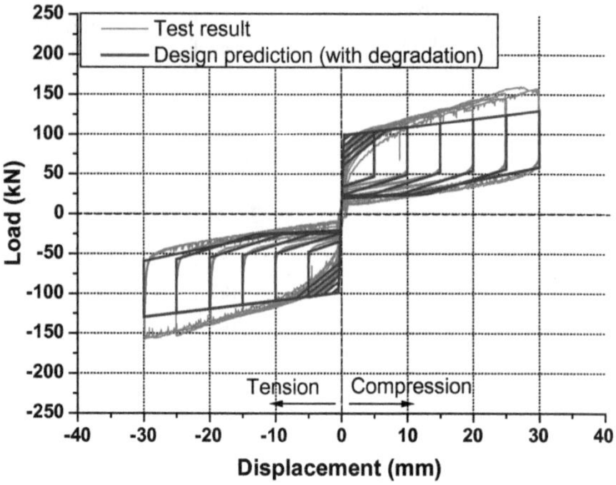

Fig. 4.30 Result from design model taking account of degradation

device should be taken into account when designing the brace, connections, beams and columns.

References

AISC (2010) Seismic provisions for structural steel buildings, (ANSI/AISC 341-10). American Institute of Steel Construction, Chicago, IL, USA

Chou CC, Chen SY (2010) Subassemblage tests and finite element analyses of sandwiched buckling-restrained braces. Eng Struct 32(8):2108–2121

Chou CC, Chen YC, Pham DH, Truong VM (2014) Steel braced frames with dual-core SCBs and sandwiched BRBs: mechanics, modeling and seismic demands. Eng Struct 72:26–40

Chou CC, Wu TH, Beato ARO, Chung PT, Chen YC (2016) Seismic design and tests of a full-scale one-story one-bay steel frame with a dual-core self-centering brace. Eng Struct 111:435–450

Christopoulos C, Tremblay R, Kim HJ, Lacerte M (2008) Self-centering energy dissipative bracing system for the seismic resistance of structures: development and validation. J Struct Eng-ASCE 134(1):96–107

DesRoches R, McCormick J, Delemont MA (2004) Cyclical properties of superelastic shape memory alloys. J Struct Eng-ASCE 130(1):38–46

Dolce M, Cardone D (2001) Mechanical behaviour of SMA elements for seismic applications—part 2 austenite NiTi wires subjected to tension. Int J Mech Sci 43(11):2657–2677

Eatherton MR, Fahnestock LA, Miller DJ (2014) Computational study of self-centering buckling-restrained braced frame seismic performance. Earthq Eng Struct D 43(13):1897–1914

Eatherton MR, Hajjar JF (2011) Residual drifts of self-centering systems including effects of ambient building resistance. Earthq Spectra 27(3):719–744

Erochko J, Christopoulos C, Tremblay R, Choi H (2011) Residual drift response of SMRFs and BRB frames in steel buildings designed according to ASCE 7-05. J Struct Eng-ASCE 137(5):589–599

Erochko J, Christopoulos C, Tremblay R (2015a) Design, testing, and detailed component modeling of a high-capacity self-centering energy-dissipative brace. J Struct Eng-ASCE 141(8):04014193

Erochko J, Christopoulos C, Tremblay R (2015b) Design and testing of an enhanced-elongation telescoping self-centering energy-dissipative brace. J Struct Eng-ASCE 141(6):04014163

Fahnestock LA, Ricles JM, Sause R (2007) Experimental evaluation of a large-scale buckling-restrained braced frame. J Struct Eng-ASCE 133(9):1205–1214

Fang C, Yam MCH, Lam ACC, Zhang YY (2015) Feasibility study of shape memory alloy ring spring systems for self-centring seismic resisting devices. Smart Mater Struct 24(7):075024

Fang C, Wang W, Zhang A, Sause R, Ricles J, Chen YY (2019) Behavior and design of self-centering energy dissipative devices equipped with superelastic SMA ring springs. In Press, J Struct Eng-ASCE

Federal Emergency Management Agency (FEMA) (2012) Seismic performance assessment of buildings, volume 1—methodology. FEMA P-58-1, prepared by the SAC Joint Venture for FEMA, Washington, DC

Hjelmstad KD, Popov EP (1984) Characteristics of eccentrically braced frames. J Struct Eng-ASCE 110(2):340–353

Kari A, Ghassemieh M, Abolmaali SA (2011) A new dual bracing system for improvingthe seismic behavior of steel structures. Smart Mater Struct 20(12):125020

Kersting RA, Fahnestock LA, López WA (2015) Seismic design of steel buckling-restrained braced frames-a guide for practicing engineers. NIST GCR 15-917-34

McCormick J, Aburano H, Ikenaga M, Nakashima M (2008) Permissible residual deformation levels for building structures considering both safety and human elements. In: Proceedings of 14th world conference on earthquake engineering, Seismological Press of China, Beijing

Miller DJ, Fahnestock LA, Eatherton MR (2012) Development and experimental validation of a nickel-titanium shape memory alloy self-centering buckling-restrained brace. Eng Struct 40:288–298

Moradi S, Alam MS, Asgarian B (2014) Incremental dynamic analysis of steel frames equipped with NiTi shape memory alloy braces. Struct Des Tall Spec 23:1406–1425

Ozbulut OE, Hurlebaus S (2012) Application of a SMA-based hybrid control device to 20-story nonlinear benchmark building. Earthq Eng Struct D 41(13):1831–1843

Piedboeuf MC, Gauvin R, Thomas M (1998) Damping behaviour of shape memory alloys: strain amplitude, frequency and temperature effects. J Sound Vib 214(5):885–901

Qiu CX (2016) Seismic-resisting self-centering structures with superelastic shape memory alloy damping devices. PhD thesis, The Hong Kong Polytechnic University

Qiu CX, Zhu SY (2016) High-mode effects on seismic performance of multi-story self-centering braced steel frames. J Constr Steel Res 119:133–143

Qiu CX, Zhu SY (2017) Shake table test and numerical study of self-centering steel frame with SMA braces. Earthq Eng Struct D 46(1):117–137

Sabelli R, Mahin SA, Chang C (2003) Seismic demands on steel braced frame buildings with buckling-restrained braces. Eng Struct 25(5):655–666

Takeuchi T, Hajjar JF, Matsui R, Nishimoto K, Aiken ID (2010) Local buckling restraint condition for core plates in buckling restrained braces. J Constr Steel Res 66(2):139–149

Tremblay R, Bolduc P, Neville R, DeVall R (2006) Seismic testing and performance of buckling-restrained bracing systems. Can J Civil Eng 33(2):183–198

Wang W, Fang C, Liu J (2016) Large size superelastic SMA bars: heat treatment strategy, mechanical property and seismic application. Smart Mater Struct 25(7):075001

Wang W, Fang C, Liu J (2017a) Self-centering beam-to-column connections with combined superelastic SMA bolts and steel angles. J Struct Eng-ASCE 143(2):04016175

Wang W, Fang C, Yang X, Chen YY, Ricles J, Sause R (2017b) Innovative use of a shape memory alloy ring spring system for self-centering connections. Eng Struct 153:503–515

Wang W, Fang C, Zhang A, Liu XS (2019) Manufacturing and performance of a novel self-centring damper with shape memory alloy ring springs for seismic resilience. Struct Control Hlth. https://doi.org/10.1002/stc.2337

Xu X, Zhang YF, Luo YZ (2016) Self-centering eccentrically braced frames using shape memory alloy bolts and post-tensioned tendons. J Constr Steel Res 125:190–204

Zhou Z, Xie Q, Lei XC, He XT, Meng SP (2015) Experimental investigation of the hysteretic performance of dual-tube self-centering buckling-restrained braces with composite tendons. J Compos Constr 19(6):04015011

Chapter 5
Structural Responses: Single-Degree-of-Freedom (SDOF) Systems

Abstract Following the discussions of the working principle and behaviour of the various SMA elements and SMA-based members and devices, it is time to move on to understand how SMA-based structural systems respond to earthquake excitations. A good starting point is to examine the fundamental dynamic responses of single-degree-of-freedom (SDOF) systems with varying parameters that characterise the basic hysteretic behaviour of the structural systems. This chapter starts with a comprehensive seismic evaluation of SDOF systems with a wide spectrum of structural parameters. The primary objective is to quantify the structural and non-structural performances of conventional and novel self-centring systems under both near-fault (NF) and far-field (FF) earthquakes. The analysis involves 5760 different SDOF models with more than 1.45 million statistical response results being processed. According to the available data, two sets of design models that provide the predictions of the inelastic displacement demand and residual displacement response for various systems are given. The results presented in this chapter can form the basis of the performance-based seismic design of SMA-based self-centring structures.

5.1 Introduction

The preceding chapters have shed considerable light on the fundamental behaviour and potential application of various new SMA elements. After showing the unique hysteretic behaviour of the novel SMA-based devices and members subjected to standardised pseudo-static loading, a further understanding of the dynamic response of the SMA-based structures under 'real' earthquake excitations is desired, and the answer to whether these new-generation structural systems are in some way superior to the conventional construction is sought in this chapter.

There are several structural responses that are of great importance to designers. Conventional structural systems designed in accordance with modern seismic codes such as Eurocode 8 (CEN 2004) and ASCE/SEI 7-10 (2010) are deemed to have satisfactory ductility, energy dissipation capacity and collapse resistance against strong earthquakes. However, the satisfactory seismic performance of these structural systems is realised at the cost of considerable residual drifts with the damage concen-

C. Fang and W. Wang, *Shape Memory Alloys for Seismic Resilience*, https://doi.org/10.1007/978-981-13-7040-3_5

trated at major structural members such as beams, connection zones and braces. A residual inter-storey drift exceeding 0.5% after earthquakes may lead to prohibitively high repair cost for the structure (McCormick et al. 2008). Unfortunately, the chance of exceeding this threshold is quite high for code-compliant conventional structural systems, as mentioned in Chap. 4. Therefore, apart from the maximum structural deformation which is directly related to the damage to both the major and affiliated structural members, residual deformation is another important metric indicating the reparability and resilience performance of the system. Moreover, past earthquakes revealed that injuries, fatalities, repair costs and disruption time caused by non-structural failure could far exceed those associated with structural damages (Perry et al. 1994). Therefore, the peak absolute acceleration response, which is associated with the damage condition of the building contents, is an essential quantity worthy of investigation.

The response of structures to earthquakes also depends on the intensity and frequency components of the ground motion. Special attention should be paid to near-fault pulse-like earthquakes. The importance of including near-fault ground motions in the analysis lies in the fact that in practice, many high-performance structural systems (e.g. SMA-based self-centring structures) are likely to be adopted in strong earthquake regions near proven faults. Near-fault ground motions can be characterised by large, long-period velocity pulses in the fault-normal direction when the fault rupture propagates towards the site, normally with a speed close to the shear wave velocity. In this case, significant seismic energy is released in a short time at the 'forward-directivity' site, causing much higher demands for engineering structures compared with the case of far-field earthquakes (Fang et al. 2018; Alavi and Krawinkler 2004; Gerami and Abdollahzadeh 2015). 'Fling step', which occurs parallel to the strike or dip directions, is another typical characteristic of near-fault ground motions, which are featured by a unidirectional large-amplitude velocity pulse with a permanent offset of the ground (Yadav and Gupta 2017). These characteristics have been identified in many historical strong earthquakes, including the 1979 Imperial Valley, 1992 Landers, 1994 Northridge, 1995 Kobe and 1999 Chi-Chi earthquakes. Although the detrimental effects of forward directivity and fling step on conventional structural systems have been recognised to a certain extent, the performance of self-centring structures under near-fault earthquakes has not been extensively investigated so far.

How self-centring structures respond to the different types of earthquake excitations is the first question to be answered before we consider using the SMA technology in our structural design. Perhaps, a good starting point is to examine the fundamental dynamic responses of single-degree-of-freedom (SDOF) systems with varying parameters that characterise the basic hysteretic behaviour. Although SDOF-based analysis has certain limitations, e.g. the higher mode of response and P-Δ effects are not considered, the tendency revealed from the analysis can at least quantify the influences of a broadened range of parameter matrix on the key dynamic performances of structures in a statistical and effective way.

This chapter concentrates on the peak/residual deformation and acceleration responses of idealised self-centring SDOF systems under suites of ground motions, covering both far-field and pulse-like near-fault earthquakes. Conventional elasto-plastic SDOF systems are also considered for comparison purpose. The SDOF analysis data form the basis of design models that estimate the inelastic displacement ratio (IDR) and residual displacement response of the considered systems. These models will be discussed in detail in Sects. 5.5 and 5.6. After identifying the key parameters that could affect the system-level responses, the next chapter will go on with a more comprehensive study on multi-storey framed buildings, including both conventional ones and those incorporating SMA members.

5.2 Fundamentals of SDOF Analysis

SDOF systems can reliably capture the fundamental dynamic response of regular structures dominated by the first mode of vibration. In fact, many design parameters stipulated in the current seismic design standards were derived based on SDOF analysis. Bilinear elasto-plastic hysteretic oscillators or multilinear ones are the most commonly considered SDOF systems in the existing analysis (Kawashima et al. 1998; Hatzigeorgiou and Beskos 2009; Dimakopoulou et al. 2013; Hou and Qu 2015; Baltzopoulos et al. 2016; Durucan and Durucan 2016). There have been several studies particularly looking into the behaviour of flag-shaped self-centring SDOF systems, with the focus on their peak and residual displacement responses under far-field earthquakes (Christopoulos et al. 2003; Karavasilis and Seo 2011). The key findings and trends revealed in those studies are the main basis of the current (and future) performance-based earthquake design framework.

5.2.1 Basic Parameters

As shown in Fig. 5.1, an idealised SDOF system consists of a lumped mass supported by a 'spring'. The governing equation of motion of an elastic SDOF system under earthquake excitations is expressed by

$$m\ddot{u} + c\dot{u} + ku = -m\ddot{u}_g(t) \tag{5.1}$$

where u is the relative displacement of the system, m is the mass, c is the damping coefficient, k is the elastic stiffness and $\ddot{u}_g(t)$, with the upper dots standing for time derivatives, is the ground acceleration with time. The inherent dynamic properties of the system include natural period of vibration T and viscous damping ratio ζ:

$$T = 2\pi\sqrt{m/k} \tag{5.2}$$

$$\zeta = c\Big/2\sqrt{mk} \tag{5.3}$$

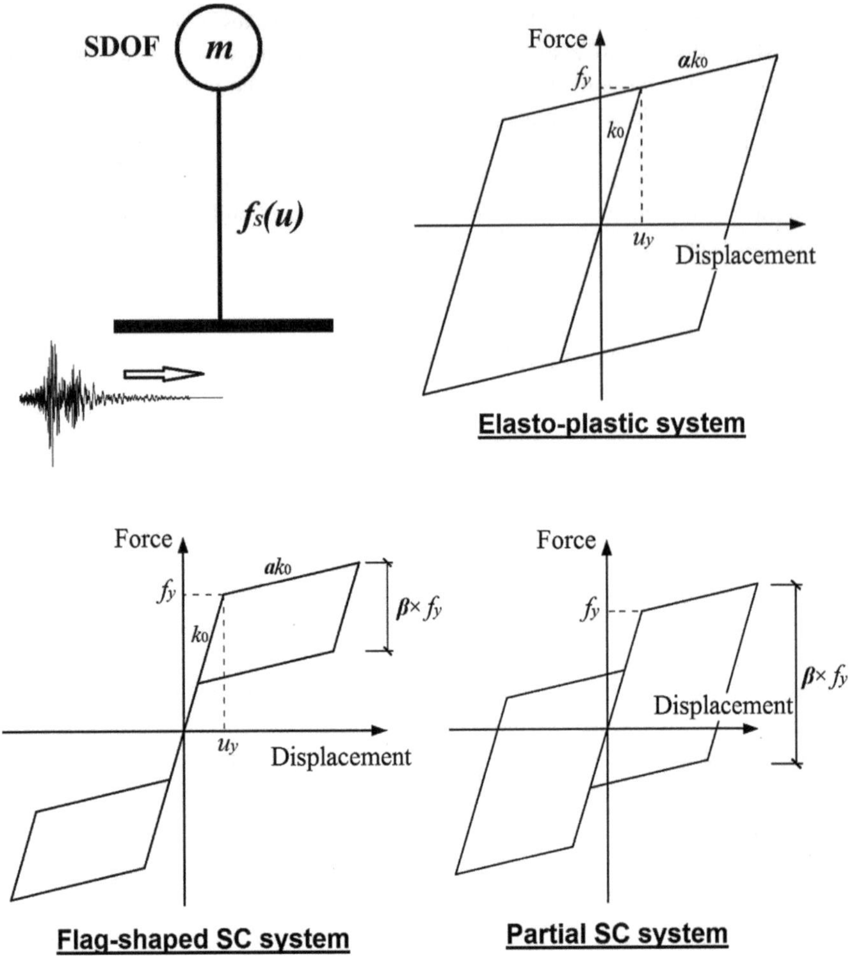

Fig. 5.1 Illustration of typical SDOF systems

Equation (5.1) can be easily extended to nonlinear SDOF systems, as expressed in a more general form by

$$m\ddot{u} + c\dot{u} + f_S(u) = -m\ddot{u}_g(t) \tag{5.4}$$

where $f_S(u)$ is the nonlinear restoring force with the relative displacement u. Figure 5.2 shows the typical force–displacement response of an elasto-plastic system, together with the response of a linear system having the same mass and initial stiffness. Assuming that the maximum force of the linear system under a certain earthquake excitation reaches f_{el}, the strength ratio R, also known as response modification coefficient in a codified term (ASCE 2010), can be defined as

Fig. 5.2 Force–displacement responses of elastic and elasto-plastic systems

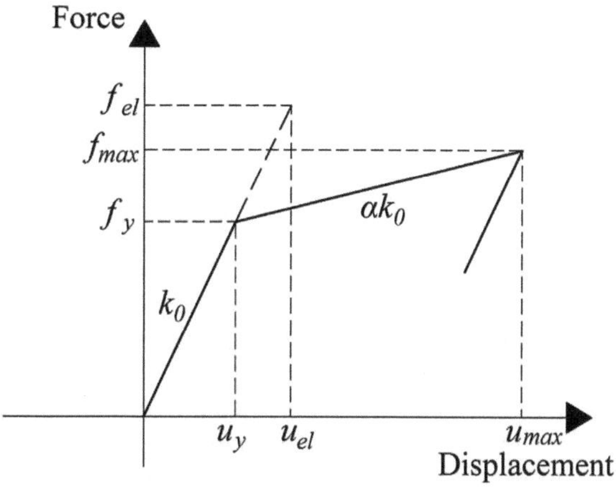

$$R = f_{el} \big/ f_y \tag{5.5}$$

where f_y is the yield strength of the corresponding elasto-plastic system. The ductility demand (μ) and inelastic displacement ratio (IDR) are defined in Eqs. (5.6) and (5.7), respectively

$$\mu = u_{max} \big/ u_y \tag{5.6}$$

$$IDR = u_{max} \big/ u_{el} = \mu \big/ R \tag{5.7}$$

As revealed in the previous chapters, SMA components exhibit typical flag-shaped hysteretic responses. It is reasonable to use an idealised flag-shaped model to mimic a SMA-based self-centring system, just like the conventional yielding systems which are often simulated by idealised elasto-plastic models. As illustrated in Fig. 5.1, the load–displacement curve of a self-centring system first follows a linear branch with an initial stiffness k_0 and then reaches the yield strength f_y before advancing into the 'inelastic' plateau. Being different from the elasto-plastic system, the yield point of the self-centring system is not induced by material yielding, but is instead attributed to forward transformation of SMA or decompression of PT elements.

For either conventional elasto-plastic systems or flag-shaped self-centring systems, the strain hardening (or softening) response in the inelastic range is depicted by a post-yield stiffness ratio α. The practical value of α for common lateral resisting systems such as steel moment resisting frames (MRFs) and braced frames typically ranges between 3 and 5%, although a larger value is possible for self-centring systems with PT-based members.

Apart from the basic parameters such as yield strength f_y, initial stiffness k_0 and post-yield stiffness ratio α, the complete force–displacement hysteretic relationship of a flag-shaped self-centring system needs to be defined by an additional parameter, namely, energy dissipation factor β. As illustrated in Fig. 5.1, this parameter describes how 'full' the hysteretic loop is. When β increases from 0.0 to 2.0, the system transforms from an elastic bilinear system (with no energy dissipation) to a full elasto-plastic system. We can define a 'partial' self-centring response when β ranges between 1.0 and 2.0, in which case a certain residual displacement is induced, but the hysteresis is not full. An identical slope of the loading and unloading plateaus are often assumed for self-centring SDOF systems, and this is generally true for typical SMA- or PT-based self-centring devices/members. Figure 5.3 further illustrate the load–displacement hysteretic curves of SDOF systems with varying α and β values.

In practice, the following structural parameters are of major interest to engineers: (1) natural period of vibration T, (2) strength ratio R, (3) post-yield stiffness ratio α, (4) energy dissipation factor β and (5) total viscous damping ratio ζ. The practical range of the T is between 0.1 and 3.0 s which adequately covers the fundamental periods of most multi-storey and high-rise building structures.

The practical strength ratio R ranges from 1.0 (elastic system) to 8.0, as stipulated in major design codes (e.g. ASCE 2010). A larger value of R is employed for structures with a higher ductility supply. For example, R is taken as 8.0 for steel special moment resisting frames, and is from 2.0 to 5.5 for reinforced masonry shear walls.

For the inelastic behaviour, the practical range of α is 0.0–0.1, and that of β is 0.5–2.0. The total viscous damping ratio ζ typically ranges from 2 to 30% of critical, where $\zeta \geq 5\%$ represents the cases where additional damping devices are used. As one of the most commonly used rate-dependent dampers, a fluid viscous damper operates on the principle of fluid (e.g. silicone oil) flow through orifices, and the output damping force (f_d) is equal to

$$f_d = c|\dot{u}|^{\upsilon}\text{sgn}(\dot{u}) \tag{5.8}$$

where sgn is the signum function and υ is a velocity exponent that is related to the damper nonlinearity. The value of υ is determined by the piston head orifice details and practically ranges from 0.3 to 1.0 for seismic applications. In order to have a

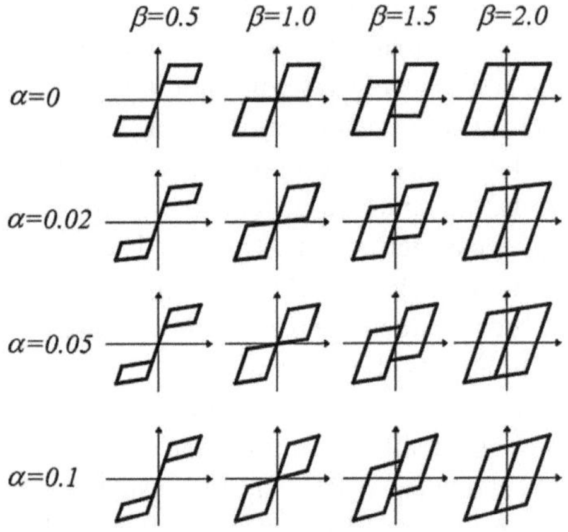

Fig. 5.3 Illustration of SDOF systems with varying α and β values

consistent comparison, υ is often taken as unity, which characterises an ideal linear viscous dashpot.

5.2.2 Engineering Demand Parameters

The key engineering demand parameters of practical interest include maximum displacement (u_{max}), residual (permanent) displacement (u_{res}) and peak absolute acceleration (a_{max}). These quantities can be obtained via numerical approaches such as the central difference method or Newmark's method along with Newton–Raphson iteration. Commercial or open source software such as Matlab (MathWorks 2010) and OpenSees (Mazzoni et al. 2006) can be conveniently used to integrate the nonlinear equation of motion. Note that, the integration time step should be sufficiently small to ensure an accurate prediction of the time-history results, and u_{res} should be obtained in each dynamic analysis by recording the displacement at a stage well beyond the main shaking to allow for vibration decay.

5.2.3 Ground Motions

Nonlinear time-history analysis is commonly used for obtaining the real-time dynamic responses of SDOF systems subjected to earthquake excitations. This procedure needs ground motion inputs, which are often prepared as data files consisting of ground acceleration time histories. These ground motions can be either recorded or artificially produced, and can be scaled where and when necessary.

As just mentioned, there is a consensus among the community of earthquake engineers that near-fault ground motions have some special characteristics which distinguish them from far-field ground motions. Figure 5.4 illustrates the typical ground acceleration and the associated ground velocity and displacement time histories of selected near-fault ground motions exhibiting either forward directivity or fling step. Forward directivity produces large velocity amplitudes with two- or multi-sided velocity pulses but with an unremarkable permanent ground displacement. On the other hand, fling step typically produces a permanent ground offset with a corresponding one-sided dominated velocity pulse, as can be seen from the velocity and displacement time histories. In general, the pulse-like nature is best observed in the velocity time history, while it can be hardly identified from the acceleration time history alone.

Peak ground velocity-to-acceleration ratio (PGV/PGA ratio) and pulse period (T_p) are two important parameters of near-fault ground motions. The PGV/PGA ratio is to some extent related to the significance of the velocity pulse, where a higher ratio indicates a more obvious pulse-like effect of the ground motion (Loh et al. 2002). Pulse period (T_p) is another important factor which affects the structural response. While consensus has not been fully reached so far for the determination of T_p, the

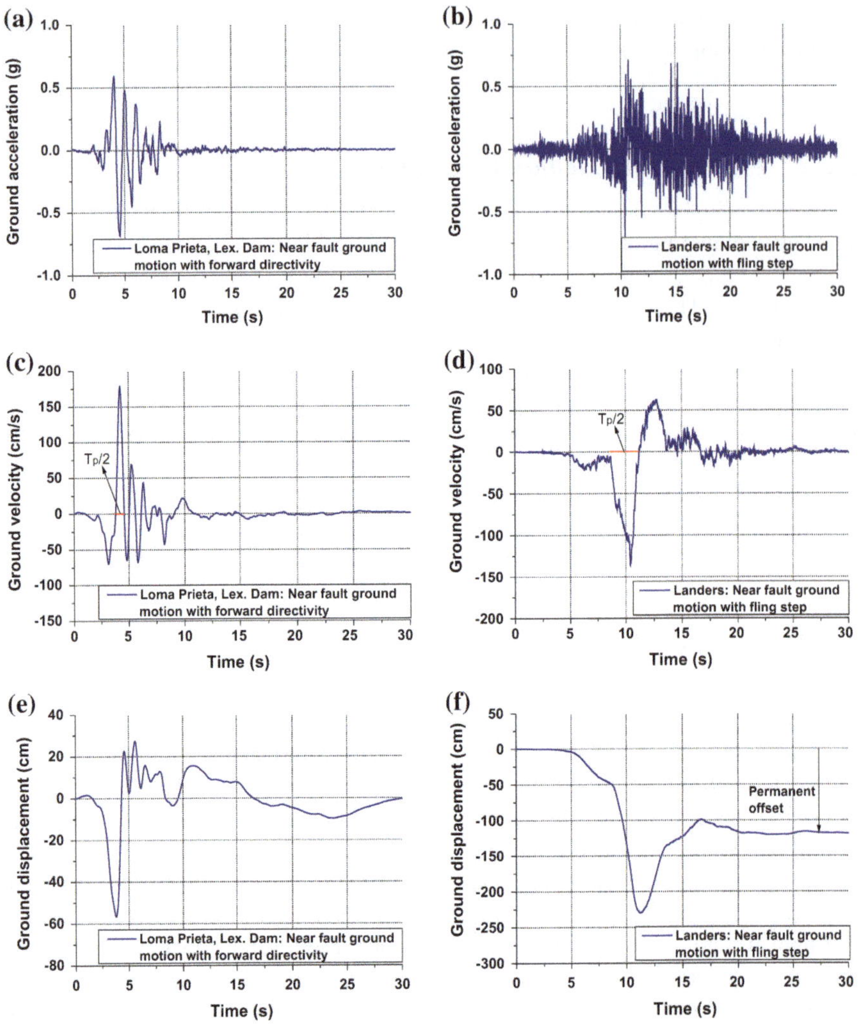

Fig. 5.4 Typical time histories of near-fault ground motions: **a** ground acceleration—forward directivity, **b** ground acceleration—fling step, **c** ground velocity—forward directivity, **d** ground velocity—fling step, **e** ground displacement—forward directivity, **f** ground displacement—fling step

most straightforward way is to estimate T_p based on the zero-crossing time of the pulse (Bray and Rodriguez-Marek 2004), as illustrated in Fig. 5.5. Certain subjectivity may be involved especially when multiple and overlapping velocity pulses exist. A more convenient approach is to identify an equivalent pulse period ($T_{p\text{-}v}$) corresponding to the peak point of the pseudo-velocity response spectrum of the ground motion (Krawinkler and Alavi 1998), as illustrated in Fig. 5.6. The two different approaches could sometimes lead to similar pulse period values, but inconsistency is

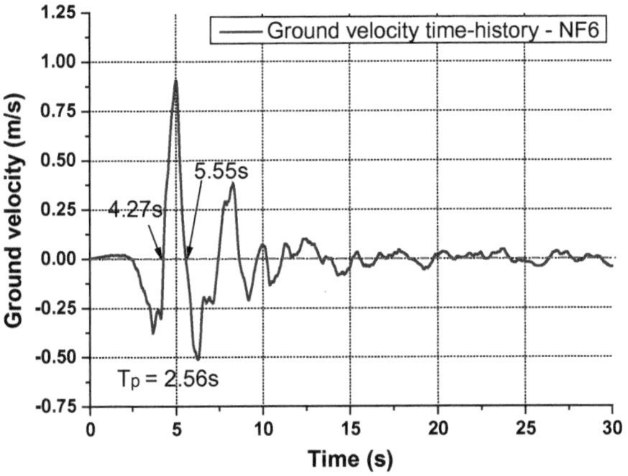

Fig. 5.5 Determination of pulse period using zero-crossing method

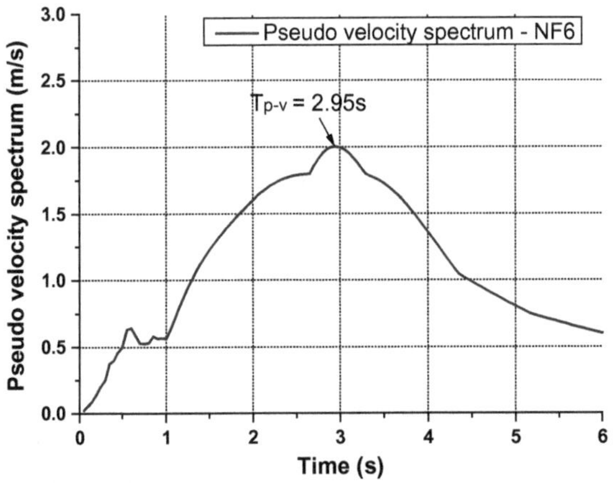

Fig. 5.6 Determination of pulse periods using pseudo-velocity response spectrum method

expected for ground motions with complex velocity pulsing characteristics. According to limited data, Rodriguez-Marek (2000) claimed that the mean ratio of the pulse periods obtained from the two approaches (i.e. zero-crossing and velocity response spectrum methods) is 0.84 with a standard deviation of 0.28.

5.3 General Behaviour of SDOF Systems

Before conducting an extensive parametric study, the dynamic responses of one representative self-centring ($\beta = 1.0$) SDOF system, as well as one elasto-plastic ($\beta = 2.0$) system subjected to single earthquake excitation, are first examined. The

SDOF systems selected in the case study are defined with the following parameters: $T = 1.0$ s, $R = 6$, $\alpha = 0.05$ and $\zeta = 5\%$. The Loma Prieta, Lex. Dam earthquake, as shown in Fig. 5.4, is considered as the ground motion input.

As can be seen in Fig. 5.7, the self-centring system has no residual displacement, compared with the residual displacement value of more than 10 cm for the elasto-plastic system; however, the former exhibits a slightly larger maximum displacement response than the latter. These are quite typical observations for self-centring and elasto-plastic SDOF systems, as can be confirmed in the later discussion. The occurrence of u_{max} seems to coincide with the occurrence of the predominant velocity pulse. Figure 5.8 shows that the two SDOF systems have comparable total acceleration responses. A close examination of the force–displacement histories of the two systems is provided in Fig. 5.9.

Further to the case study, Fig. 5.10 shows the normalised engineering demand parameters u_{max}, u_{res} and a_{max} of typical elasto-plastic SDOF systems with natural

Fig. 5.7 Case study: displacement time-history responses

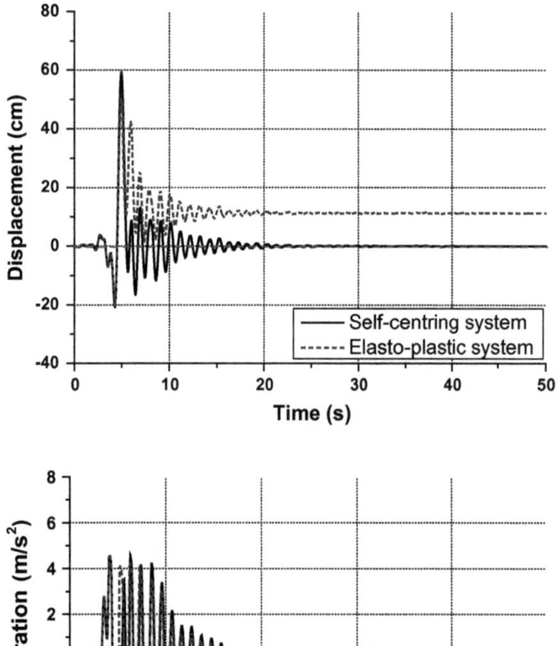

Fig. 5.8 Case study: total acceleration time-history responses

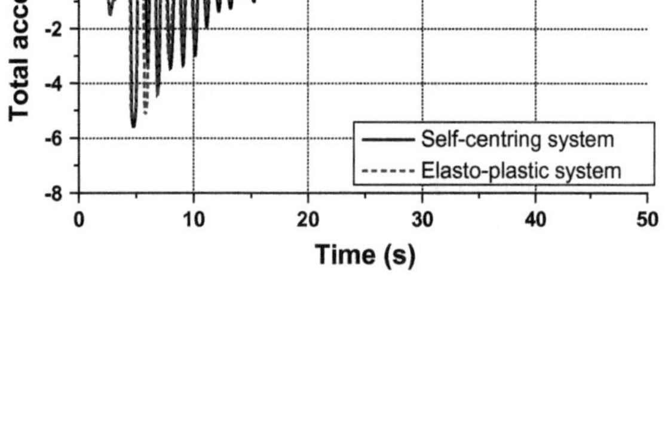

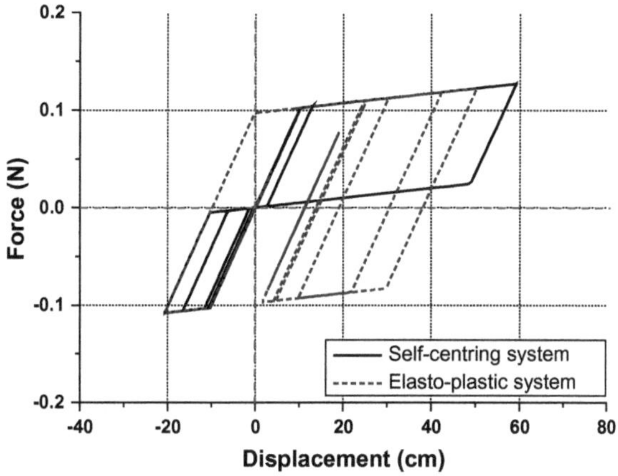

Fig. 5.9 Case study: force–displacement responses

periods from 0.1 to 3.0 s under suites of near-fault and far-field ground motions. The values of u_{max} and u_{res} are normalised with respect to the maximum displacement of the corresponding elastic system with damping ratio $\zeta = 5\%$, i.e. u_{max} $(R = 1, \zeta = 5\%)$. The peak absolute acceleration a_{max} is normalised with respect to the PGA of the ground motion (Fang et al. 2019).

The basic information of the ground motions is provided in Tables 5.1 and 5.2. The 40 pulse-like near-fault ground motion records given in Table 5.1 are taken from 13 events (including 2 simulated events) in different countries and regions, where both forward directivity and fling step are considered. Most of these ground motions are recorded at sites within a distance of 15 km from the rupture fault, and the event magnitudes M_w ranges from 6.5 to 7.6. Table 5.2 summarises the 44 far-field ground motions provided by FEMA (2009). These records are taken from 14 events, 8 in California and 6 in other countries. The event magnitude M_w also ranges from 6.5 to 7.6. These far-field ground motions exhibit no evident velocity pulses. The acceleration response spectra of the recorded near-fault and far-field ground motions are given in Fig. 5.11. It can be seen that the near-fault records are likely to impose larger response demands at longer periods, a case which is related to the long-period pulse characteristic of these ground motions.

The degree of dispersion is highlighted in Fig. 5.10 by giving both the median responses (thick line) and those from individual earthquake excitations (thin lines). The dispersion is attributed to the record-to-record (RTR) variability of the ground motions. The dispersions of the responses under the near-fault earthquakes seem to be larger than far-field earthquakes, which is likely due to the large variability of the pulse characteristics of the former. With an increase in period, the normalised u_{max} approaches unity, which conforms to the well-known ‘equal displacement rule’. The near-fault ground motions cause increased maximum and residual displacement demands compared with the far-field ground motions over the considered range of period. The normalised acceleration response tends towards unity when the period approaches 0.0 s, and decreases with an increase in period.

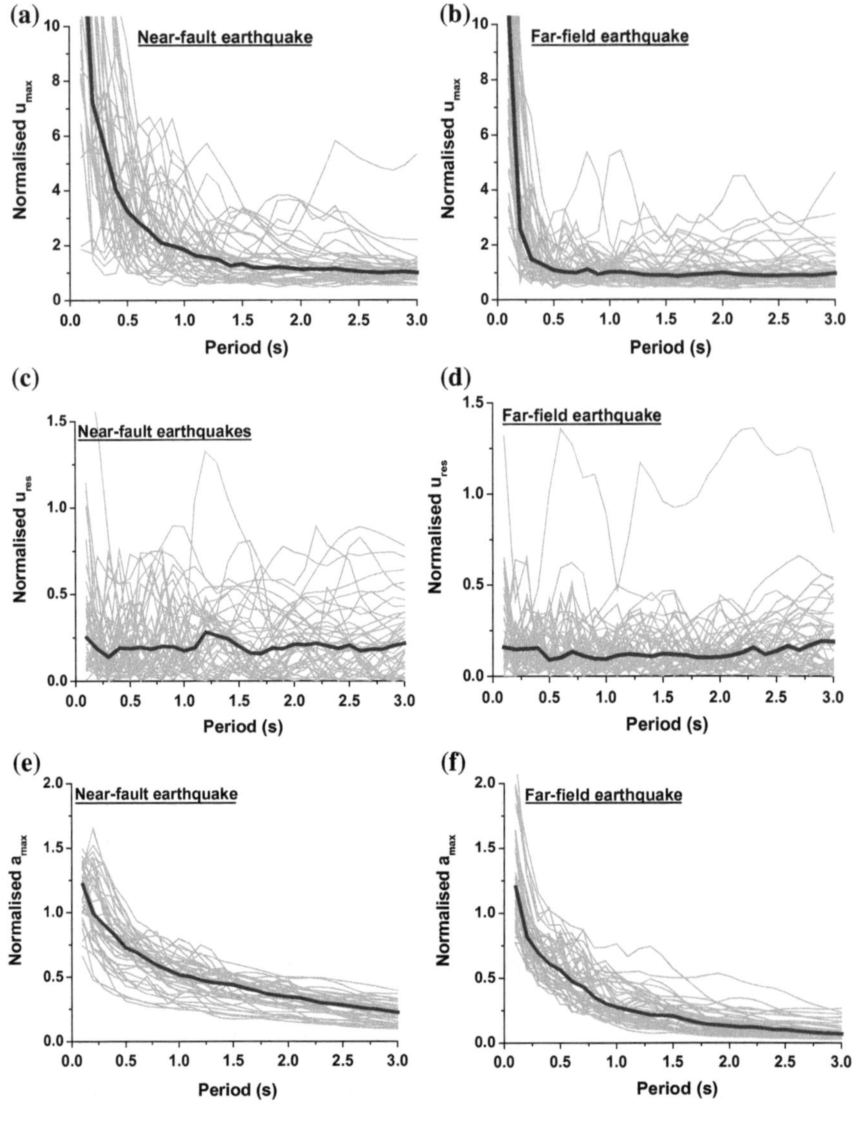

Fig. 5.10 Dispersions of normalised engineering demand parameters for SDOF systems with $R = 6$, $\alpha = 0.05$, $\beta = 2.0$ and $\zeta = 5\%$: **a** normalised u_{max} under near-fault earthquake, **b** normalised u_{max} under far-field earthquake, **c** normalised u_{res} under near-fault earthquake, **d** normalised u_{res} under far-field earthquake, **e** normalised a_{max} under near-fault earthquake, **f** normalised a_{max} under far-field earthquake

Table 5.1 Basic information of selected near-fault ground motions

Event no.	Earthquake	Station	M_w	PGA (g)	PGV (cm/s)	PGV/PGA	Type	T_p (s)	$T_{p\text{-}v}$ (s)
NF1	Chi-Chi	NSY	7.6	0.145	47.46	0.334	FD	6.61	5.45
NF2	Chi-Chi	NSY	7.6	0.128	41.81	0.334	FD	4.01	4.50
NF3	Chi-Chi	TCU	7.6	0.192	34.42	0.183	FD	2.64	3.50
NF4	Chi-Chi	TCU120	7.6	0.225	63.10	0.286	FD	2.87	2.00
NF5	Imperial Valley	5155 EC Meloland	6.5	0.314	71.65	0.233	FD	1.85	2.65
NF6	Imperial Valley	5155 EC Meloland	6.5	0.296	90.34	0.311	FD	2.56	2.95
NF7	Imperial Valley	5165 EI Centro	6.5	0.352	71.19	0.206	FD	2.88	2.60
NF8	Kobe	Takatori	6.9	0.611	127.06	0.212	FD	1.62	1.25
NF9	Kobe	Takatori	6.9	0.616	120.61	0.200	FD	1.50	1.25
NF10	Kocaeli	Sakarya	7.5	0.376	79.40	0.215	FD	5.10	5.70
NF11	Kocaeli	Yarimca	7.5	0.268	65.68	0.250	FD	4.48	4.35
NF12	Kocaeli	Yarimca	7.5	0.349	62.12	0.182	FD	3.28	3.65
NF13	Superstitn Hills	5051	6.6	0.455	111.97	0.251	FD	2.06	1.85
NF14	Loma Prieta	Los Gatos	7.0	0.718	172.84	0.246	FD	2.32	3.25
NF15	Loma Prieta	Los Gatos	7.0	0.459	91.06	0.203	FD	2.02	1.45
NF16	Loma Prieta	Lex. Dam	7.0	0.687	178.66	0.266	FD	1.58	1.10
NF17	Loma Prieta	Lex. Dam	7.0	0.370	68.65	0.189	FD	2.54	0.70
NF18	C. Mendocino	Petrolia	7.1	0.638	125.79	0.201	FD	2.32	0.75
NF19	Erzincan	95 Erzincan	6.7	0.433	99.41	0.235	FD	1.55	1.95
NF20	Landers	Landers	7.3	0.714	136.05	0.194	FS	4.39	4.00

(continued)

Table 5.1 (continued)

Event no.	Earthquake	Station	M_w	PGA (g)	PGV (cm/s)	PGV/PGA	Type	T_p (s)	$T_{p\text{-}v}$ (s)
NF21	Nothridge	Rinaldi	6.7	0.891	174.22	0.200	FD	1.39	1.05
NF22	Kobe	Takatori	6.9	0.787	144.73	0.188	FD	1.52	1.05
NF23	Elysian Park 4	–	7.1	0.923	286.93	0.317	FD	2.32	1.90
NF24	Elysian Park 5	–	7.1	1.162	250.95	0.220	FD	2.56	2.20
NF25	Palos Verdes 1	–	7.1	0.974	273.92	0.287	FD	2.72	2.45
NF26	Palos Verdes 2	–	7.1	0.969	287.86	0.303	FD	3.08	2.60
NF27	Palos Verdes 3	–	7.1	0.874	263.16	0.307	FD	1.84	1.60
NF28	Palos Verdes 4	–	7.1	0.794	195.32	0.251	FD	2.04	1.65
NF29	Palos Verdes 5	–	7.1	0.916	264.49	0.295	FD	2.24	2.10
NF30	Palos Verdes 5	–	7.1	0.601	139.45	0.237	FD	3.04	2.65
NF31	Chi-Chi	TCU52	7.6	0.448	220.68	0.503	FS	9.08	1.75
NF32	Chi-Chi	TCU52	7.6	0.356	183.85	0.527	FS	8.05	2.40
NF33	Chi-Chi	TCU65	7.6	0.790	132.48	0.171	FS	4.12	4.65
NF34	Chi-Chi	TCU67	7.6	0.499	97.47	0.199	FS	3.94	2.20
NF35	Chi-Chi	TCU68	7.6	0.369	305.59	0.844	FS	13.0	8.50
NF36	Chi-Chi	TCU68	7.6	0.512	280.28	0.559	FS	8.17	9.50
NF37	Chi-Chi	TCU75	7.6	0.332	116.43	0.358	FS	6.09	4.20
NF38	Chi-Chi	TCU76	7.6	0.347	69.34	0.204	FS	6.99	4.25
NF39	Chi-Chi	TCU87	7.6	0.114	45.26	0.406	FS	4.54	3.90
NF40	Chi-Chi	TCU120	7.6	0.228	62.55	0.280	FS	4.52	2.00

Note FD = forward directivity, FS = fling step

Table 5.2 Basic information of selected far-field ground motions

Event no.	Earthquake name	Station	M_w	PGA (g)	PGV (cm/s)	PGV/PGA
FF1	San Fernando	LA-Hollywood_Stor_FF	6.6	0.210	18.86	0.092
FF2	San Fernando	LA-Hollywood_Stor_FF	6.6	0.174	14.84	0.087
FF3	Friuli Italy	Tolmezzo	6.5	0.351	22.02	0.064
FF4	Friuli Italy	Tolmezzo	6.5	0.315	30.76	0.100
FF5	Imperial Valley	Delta	6.5	0.238	25.97	0.111
FF6	Imperial Valley	Delta	6.5	0.351	32.97	0.096
FF7	Imperial Valley	El_Centro_Array_#11	6.5	0.364	34.42	0.096
FF8	Imperial Valley	El_Centro_Array_#11	6.5	0.380	42.10	0.113
FF9	Superstition Hills	El_Centro_Imp._Co._C	6.5	0.358	46.31	0.132
FF10	Superstition Hills	El_Centro_Imp._Co._C	6.5	0.258	40.83	0.161
FF11	Superstition Hills	Poe Road (temp)	6.5	0.446	35.67	0.082
FF12	Superstition Hills	Poe Road (temp)	6.5	0.300	32.75	0.111
FF13	Loma Prieta	Capitola	6.9	0.529	34.98	0.068
FF14	Loma Prieta	Capitola	6.9	0.443	29.20	0.067
FF15	Loma Prieta	Gilroy Array #3	6.9	0.555	35.65	0.066
FF16	Loma Prieta	Gilroy Array #3	6.9	0.367	44.62	0.124
FF17	Cape Mendocino	Rio_Dell_Overpass-FF	7.0	0.385	43.72	0.116
FF18	Cape Mendocino	Rio_Dell_Overpass-FF	7.0	0.549	41.90	0.078
FF19	Landers	Coolwater	7.3	0.284	28.29	0.102
FF20	Landers	Coolwater	7.3	0.417	44.49	0.109
FF21	Landers	Yermo_Fire_Station	7.3	0.245	51.36	0.214
FF22	Landers	Yermo_Fire_Station	7.3	0.152	29.69	0.200
FF23	Northridge	Beverly_Hills-14145_Mulhol	6.7	0.416	58.89	0.145
FF24	Northridge	Beverly_Hills-14145_Mulhol	6.7	0.516	62.71	0.124

(continued)

The above observations are generally true for most SDOF systems, although different structural parameters can lead to different 'shapes' of the spectra. Given the RTR variability of the ground motions, the median responses from the SDOF analysis are mainly used in the subsequent parametric study, in which case the basic trends can be more clearly revealed.

Table 5.2 (continued)

Event no.	Earthquake name	Station	M_w	PGA (g)	PGV (cm/s)	PGV/PGA
FF25	Northridge	Canyon_Country-W_Lost_C	6.7	0.410	42.93	0.107
FF26	Northridge	Canyon_Country-W_Lost_C	6.7	0.482	44.87	0.095
FF27	Kobe—Japan	Nishi-Akashi	6.9	0.509	37.25	0.075
FF28	Kobe—Japan	Nishi-Akashi	6.9	0.503	36.59	0.074
FF29	Kobe—Japan	Shin-Osaka	6.9	0.243	37.76	0.158
FF30	Kobe—Japan	Shin-Osaka	6.9	0.212	27.87	0.134
FF31	Kocaeli—Turkey	Arcelik	7.5	0.219	17.68	0.082
FF32	Kocaeli—Turkey	Arcelik	7.5	0.150	39.53	0.269
FF33	Kocaeli—Turkey	Duzce	7.5	0.312	58.79	0.192
FF34	Kocaeli—Turkey	Duzce	7.5	0.358	46.35	0.132
FF35	Chi-Chi—Taiwan	CHY101	7.6	0.353	70.58	0.204
FF36	Chi-Chi—Taiwan	CHY101	7.6	0.440	114.92	0.266
FF37	Chi-Chi—Taiwan	TCU045	7.6	0.473	50.03	0.108
FF38	Chi-Chi—Taiwan	TCU046	7.6	0.507	46.33	0.093
FF39	Duzce—Turkey	Bolu	7.1	0.728	56.39	0.079
FF40	Duzce—Turkey	Bolu	7.1	0.822	62.04	0.077
FF41	Manjil Iran	Abbar	7.4	0.515	42.40	0.084
FF42	Manjil Iran	Abbar	7.4	0.497	5.05	0.010
FF43	Hector_Mine	Hector	7.1	0.266	28.53	0.110
FF44	Hector_Mine	Hector	7.1	0.337	41.70	0.126

5.4 Factors Influencing Behaviour of SDOF Systems

5.4.1 *Yield Strength*

There are a number of factors which can affect the engineering demand parameters of SDOF systems, and yield strength is one of them. The variations of the maximum displacements, residual displacements and peak absolute accelerations of typical SDOF systems with different strength ratios R are illustrated in Figs. 5.12 through 5.14. In these figures, $\beta = 1.0$, 1.5 and 2.0 correspond to the representative self-centring, 'partial' self-centring and elasto-plastic systems, respectively. All the SDOF systems shown in Figs. 5.12 through 5.14 have $\alpha = 0.05$ and $\zeta = 5\%$.

It can be clearly seen from Fig. 5.12 that the normalised maximum displacement increases as the yield strength decreases, a trend which is particularly evident at short periods. Compared with the far-field ground motions, the near-fault ones tend to increase the maximum displacement demands. It is worth noting that the equal displacement rule provides unsafe results for the systems with $T < 2.0$ s under the

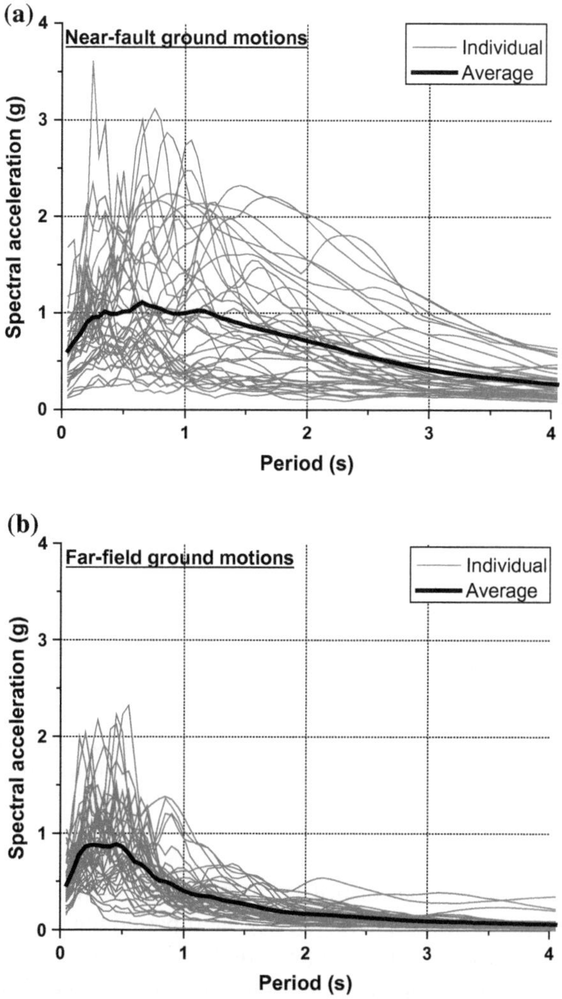

Fig. 5.11 Spectral accelerations of considered ground motions: **a** near-fault earthquakes, **b** far-field earthquakes

near-fault earthquakes. The difference in the maximum displacement response caused by the different earthquake types is related to the long-period pulse exhibited by the near-fault ground motions. It is not difficult to imagine that when the pulse period is close to the effective period of the system, the displacement demand would be amplified due to resonance. Table 5.2 has shown that the average pulse period of the 40 near-fault ground motions is around 3.0 s, so it is expected that these long-period pulses raise the maximum displacement demands at the long-period region. Another finding from Fig. 5.12 is that the self-centring systems exhibit slightly larger maximum displacements than their elasto-plastic counterparts, which is probably due to the less energy dissipation capacity provided by the former.

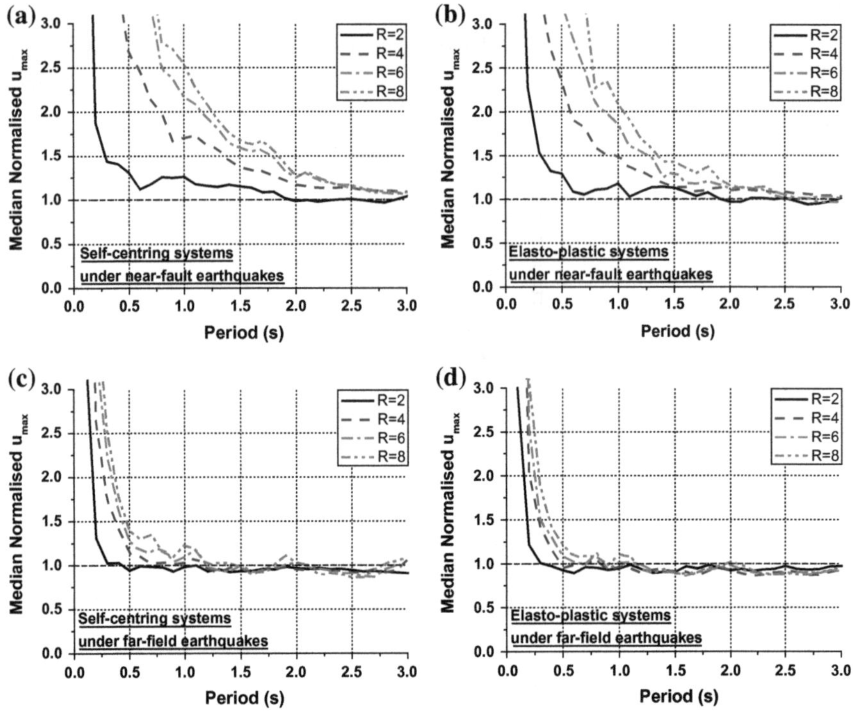

Fig. 5.12 Influence of strength ratio on normalised maximum displacement of SDOF systems with $\alpha = 0.05$, $\zeta = 5\%$: **a** $\beta = 1.0$ under near-fault earthquake, **b** $\beta = 2.0$ under near-fault earthquake, **c** $\beta = 1.0$ under far-field earthquake, **d** $\beta = 2.0$ under far-field earthquake

The residual displacement responses of the considered systems are shown in Fig. 5.13. Except for the very short period region, the normalised u_{res} exhibits less clear trends with varying periods. This is quite different from the maximum displacement responses shown in Fig. 5.12. For the elasto-plastic systems, decreasing the yield strength decreases the residual displacement, although an opposite trend may be seen at a certain range of period. What we can at least deduce from this tendency is that decreasing the yield strength of conventional elasto-plastic systems leads to practically no risk of residual displacement amplification.

The results also show that near-fault earthquakes lead to larger residual displacement responses than far-field earthquakes. This warns that the seismic resilience of conventional structures located near proven faults should receive particular attention in the design. For the partial self-centring systems with $\beta = 1.5$, the residual displacement is much smaller. A detailed discussion on the influence of β is reserved for Sect. 5.4.3.

The peak absolute acceleration responses of both the elasto-plastic and self-centring systems, as shown in Fig. 5.14, consistently increase with an increase in their yield strength. The change of the acceleration response is particularly remark-

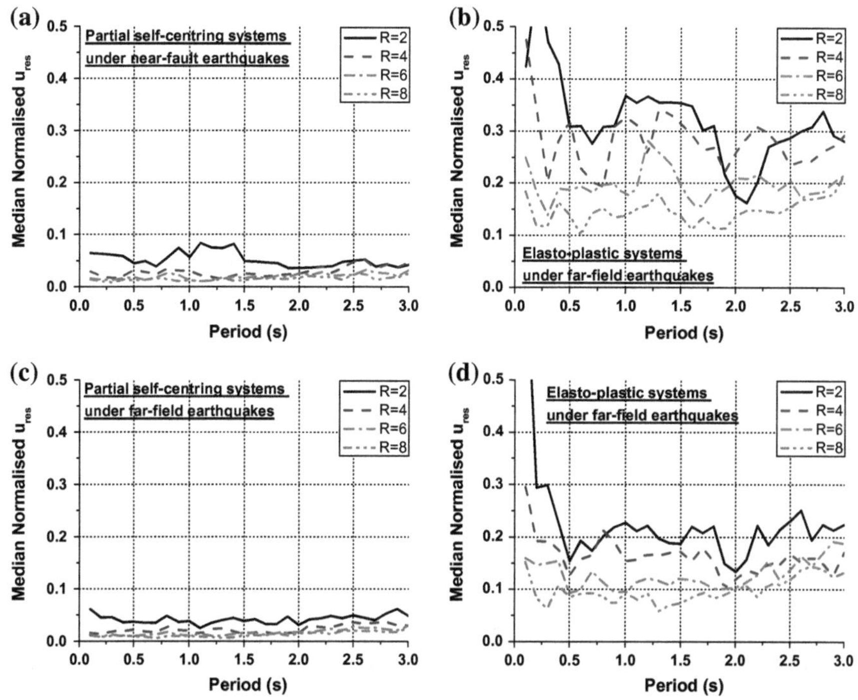

Fig. 5.13 Influence of strength ratio on normalised residual displacement of SDOF systems with $\alpha = 0.05$, $\zeta = 5\%$: **a** $\beta = 1.5$ under near-fault earthquake, **b** $\beta = 2.0$ under near-fault earthquake, **c** $\beta = 1.5$ under far-field earthquake, **d** $\beta = 2.0$ under far-field earthquake

able when *R* increases from 2 to 4, whereas a further increase in *R* leads to less remarkable changes. The relatively large acceleration response for the case of *R* = 2 is attributed to the mild inelastic behaviour of the system, noting that a more pronounced inelastic behaviour often reduces acceleration demands (Ray-Chaudhuri and Hutchinson 2011). As we know that the acceleration response of an infinitely stiff system is identical to the PGA of the ground motion, the normalised peak absolute acceleration tends towards unity when the period moves towards 0.0 s.

The SDOF analysis shows no obvious difference in the acceleration demands between the self-centring and elasto-plastic systems. It should be kept in mind that this observation is not applicable to the case of multi-degree-of-freedom (MDOF) systems. Previous studies, e.g. Tremblay et al. (2008) and Fang et al. (2018), have shown that the floor acceleration response of multi-storey self-centring building frames is amplified compared with the conventional systems exhibiting typical yielding behaviour. This is because of the more complex unloading path of the flag-shaped hysteresis. The unique behaviour leads to larger differences in the shear force between adjacent stories, a case which produces a transient large floor acceleration. We will discuss this issue in more in detail in the next chapter.

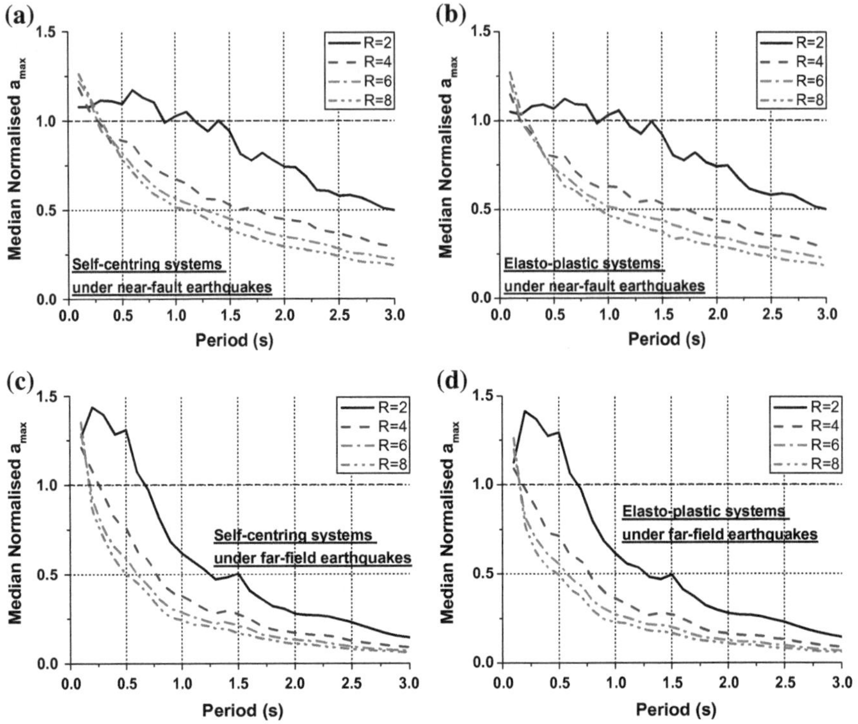

Fig. 5.14 Influence of strength ratio on normalised peak absolute acceleration of SDOF systems with $\alpha = 0.05$, $\zeta = 5\%$: **a** $\beta = 1.0$ under near-fault earthquake, **b** $\beta = 2.0$ under near-fault earthquake, **c** $\beta = 1.0$ under far-field earthquake, **d** $\beta = 2.0$ under far-field earthquake

5.4.2 *Post-yield Stiffness*

The influences of post-yield stiffness ratio α on the maximum displacement, residual displacement and peak absolute acceleration responses of typical ductile SDOF systems with $R = 6$ and $\zeta = 5\%$ are shown in Figs. 5.15 through 5.17.

For both the self-centring and elasto-plastic systems, the maximum displacement demand exhibits a certain sensitivity to the post-yield stiffness. Such sensitivity is slightly more obvious for the case of near-fault earthquakes. The SDOF analysis results are generally consistent with the existing studies on bridges and building structures. For example, Alhan et al. (2016) showed that the excessive displacement of bridge girders due to the pulsing effect of near-fault ground motions can be effectively controlled through increasing the hardening stiffness of the bearings in the inelastic range. Such a conclusion can also be extended to MDOF systems, for example, Moradi et al. (2014) confirmed that the risk of 'soft storey' in building frames is effectively mitigated by enhancing the hardening stiffness of the braces.

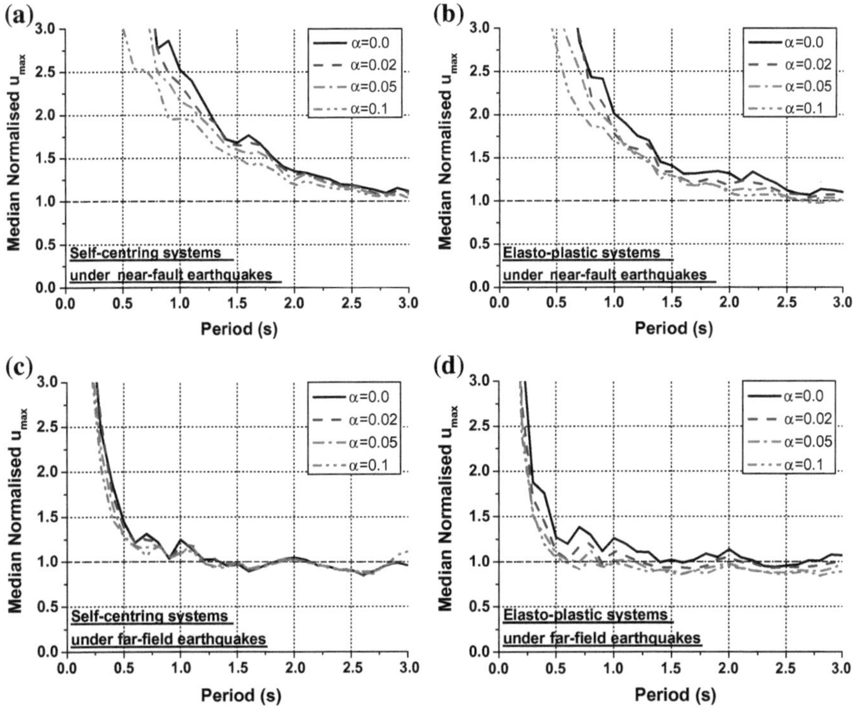

Fig. 5.15 Influence of post-yield stiffness on normalised maximum displacement of SDOF systems with $R = 6$, $\zeta = 5\%$: **a** $\beta = 1.0$ under near-fault earthquake, **b** $\beta = 2.0$ under near-fault earthquake, **c** $\beta = 1.0$ under far-field earthquake, **d** $\beta = 2.0$ under far-field earthquake

Consensus has been reached that the post-yield stiffness tremendously affects the residual displacement of conventional elasto-plastic systems (Eatherton and Hajjar 2011). As shown in Fig. 5.16, compared with the elastic-perfectly plastic systems ($\alpha = 0.0$), the residual displacements of the elasto-plastic systems with $\alpha = 0.1$ are reduced by more than 80% when $T > 1.0$ s. The reduction is even more obvious at shorter periods. As emphasised in the previous chapters, the benefits from increasing the post-yield stiffness for residual displacement control is related to the concept of 'dynamic shakedown' or 'probabilistic self-centring', based on the fact that the actual residual displacement after dynamic shaking is normally smaller than the maximum possible residual displacement derived from 'slow' unloading (MacRae and Kawashima 1997; Eatherton and Hajjar 2011). Increasing the post-yield stiffness has been found to increase the probability of self-centring (Eatherton and Hajjar 2011). On the other hand, for the partial self-centring systems, the change of post-yield stiffness has a less significant influence on the residual displacement response, because the u_{res} value has already been reduced to practically negligible levels.

Reducing displacement and acceleration responses are often contradictory objectives (Karavasilis and Seo 2011). Figure 5.17 clearly demonstrates that increasing

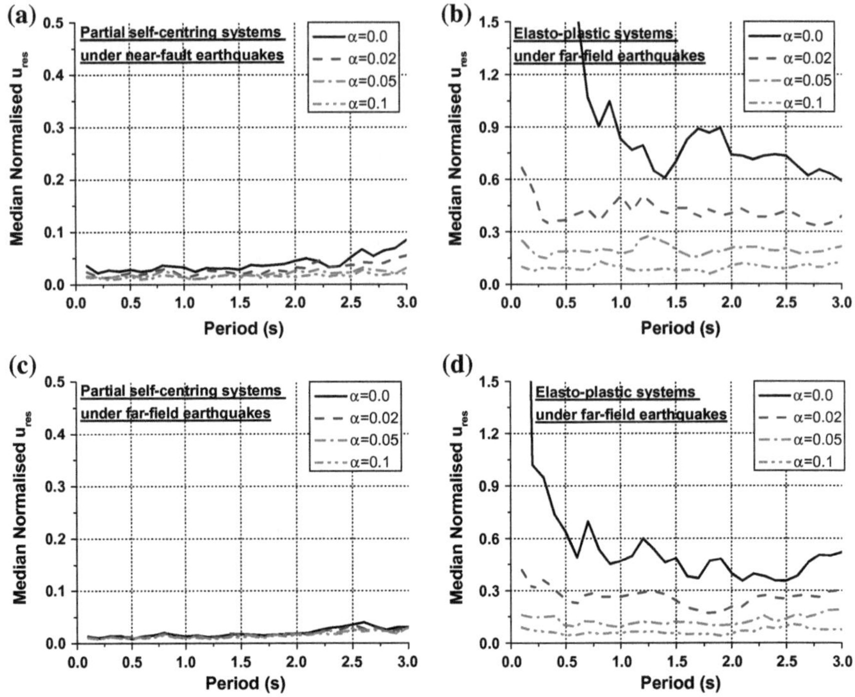

Fig. 5.16 Influence of post-yield stiffness on normalised residual displacement of SDOF systems with $R = 6$, $\zeta = 5\%$: **a** $\beta = 1.5$ under near-fault earthquake, **b** $\beta = 2.0$ under near-fault earthquake, **c** $\beta = 1.5$ under far-field earthquake, **d** $\beta = 2.0$ under far-field earthquake

the post-yield stiffness increases the acceleration demand. The acceleration demand of the systems is quite sensitive to the post-yield stiffness in the short and intermediate period regions when subjected to near-fault earthquakes, while for the case of far-field earthquakes, the sensitivity is remarkable only in the short period region. Again, the difference is somehow linked to the pulsing nature of the near-fault ground motions.

5.4.3 *Energy Dissipation Ratio*

One common concern regarding the design of SMA-based structures is whether the energy dissipation provided by the SMA devices is sufficient or not. To seek answers to this question, the influence of different hysteretic shapes on the system response should be examined, starting from SDOF analysis. Figures 5.18 through 5.20 shows the influence of energy dissipation ratio β on the maximum displacement, residual displacement and peak absolute acceleration responses of typical ductile

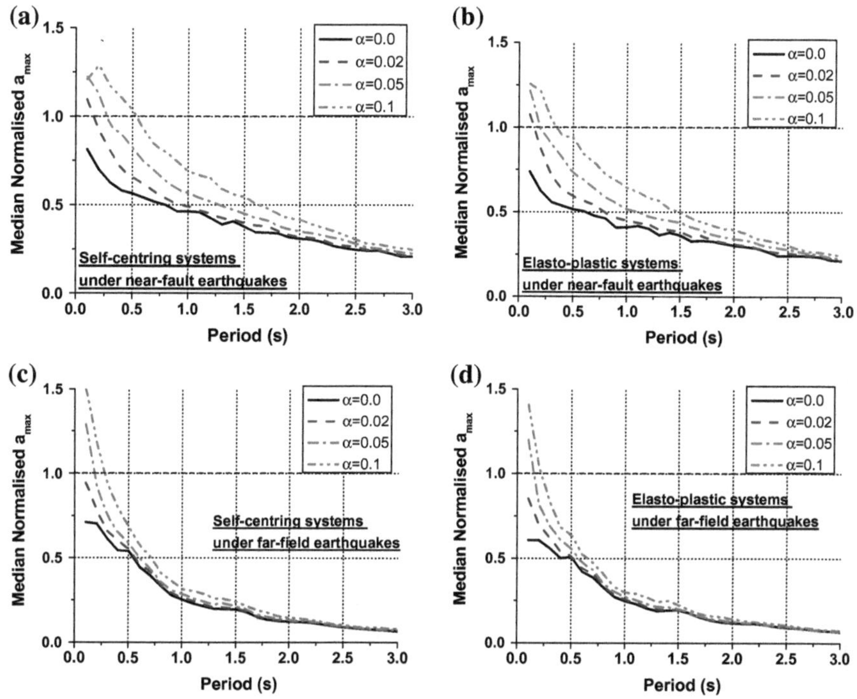

Fig. 5.17 Influence of post-yield stiffness on normalised peak absolute acceleration of SDOF systems with $R = 6$, $\zeta = 5\%$: **a** $\beta = 1.0$ under near-fault earthquake, **b** $\beta = 2.0$ under near-fault earthquake, **c** $\beta = 1.0$ under far-field earthquake, **d** $\beta = 2.0$ under far-field earthquake

SDOF systems with $R = 6$, $\alpha = 0.05$ and $\zeta = 5\%$. The maximum displacement response shows a moderate sensitivity to the energy dissipation ratio, where $\beta = 2.0$ leads to the smallest maximum displacement. This means that although the influence of energy dissipation ratio is generally not critical, decreasing the energy dissipation capacity of the system can indeed increase the maximum displacement demand. This observation applies to both the near-fault and far-field earthquakes.

As expected, the elasto-plastic systems exhibit the largest residual displacement, where the u_{res} values decrease as the energy dissipation ratio decreases. Perhaps, the most encouraging finding from Fig. 5.19 is that when the energy dissipation ratio decreases from 2.0 to 1.5, the residual displacement demand drops dramatically to practically negligible levels. This echoes the aforementioned 'probabilistic self-centring' concept. Figure 5.20 shows that the acceleration response is not sensitive to the energy dissipation ratio.

Judging from the trends revealed above, we have reason to believe that partial self-centring systems (e.g. with $\beta = 1.5$) could be a desirable option which can effectively control the residual displacement and peak acceleration responses without inducing a notable increase in the maximum displacement. In other words, well-balanced

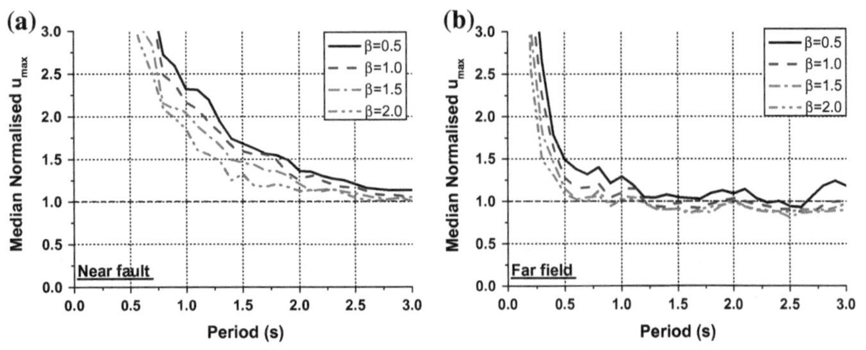

Fig. 5.18 Influence of energy dissipation ratio on normalised maximum displacement of SDOF systems with $R = 6$, $\alpha = 0.05$, $\zeta = 5\%$: **a** near-fault earthquake, **b** far-field earthquake

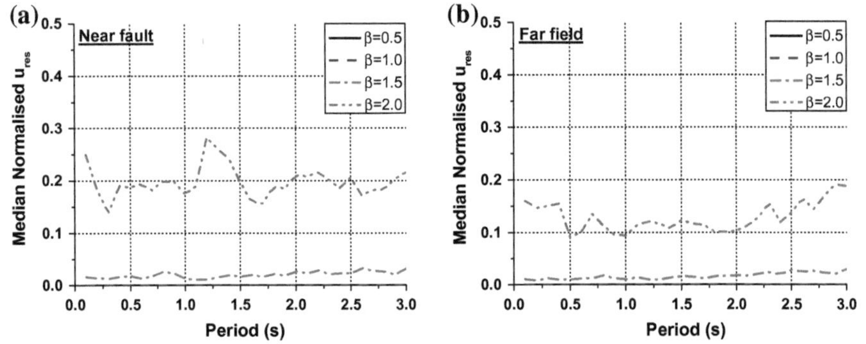

Fig. 5.19 Influence of energy dissipation ratio on normalised residual displacement of SDOF systems with $R = 6$, $\alpha = 0.05$, $\zeta = 5\%$: **a** near-fault earthquake, **b** far-field earthquake

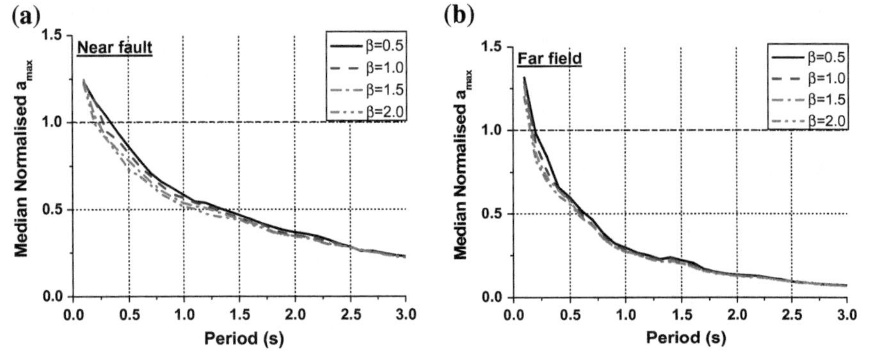

Fig. 5.20 Influence of energy dissipation ratio on normalised peak absolute acceleration of SDOF systems with $R = 6$, $\alpha = 0.05$, $\zeta = 5\%$: **a** near-fault earthquake, **b** far-field earthquake

maximum displacement, residual displacement and peak acceleration demands can be attained when partial self-centring systems are used. As a result, a harmonisation of structural and non-structural damage control can be achieved in structural seismic design.

5.4.4 Damping Ratio

One common strategy for peak response control against earthquakes is to add extra damping to the structural systems. As the energy dissipation provided by superelasticity is often less than that offered by material yielding, friction action and other reliable energy dissipation mechanisms, a combined use of SMA devices and extra sources of energy dissipation may be a good idea. Fluid viscous dampers can be viable candidates for such purpose. A possible advantage of such combination is that fluid viscous dampers are only 'active' under dynamic (high frequency) loading cases whilst providing little resisting force under quasi-static loading conditions. This enables the use of fluid viscous dampers without compromising the self-centring capability of the system. From an economic point of view, the required number of fluid viscous dampers in a self-centring structure can be kept small, depending on the specific design target.

Figures 5.21 through 5.23 shows the influence of damping ratio ζ on the maximum displacement, residual displacement and peak absolute acceleration responses of the SDOF systems with $R = 6$ and $\alpha = 0.05$. As anticipated, increasing the viscous damping ratio is very effective in suppressing the maximum displacement response of the systems under both the near-fault and far-field earthquakes (Fig. 5.21). Taking $T = 1.0$ s for instance, when the total damping ratio increases from 5% to 25%, the maximum displacement of the elasto-plastic and self-centring systems is decreased by 37% and 42%, respectively, under the far-field earthquakes. For the two systems with the same period under the near-fault earthquakes, the maximum displacement is decreased by 50% and 53%, respectively. We can deduce from the results that fluid viscous dampers (or other similar rate-dependent dampers), which are in practice popular in peak deformation and vibration control for conventional yielding structures, can be equally (and perhaps more) effective for self-centring structural systems. From a residual displacement point of view, the variation in damping is not influential. It is noted that in some cases, the added damping can even increase the residual displacement, but generally speaking, the detrimental effect is not critical (Fig. 5.22).

In contrast to the maximum displacement response, the acceleration demand increases as the damping ratio increases, with the exception of the very short period. For the systems with $T = 1.0$ s, when the total damping ratio increases from 5% to 25%, the peak absolute acceleration is increased by 57.7% and 67.9% for the elasto-plastic and self-centring systems, respectively, under the far-field earthquakes. For the case of near-fault earthquakes, the increases are 40.4% and 36.5%, respectively.

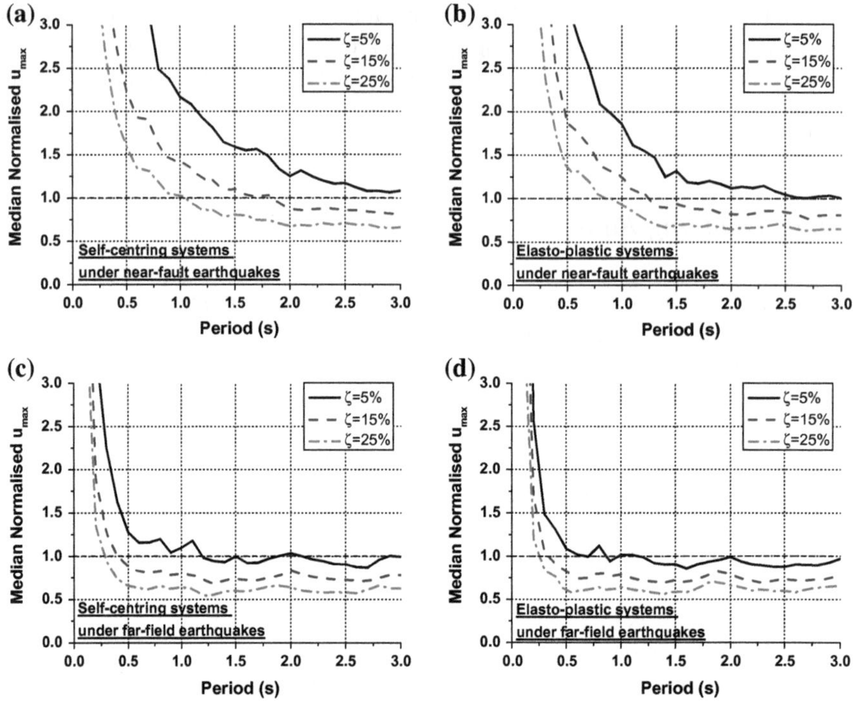

Fig. 5.21 Influence of viscous damping ratio on normalised maximum displacement of SDOF systems with $R = 6$, $\alpha = 0.05$: **a** $\beta = 1.0$ under near-fault earthquake, **b** $\beta = 2.0$ under near-fault earthquake, **c** $\beta = 1.0$ under far-field earthquake, **d** $\beta = 2.0$ under far-field earthquake

It is worth noting that this observation is based on the case of relatively ductile systems ($R = 6$) with significant inelastic behaviour expected. For mildly inelastic systems (e.g. $R = 2$), the effect of damping ratio is not that obvious. This can be explained by rearranging Eq. (5.4) into

$$\ddot{u}(t) + \ddot{u}_g(t) = a_t(t) = -[c\dot{u}(t) + f_S(t)]/m \tag{5.9}$$

where a_t is the total acceleration. For purely elastic or mildly inelastic structures, the peaks of the damping force are out of phase with those of the restoring force, and as a result, the effect of damping on the total acceleration can be less obvious. For damped systems with significant inelastic behaviour, the phase difference between the peaks of the restoring and damping forces is small, and therefore increasing the damping correspondingly leads to an increase in total acceleration (Karavasilis and Seo 2011).

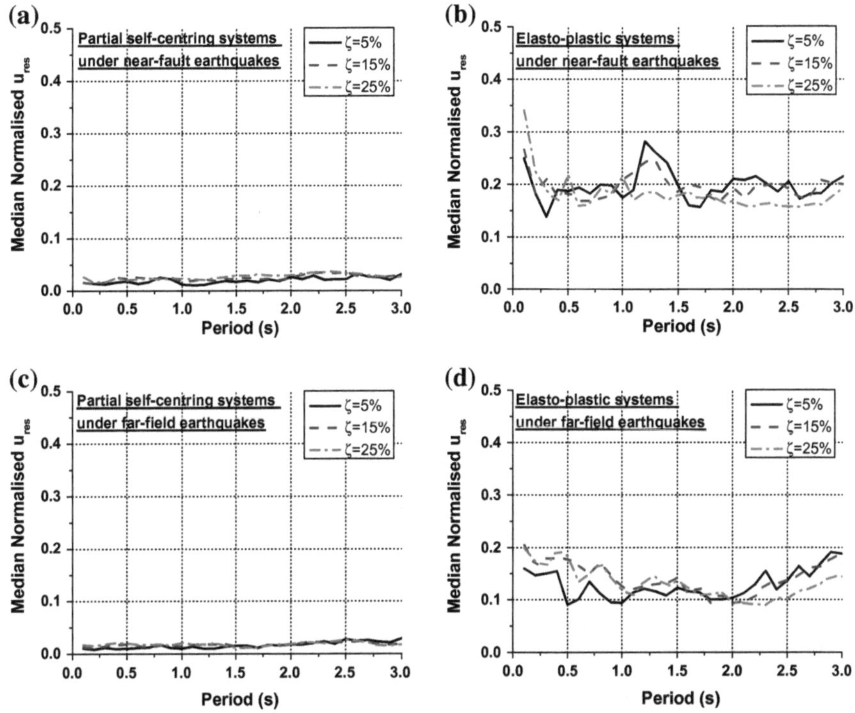

Fig. 5.22 Influence of viscous damping ratio on normalised residual displacement of SDOF systems with $R = 6$, $\alpha = 0.05$: **a** $\beta = 1.5$ under near-fault earthquake, **b** $\beta = 2.0$ under near-fault earthquake, **c** $\beta = 1.5$ under far-field earthquake, **d** $\beta = 2.0$ under far-field earthquake

5.5 Inelastic Displacement Ratio

The IDR, short for inelastic displacement ratio, is the ratio of the maximum inelastic displacement of a nonlinear SDOF system to its elastic spectral displacement. It is an important indicator of structural damage under the framework of performance-based seismic design. IDR forms the basis of the mainstream nonlinear static approaches, such as pushover analysis, for predicting the target inter-storey drift demand of building frames. This concept has been adopted in many modern design guidelines (FEMA 2000, 2005).

The typical median IDR responses of the SDOF systems under the near-fault and far-field earthquakes are presented in Figs. 5.24 and 5.25, respectively. Power function forms are suitable for describing the relationship between the IDR and the period (Durucan and Durucan 2016). It is also noticed that the shapes of the IDR curve as a function of period are quite different for near-fault and far-field earthquakes. Building on the basic power function form recommended in FEMA (2005) and ASCE (2007) for estimating the target roof drift of building frames, the

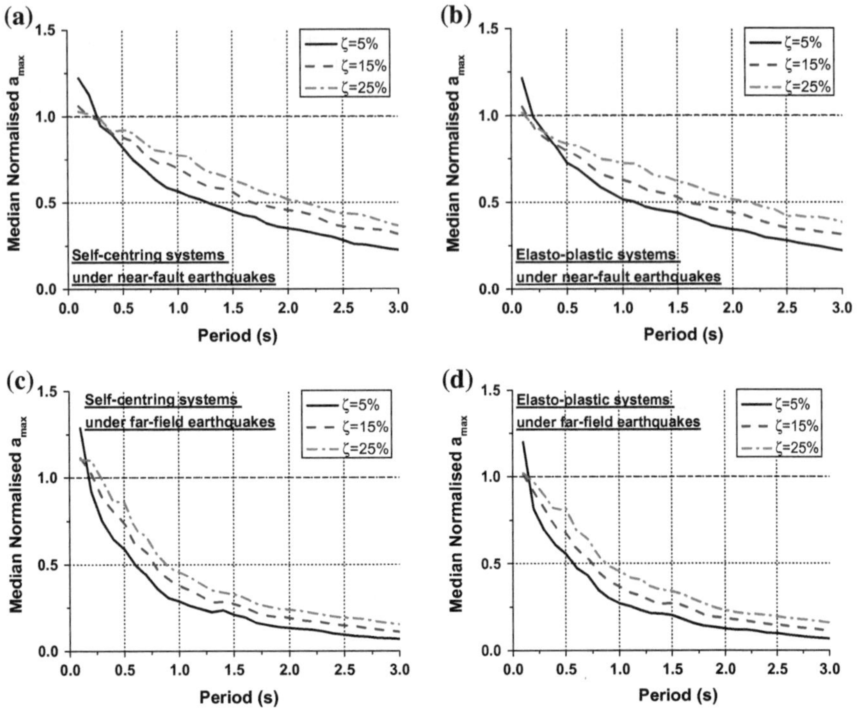

Fig. 5.23 Influence of viscous damping ratio on normalised peak absolute acceleration of SDOF systems with $R = 6$, $\alpha = 0.05$: **a** $\beta = 1.0$ under near-fault earthquake, **b** $\beta = 2.0$ under near-fault earthquake, **c** $\beta = 1.0$ under far-field earthquake, **d** $\beta = 2.0$ under far-field earthquake

following basic functional forms are suitable for predicting the IDR of the systems with an extended spectrum of parameters (Fang et al. 2019):

$$IDR_{NF} = 1 + \frac{(R-1)^{a_1}(1-\alpha)^{a_6}}{a_2 T^{a_3} \zeta^{a_4} \beta^{a_5}} \tag{5.10}$$

$$IDR_{FF} = 1 + \frac{(R-1)^{a_1}(1-\alpha)^{a_6}}{a_2\left(T^{b_1} e^{b_2 T}\right) \zeta^{a_4} \beta^{a_5}} \tag{5.11}$$

In Eqs. (5.10) and (5.11), which are applicable to the near-fault and far-field earthquakes, respectively, the constants a_1 through a_6 as well as b_1 and b_2 are determined through nonlinear regression. Notice that an additional exponential term with constant b_2 is included in the equation for the far-field earthquakes to better capture the local 'sharp' turning of the IDR curves as a function of period.

Fig. 5.24 Comparisons between median time-history analysis (THA) and fitted results for selected systems under near-fault earthquakes: **a** influence of R, **b** influence of α, **c** influence of ζ

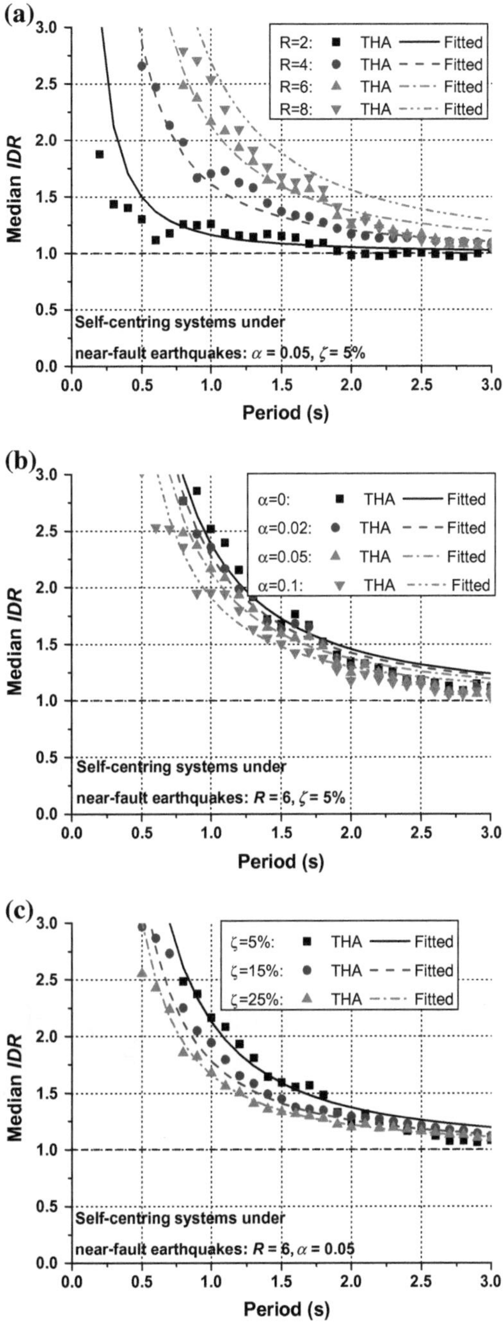

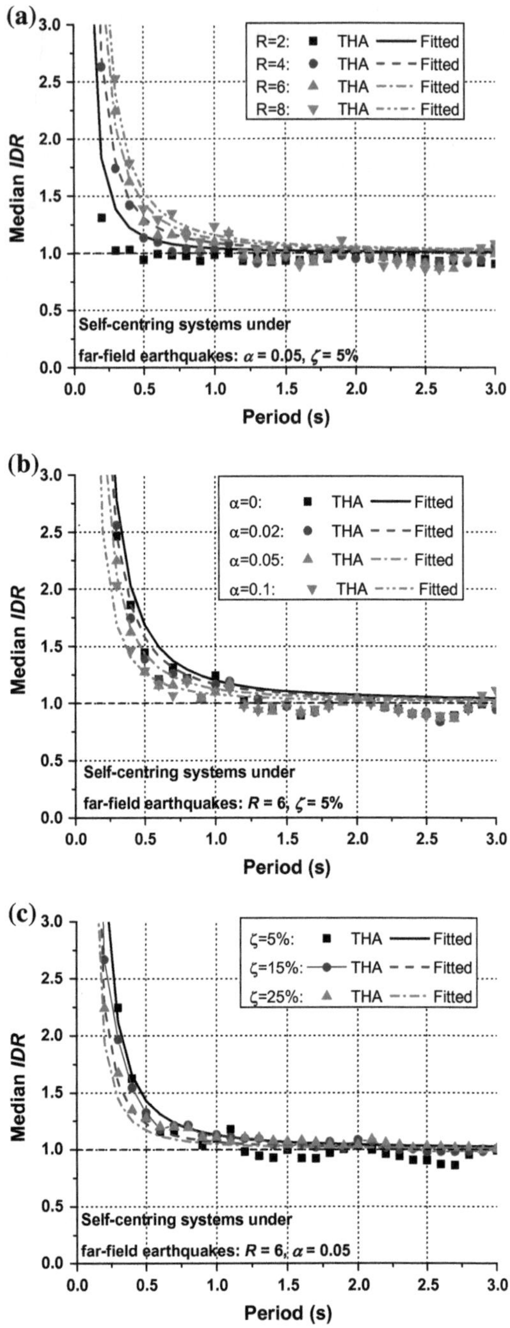

Fig. 5.25 Comparisons between median time-history analysis (THA) and fitted results for selected systems under far-field earthquakes: **a** influence of R, **b** influence of α, **c** influence of ζ

Based on the median response of the considered SDOF systems and using nonlinear regression, the IDR prediction models for the cases of near-fault and far-field earthquakes are given in Eqs. (5.12) and (5.13), respectively.

$$IDR_{NF-median} = 1 + \frac{(R-1)^{6/5}(1-\alpha)^4}{13.5T^{8/5}\varsigma^{1/3}\beta^{1/3}} \quad (5.12)$$

$$IDR_{FF-median} = 1 + \frac{(R-1)^{2/3}(1-\alpha)^9}{122\left(T^2 e^{(-1/3)T}\right)\varsigma^{3/5}\beta^{3/5}} \quad (5.13)$$

The equations are applicable to SDOF systems with varying strengths, periods, post-yield stiffness, energy dissipation ratios and damping ratios. The comparisons between the predicted and 'exact' time-history analysis (THA) results are typically shown in Figs. 5.24 and 5.25, where good agreements are observed. The trends of the IDR responses with varying parameters are also well captured.

The fitted results based on the 84th percentile data are provided in Eqs. (5.14) and (5.15). Depending on the engineers' judgment, the 84th percentile design models can be alternatively used to ensure a more conservative IDR prediction (Fang et al. 2019).

$$IDR_{NF-84th} = 1 + \frac{(R-1)^{4/5}(1-\alpha)^4}{4.5T^{6/5}\varsigma^{1/2}\beta^{2/5}} \quad (5.14)$$

$$IDR_{FF-84th} = 1 + \frac{(R-1)^{3/5}(1-\alpha)^5}{14.3\left(T^{7/5}e^{-3/5T}\right)\varsigma^{1/2}\beta^{1/3}} \quad (5.15)$$

5.6 Residual Displacement

Residual displacement is another critical index that determines whether the structure is repairable or not after earthquakes. Under the performance-based seismic design framework, engineers are often more interested in understanding the relationship between the residual and maximum displacements. Such a relationship is particularly useful when they want to check the resilience category of the designed structure based on the expected maximum displacement. Alternatively, if the permissible residual displacement level (e.g. 0.5% inter-storey drift) is predetermined, based on which a target maximum displacement can be set to restrict the structural design.

In theory, the maximum possible residual displacement (u_{res}) would never exceed the maximum displacement (u_{max}) minus the yield displacement (u_y). The relationship between u_{res} and u_{max} can be expressed by a residual displacement ratio (RDR), and it is expressed by

$$RDR = \frac{u_{res}}{u_{max} - u_y} = \frac{u_{res}}{u_{max} - \frac{u_{el}}{R}} \tag{5.16}$$

where the denominator $u_{max} - u_y$ is an upper bound for u_{res}, and therefore, the RDR always ranges from 0.0 to 1.0.

While dynamic nonlinear THA can give a definite prediction of the residual displacement response of structures under assigned ground motions, THA is generally computational expensive and is not commonly adopted in daily design. Many researchers have attempted to derive simplified residual displacement prediction models (MacRae and Kawashima 1997; Christopoulos et al. 2003; Erochko et al. 2011), either at SDOF or MDOF levels. At present, one of the most recognised residual displacement prediction models is that given in FEMA-P58 (2012) for elasto-plastic yielding systems, where the following simple equations, based on SODF analysis, are recommended for estimating the median residual displacement u_{res} as a function of the peak transient response u_{max} of the structure:

$$u_{res} = 0.3(u_{max} - u_y) \text{ for } u_y < u_{max} < 4u_y \tag{5.17}$$

$$u_{res} = u_{max} - 3u_y \text{ for } u_{max} \geq 4u_y \tag{5.18}$$

The existing models presented in Eqs. (5.17) and (5.18) need to be extended when more diverse structural parameters, as well as different earthquake types, are to be considered. By dividing the SDOF analysis data pool into the following four groups, i.e. partial self-centring systems ($\beta = 1.5$) under near-fault earthquakes, partial self-centring systems ($\beta = 1.5$) under far-field earthquakes, elasto-plastic systems ($\beta = 2.0$) under near-fault earthquakes and elasto-plastic systems ($\beta = 2.0$) under far-field earthquakes, the following quadratic polynomial format can be used for the RDR prediction model (Fang et al. 2019):

$$RDR = A\alpha^2 + B\alpha + C \tag{5.19}$$

where A, B and C are constants derived based on the median responses, as listed in Table 5.3 for the different cases. It should be noted that the case of $\beta \leq 1.0$ always leads to zero RDR, and is therefore not included in the model. The comparisons between the THA and predicted RDR results are shown in Fig. 5.26.

These models are for engineers to quickly evaluate the relationship between the residual and peak displacements (e.g. residual and peak inter-storey drifts) for different types of structures, at least for preliminary design stages. Either the median or the 84th percentile predictions can be used, depending on the design intention.

Table 5.3 Constants for RDR prediction models (Fang et al. 2019)

Cases	A	B	C
Near-fault, $\beta = 1.5$, median	5.0	−0.9	0.1
Near-fault, $\beta = 1.5$, 84th percentile	17.5	−3.3	0.3
Near-fault, $\beta = 2.0$, median	36.6	−7.6	0.6
Near-fault, $\beta = 2.0$, 84th percentile	27.9	−6.5	0.9
Far-field, $\beta = 1.5$, median	0.7	−0.2	0.0
Far-field, $\beta = 1.5$, 84th percentile	2.3	−0.7	0.2
Far-field, $\beta = 2.0$, median	28.8	−5.8	0.4
Far-field, $\beta = 2.0$, 84th percentile	25.6	−6.1	0.8

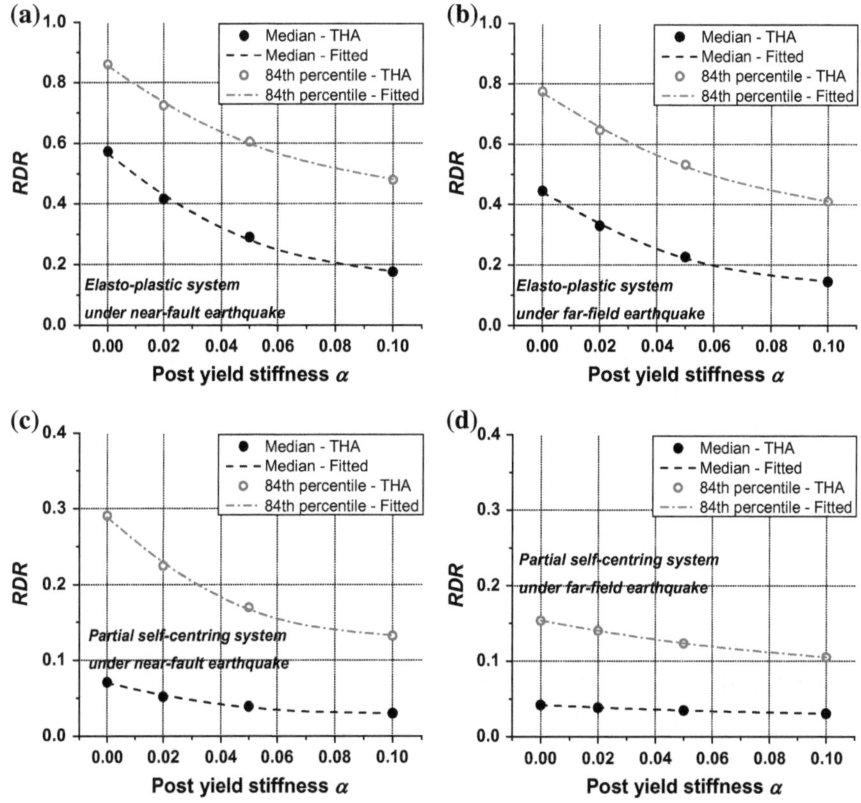

Fig. 5.26 Comparisons between THA and fitted RDR responses: **a** $\beta = 2.0$ under near-fault earthquake, **b** $\beta = 2.0$ under far-field earthquake, **c** $\beta = 1.5$ under near-fault earthquake, **d** $\beta = 1.5$ under far-field earthquake

5.7 Use of SDOF Analysis Data

The SDOF analysis data and the corresponding IDR and RDR prediction models can be employed to give a quick estimation of structural and non-structural damages to structures of which the behaviour is mainly governed by the first vibration mode. To illustrate the application of the SDOF analysis results, six different 2-storey steel frames with the basic information given in Table 5.4 and Fig. 5.27 are considered. The MRFs, both the 5% damped and highly damped versions, were originally designed by Karavasilis et al. (2011) and Karavasilis and Seo (2011), while the remaining structures are modified from the existing MRFs with the same layout of the frame members but using adjusted member dimensions. The structures are located on a stiff soil site, and the design-based earthquake (DBE) response spectrum parameters are: $S_{DS} = 1.0$ g, $S_{D1} = 0.6$ g, $T_0 = 0.12$ s and $T_s = 0.6$ s (see Fig. 5.28). The site is assumed to be sufficiently far away from proven faults, i.e. far-field earthquakes are considered. All the discussions in this section are based on the equivalent nonlinear SDOF representations of these 2-storey buildings, an assumption which is reasonable as these structures mainly vibrate on their fundamental mode.

In Table 5.4, the building codes MRF, BRBF and SMAF represent moment resisting frame, buckling restrained braced frame and frame with SMA-based braces, respectively. The subsequent numbers '100' and '50' indicate the percentage of the base shear compared with the non-reduced, 5% damped structures. D5% and D25% represent 5% damping ratio and 25% damping ratio, respectively. In addition, T_1 is the fundamental period of vibration; V/W is the ratio of the base shear strength (the same as yield strength f_y) to the seismic weight; S_a is the 5% damped spectral response acceleration at the fundamental period of vibration; R is the strength ratio which can be expressed by

$$R = f_{el}/f_y = mS_a/V = S_a\Big/\frac{gV}{W} \tag{5.20}$$

Table 5.4 Basic information of example buildings

Building code	T_1 (s)	V/W	Sa (g)	R	α	β	u_y (m)
MRF100-D5%	1.08	0.27	0.56	2.07	0.05	2.0	0.078
MRF50-D25%	1.48	0.14	0.41	2.93	0.05	2.0	0.076
BRBF100-D5%	0.76	0.27	0.79	2.93	0.05	2.0	0.039
BRBF50-D25%	1.04	0.14	0.58	4.14	0.05	2.0	0.038
SMAF100-D5%	0.76	0.27	0.79	2.93	0.05	1.5	0.039
SMAF50-D25%	1.04	0.14	0.58	4.14	0.05	1.5	0.038

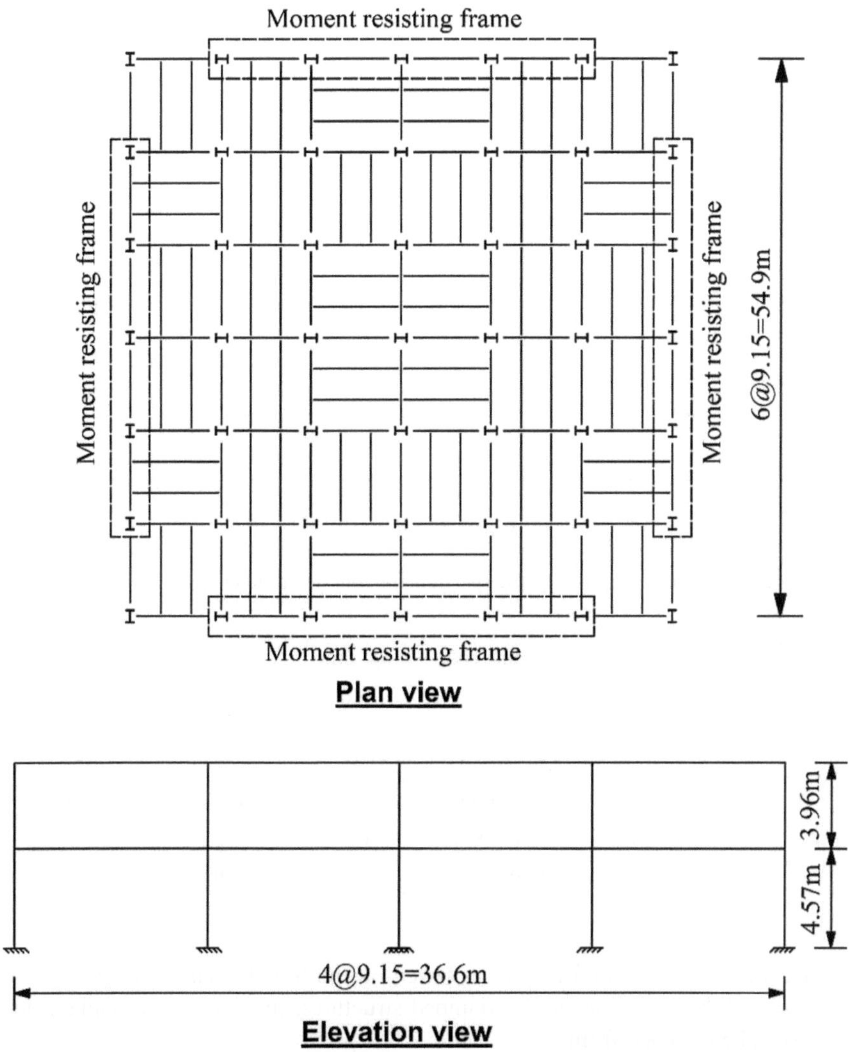

Fig. 5.27 Prototype of 2-storey steel frames

Moreover, for all the considered structures, the post-yield stiffness ratio, which may be obtained from pushover analysis, is 0.05. A full hysteretic behaviour, i.e. $\beta = 2.0$, is assumed for the MRFs and BRBFs. A partial self-centring behaviour with $\beta = 1.5$ is assumed for the SMAFs.

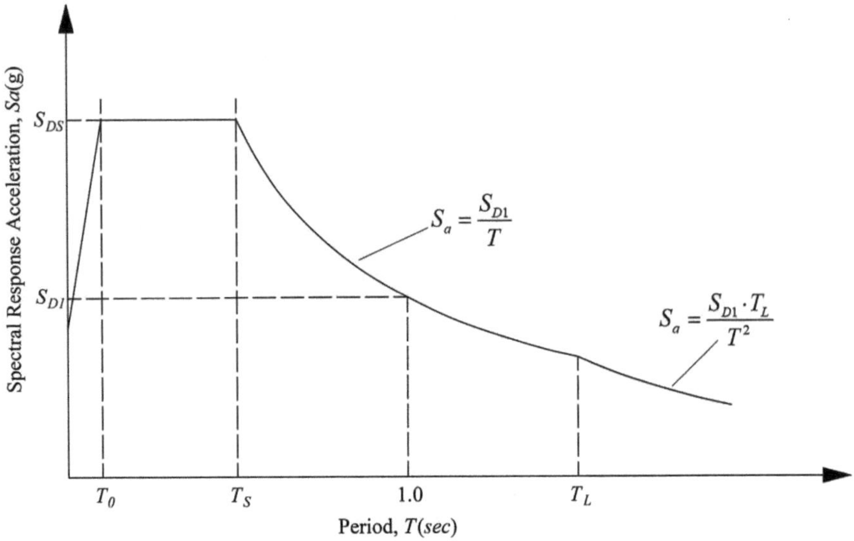

Fig. 5.28 Illustration of design response spectrum for considered structures

Table 5.5 Predicted median responses of example buildings

Building code	u_{max} (m)	u_{res} (m)	a_{max} (g)
MRF100-D5%	0.166	0.019	0.230
MRF50-D25%	0.137	0.013	0.185
BRBF100-D5%	0.122	0.018	0.278
BRBF50-D25%	0.099	0.014	0.215
SMAF100-D5%	0.123	0.003	0.274
SMAF50-D25%	0.103	0.002	0.211

The predicted structural responses based on the SDOF analysis data are summarised in Table 5.5. For the 5% damped structures, the maximum displacement (u_{max}) can be obtained from:

$$u_{max} = IDR \times u_{el} = IDR \times \frac{mS_a}{k} = IDR \times \frac{S_a T_1^2}{(2\pi)^2} \tag{5.21}$$

where IDR is calculated from Eq. (5.13), with all the necessary parameters known.

For the 25% damped systems, the maximum displacement can be first estimated using the same method presented above, assuming 5% damping. Then, the 25% damped response can be predicted via the SODF analysis data, as typically shown in Fig. 5.29. Taking MRF50-D25%, for example the 5% damped maximum displacement, according to Eq. (5.21), is 0.229 m. The natural period is 1.48 s, and $R = 2.93$.

According to Fig. 5.29, for both $R = 2$ and $R = 4$, the ratio of u_{max} ($\zeta = 25\%$) to u_{max} ($\zeta = 5\%$) for $T = 1.48$ s is around 0.6 (alternatively, a linear interpolated value based on the results from the different Rs can be used). Therefore, the calculated u_{max} for MRF50-D25% is equal to 0.229 m $\times$ 0.6 = 0.137 m. Using a similar approach, the u_{max} results for BRBF50-D25% and SMAF50-D25% are 0.099 m and 0.103 m, respectively.

It is shown that, among the six structural systems, the original 5% damped MRF exhibits the largest maximum displacement, while the displacement demands of the 5% damped BRB and SMA-based frames are decreased by more than 25% due to the larger stiffness of the braced frames. For each structural type, reducing the system strength by 50% but increasing the total damping ratio to 25% leads to an evident decrease in the maximum displacement. The reduction rate ranges from 16 to 19%. The BRBFs and SMAFs, either 5% or 25% damped, have comparable maximum displacements.

The residual displacement may be calculated from the median RDR prediction models, where $u_y = u_{el}/R$. Taking MRF50-D25% for example, the residual displacement u_{res} is equal to $(u_{max} - u_y) \times$ RDR, where $u_{max} = 0.137$ m, $u_y = 0.076$ m and RDR = 0.222 according to Fig. 5.26 (elasto-plastic systems under far-field earthquakes). The estimated residual displacements for all the considered systems are summarised in Table 5.5. It is found that a significant reduction in the system strength accompanied by a significant increase in the damping ratio could lead to a reduction of the residual displacement. Moreover, the partial self-centring SMAFs have negligible residual displacements.

The maximum acceleration a_{max} can be obtained from Fig. 5.30. According to the design response spectrum, the PGA, i.e. the spectral acceleration at $T = 0.0$ s, is equal to 0.4 g. The peak acceleration is simply calculated by multiplying the PGA by the normalised a_{max}, as shown in Fig. 5.30. Linear interpolations can be used when R is not exactly equal to the considered values (e.g. 2 or 4). For the cases of $R = 2.07$ and 4.14, the results from $R = 2$ and 4 can be used, respectively, without compromising the accuracy. The estimated maximum accelerations are summarised in Table 5.5. It is emphasised again that the acceleration responses obtained from SDOF analysis are only indicative, and may not be applicable to MDOF systems where more complex floor acceleration responses are expected. Nevertheless, the results given in Table 5.5 indicate that reducing the system strength by 50% and increasing the total damping ratio to 25% is a design case with the potential to achieve a simultaneous reduction of structural (maximum drift, residual drift) and non-structural (acceleration) damages (Karavasilis and Seo 2011).

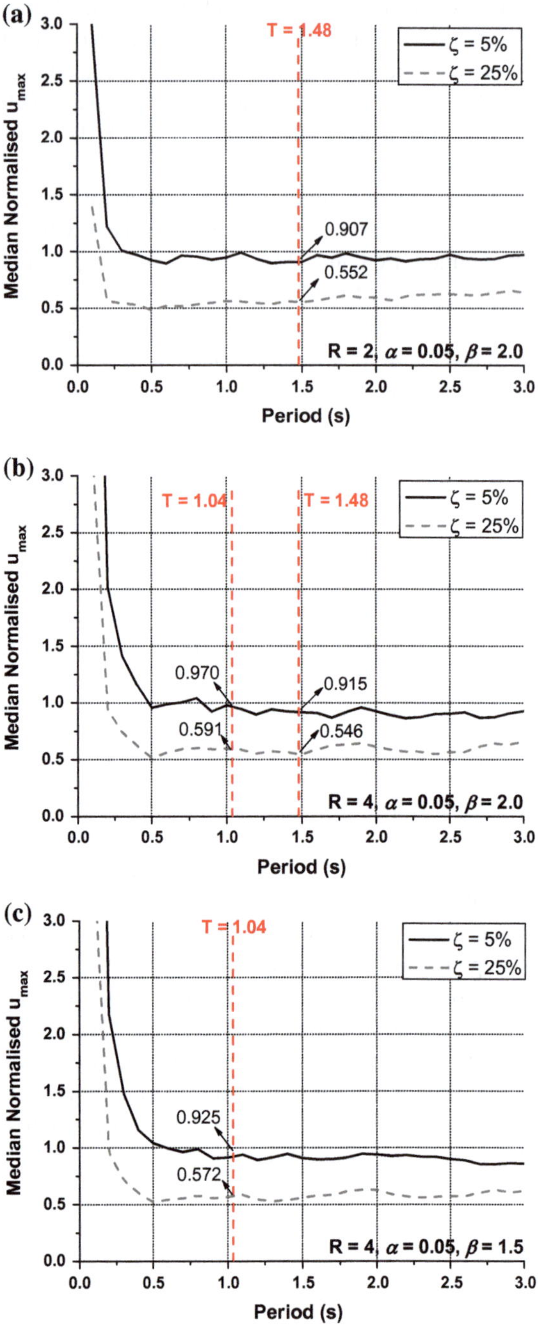

Fig. 5.29 Influence of damping ratio on maximum displacement response based on SDOF analysis data: **a** $R = 2, \alpha = 0.05, \beta = 2.0$, **b** $R = 4, \alpha = 0.05, \beta = 2.0$, **c** $R = 4, \alpha = 0.05, \beta = 1.5$

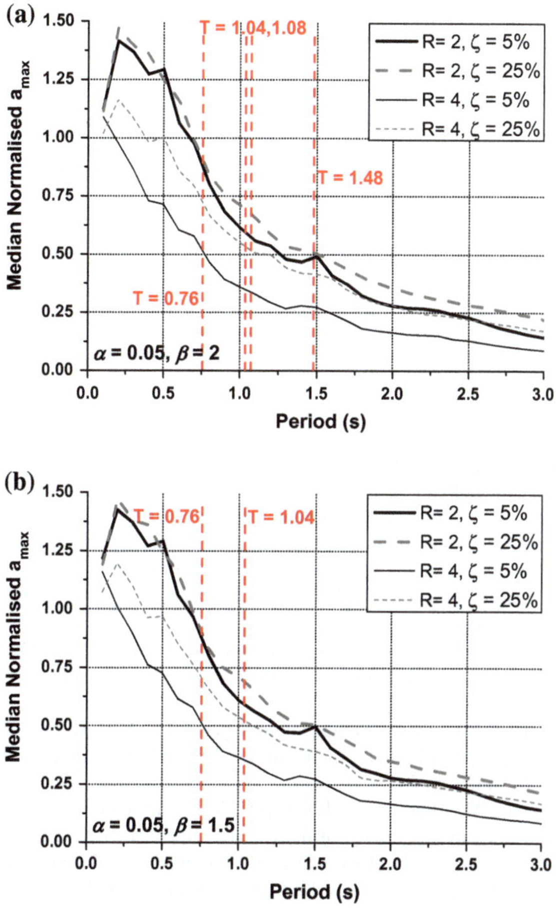

Fig. 5.30 Influences of damping ratio and strength ratio on maximum acceleration response based on SDOF analysis data: **a** $\alpha = 0.05$, $\beta = 2.0$, **b** $\alpha = 0.05$, $\beta = 1.5$

References

Alavi B, Krawinkler H (2004) Strengthening of moment-resisting frame structures against near-fault ground motion effects. Earthq Eng Struct D 33(6):707–722

Alhan C, Gazi H, Kurtuluş H (2016) Significance of stiffening of high damping rubber bearings on the response of base-isolated buildings under near-fault earthquakes. Mech Syst Signal Pr 79:297–313

American Society of Civil Engineers (ASCE) (2007) Seismic rehabilitation of existing buildings. ASCE/SEI 41-06, Reston, VA

American Society of Civil Engineers (ASCE) (2010) Minimum design loads for buildings and other structures. ASCE/SEI 7-10, Reston, VA

Baltzopoulos G, Vamvatsikos D, Iervolino I (2016) Analytical modelling of near-source pulse-like seismic demand for multi-linear backbone oscillators. Earthq Eng Struct D 45(11):1797–1815

Bray JD, Rodriguez-Marek A (2004) Characterization of forward-directivity ground motions in the near-fault region. Soil Dyn Earthq Eng 24(11):815–828

Christopoulos C, Pampanin S, Priestley MJN (2003) Performance-based seismic response of frame structures including residual deformations—part I: single-degree of freedom system. J Earthq Eng 7(1):97–118

Dimakopoulou V, Fragiadakis M, Constantine S (2013) Influence of modeling parameters on the response of degrading systems to near-field ground motions. Eng Struct 53:10–24

Durucan C, Durucan AR (2016) Ap/Vp specific inelastic displacement ratio for the seismic response estimation of SDOF structures subjected to sequential near fault pulse type ground motion records. Soil Dyn Earthq Eng 89:163–170

Eatherton MR, Hajjar JF (2011) Residual drifts of self-centering systems including effects of ambient building resistance. Earthq Spectra 27(3):719–744

Erochko J, Christopoulos C, Tremblay R, Choi H (2011) Residual drift response of SMRFs and BRB frames in steel buildings designed according to ASCE 7-05. J Struct Eng-ASCE 137(5):589–599

European Committee for Standardization (CEN) (2004) EN 1998-1, Eurocode 8: design of structures for earthquake resistance—part 1: general rules, seismic actions and rules for buildings. Belgium, Brussels

Fang C, Zhong QM, Wang W, Hu SL, Qiu CX (2018) Peak and residual responses of steel moment-resisting and braced frames under pulse-like near-fault earthquakes. Eng Struct 177:579–597

Fang C, Qiu CX, Huang LT (2019) Performance-based evaluation of seismic resilient systems under near-fault and far-field earthquakes. Eng Struct Under Rev

Federal Emergency Management Agency (FEMA) (2000) Prestandard and commentary for the seismic rehabilitation of buildings: rehabilitation requirements. FEMA 356, Washington, DC

Federal Emergency Management Agency (FEMA) (2005) Improvement of nonlinear static seismic analysis procedures. FEMA 440, prepared by Applied Technology Council for FEMA, Washington, DC

Federal Emergency Management Agency (FEMA) (2009) Quantification of building seismic performance factors. FEMA P695, prepared by the Applied Technology Council for the FEMA, Washington, DC

Federal Emergency Management Agency (FEMA) (2012) Seismic performance assessment of buildings, volume 1—methodology. FEMA P-58-1, prepared by the SAC Joint Venture for FEMA, Washington, DC

Gerami M, Abdollahzadeh D (2015) Vulnerability of steel moment-resisting frames under effects of forward directivity. Struct Des Tall Spec 24(2):97–122

Hatzigeorgiou GD, Beskos DE (2009) Inelastic displacement ratios for SDOF structures subjected to repeated earthquakes. Eng Struct 31(11):2744–2755

Hou HT, Qu B (2015) Duration effect of spectrally matched ground motions on seismic demands of elastic perfectly plastic SDOFS. Eng Struct 90:48–60

Karavasilis TL, Ricles JM, Sause R, Chen C (2011) Experimental evaluation of the seismic performance of steel MRFs with compressed elastomer dampers using large-scale real-time hybrid simulation. Eng Struct 33(6):1859–1869

Karavasilis TL, Seo CY (2011) Seismic structural and non-structural performance evaluation of highly damped self-centering and conventional systems. Eng Struct 33(8):2248–2258

Kawashima K, Macrae R, Hoshikuma JI, Nagaya K (1998) Residual displacement response spectrum. J Struct Eng-ASCE 124(5):523–530

Krawinkler H, Alavi B (1998) Development of improved design procedures for near-fault ground motions. In: SMIP 98: seminar on utilization of strong motion data, Oakland, CA

Loh CH, Wan S, Liao WI (2002) Effects of hysteretic model on seismic demands: consideration of near-fault ground motions. Struct Des Tall Buil 11(3):155–169

MacRae GA, Kawashima K (1997) Post-earthquake residual displacements of bilinear oscillators. Earthq Eng Struct D 26(7):701–716

MathWorks (2010) MATLAB 7.10.0. The MathWorks Inc., Natick, MA

Mazzoni S, McKenna F, Scott MH, Fenves GL (2006) Open system for earthquake engineering simulation (OpenSees), OpenSees Command Language Manual. Pacific Earthquake Engineering Research Center, University of California, Berkeley

McCormick J, Aburano H, Ikenaga M, Nakashima M (2008) Permissible residual deformation levels for building structures considering both safety and human elements. In: Proceedings of 14th world conference on earthquake engineering, Seismological Press of China, Beijing
Moradi S, Alam MS, Asgarian B (2014) Incremental dynamic analysis of steel frames equipped with NiTi shape memory alloy braces. Struct Des Tall Spec 23:1406–1425
Perry C, PhippsM, Hortacsu A (1994) Reducing the risks of nonstructural earthquake damage—a practical guide. In: ATC and SEI conference on improving the seismic performance of existing buildings and other structures
Ray-Chaudhuri S, Hutchinson TC (2011) Effect of nonlinearity of frame buildings on peak horizontal floor acceleration. J Earthq Eng 15(1):124–142
Rodriguez-Marek A (2000) Near-fault seismic site response. PhD dissertation, University of California at Berkeley
Tremblay R, Lacerte M, Christopoulos C (2008) Seismic response of multistory buildings with self-centering energy dissipative steel braces. J Struct Eng-ASCE 134(1):108–120
Yadav KK, Gupta VK (2017) Near-fault fling-step ground motions: characteristics and simulation. Soil Dyn Earthq Eng 101:90–104

Chapter 6
Structural Responses: Multi-storey Building Frames

Abstract This chapter extends to multi-degree-of-freedom (MDOF) systems and aims to demonstrate in more detail the effectiveness of various self-centring solutions in controlling the residual deformation and other critical responses of multi-storey building frames. Two major types of self-centring steel structures, namely self-centring braced frames and self-centring moment resisting frames (MRFs), are presented. The considered self-centring braced frames are in fact generic buildings with braces exhibiting idealised flag-shaped hysteretic behaviour, where the influences of a spectrum of brace parameters on the peak and residual responses of the buildings are examined. The self-centring MRFs are those with specifically designed beam-to-column connections employing SMA ring springs. Other commonly used structural forms, such as conventional MRFs and BRB frames, are also analysed for comparison purpose. Both far-field and pulse-like near-fault earthquakes are considered in this chapter.

6.1 Introduction

The behaviour of idealised SDOF systems under different types of earthquakes has been discussed in the previous chapter, and in particular, the influences of various structural parameters on the key engineering demand parameters of the SDOF systems have been revealed. Some performance-based design considerations in terms of inelastic displacement ratio and residual displacement were also outlined. From the SDOF analysis, we now understand that natural period, yield strength, post-yield stiffness, energy dissipation ratio and damping ratio are factors that can affect the displacement and acceleration responses of systems to varying extents, but caution should be exercised that some observations from the SODF analysis may not be applicable to multi-degree-of-freedom (MDOF) systems. Some studies (e.g. Qiu and Zhu 2016) revealed that the high-mode effect is much more pronounced in building frames with SMA-based self-centring braces than conventional structural systems such as BRB frames and steel MRFs. In addition, the maximum total floor acceleration can be significantly amplified in self-centring structures. These issues can hardly be identified from SDOF analysis only.

C. Fang and W. Wang, *Shape Memory Alloys for Seismic Resilience*, https://doi.org/10.1007/978-981-13-7040-3_6

This chapter proceeds with a further investigation into the seismic behaviour of MDOF systems. The designs of self-centring braced frames and BRB frames are introduced first. These structures employ simplified material and member models, and are designed based on design-compatible response spectrums and satisfy the minimum codified requirements, with no particular consideration for near-fault pulse-like effects. They are then assessed in terms of maximum inter-storey drift (MID), residual inter-storey drift (RID) and peak absolute floor acceleration (PFA) responses, using suites of near-fault ground motion records. Idealised flag-shaped and elasto-plastic behaviours are considered for the self-centring braces and BRBs, respectively. Such idealisation is adequate in practical design, as we can confirm from Chaps. 2 through 4 that SMA-based devices do exhibit typical flag-shaped hysteretic responses. Degradation of SMA is ignored in the considered buildings, as it is assumed that all the SMA-based braces have been appropriately trained. While certain SMA components may exhibit mechanical behaviour being slightly different from a 'regular' flag-shaped hysteresis, these can be studied via a parametric study which is also within the scope of this chapter.

After gaining fundamental understandings of the behaviour of the simplified building frames, as well as the factors that may affect the building behaviour, a further case study on more comprehensively designed SMA-based self-centring building frames is presented. Recalling that the SMA ring spring is one of the most cost-effective SMA elements that provide large load resistance, flexible deformability and good energy dissipation capability, the application of these elements in the connection zones of MRFs is demonstrated in detail.

Based on the illustrative examples, the advantage of the self-centring technology is confirmed. Some possible shortcomings and limitations of these emerging systems are also noticed, and solutions to these issues are discussed.

6.2 Simplified Multi-storey Building Frames

6.2.1 Design and Modelling

6.2.1.1 Prototype MRFs

Code-compliant 3-storey and 9-storey steel MRFs, designed for the famous SAC steel project (Gupta and Krawinkler 1999), are considered as reference cases which are compared later with self-centring and BRB frames. These MRFs were designed as office buildings located on a stiff soil site in Los Angeles, USA. The following design response spectral values are assigned to the site: $S_{DS} = 2/3S_{MS} = 1.376g$ and $S_{D1} = 2/3S_{M1} = 0.707g$. The lateral load resisting system in each of the two orthogonal directions consists of two special MRFs on the perimeter of the building. In this section, we only focus on 2D frames that represent half of the structures in the north–south (NS) direction, as shown in Fig. 6.1.

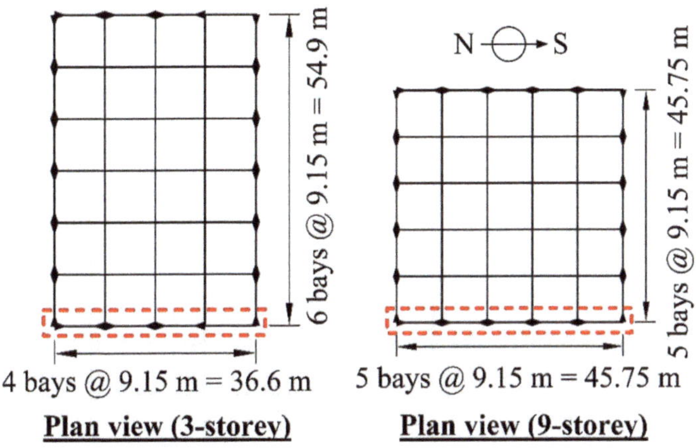

Fig. 6.1 Plan view of prototype MRFs

Figures 6.2 and 6.3 show the elevation views of the two buildings. The 3-storey building has four bays in the NS direction, including three MRF bays and one bay for resisting gravity load only. Fixed column feet are designed for the 3-storey building. The 9-storey building includes nine stories above the ground level in addition to a basement level. Each lateral resisting frame has five MRF bays in the NS direction. Note that there is one external connection which is designed as pinned, such that biaxial bending in the corner column is avoided. For both buildings, the design yield strengths of the steel beams and columns, according to the SAC report (Gupta and Krawinkler 1999), are 248 MPa and 345 MPa, respectively. The member sizes are marked in the figures, with the US practice followed. The buildings are required to conform to an inter-storey drift limit of $h/400$ (h = storey height) under regular load combinations.

Either open-source or commercial numerical programmes can be readily employed for the simulation of the structures. Taking the open-source program OpenSees (Mazzoni et al. 2006) for example, the beam and column members are adequately simulated via nonlinear force-based beam-column elements with fibre sections, such that the development and distribution of plasticity along the length

Fig. 6.2 Elevation view of 3-storey prototype MRFs

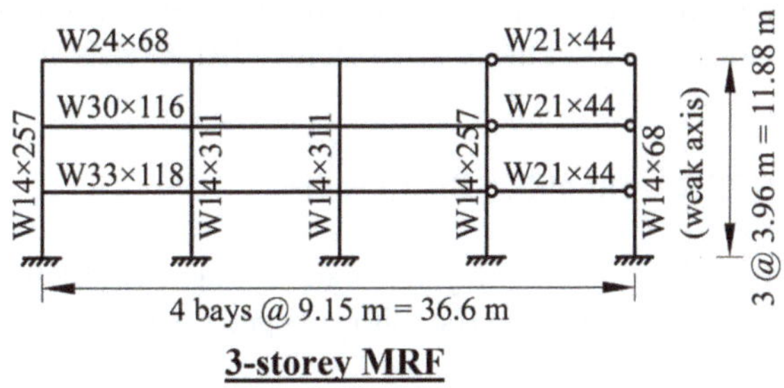

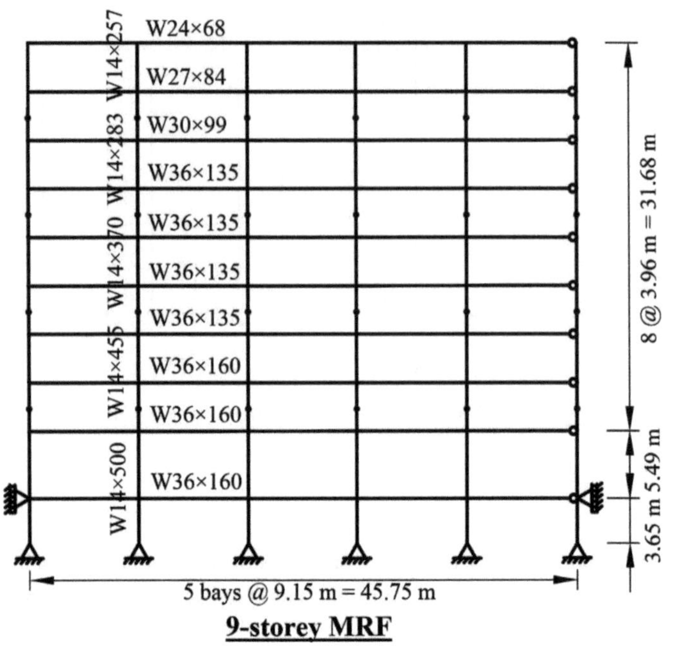

Fig. 6.3 Elevation view of 9-storey prototype MRFs

and depth of the members are taken into account. An elasto-plastic material model can be assigned to each fibre. To simplify modelling, one may choose to ignore the shear deformation of the panel zones, and hence basic 2D centreline models of the MRFs can be built. This is an acceptable simplification from a parametric study point of view, provided that all the buildings are modelled with a consistent modelling approach. The seismic mass should be taken as half the total structural mass in the 2D model. The weight from the tributary area should be directly applied to the 2D frame, while the remaining weight should be applied to adjacent lean columns, such that the P-Δ effect is fully considered. This modelling approach enables a conservative prediction of structural deformation, as the contributions from the gravity frame and infill walls in resisting the lateral load are ignored. For dynamic analysis of typical steel structures, one can assume a Rayleigh damping with a 2% damping ratio for the concerned modes of vibration. According to modal analysis, the first mode periods of the 3- and 9-storey MRF models are 1.01 s and 2.27 s, respectively.

6.2.1.2 Simplified BRB Frames

Both the 3- and 9-storey MRFs are redesigned as BRB frames for comparison purpose (Fang et al. 2018a). The following process is adopted for the BRB frame design:

A reduced-strength version of the boundary frame is first designed that satisfies the basic gravity load resistance requirement. BRBs, preferably with an inverted-V configuration according to the layout of the frame, are then added to provide seismic load resistance. These BRBs are mainly responsible for storey drift control. If the BRB frames are designed as dual systems, then the boundary frame, with rigid beam-to-column connections, is supposed to make contributes to the overall lateral resistance. Pin connections, which lead to reduced costs, are also allowed according to the design standards such as ASCE/SEI 7-10 (2010). In fact, pin connections are quite popular for BRB frames in the US practice. Nevertheless, dual systems are consistently considered for all the braced frames in this chapter.

A standard equivalent lateral force or modal response spectrum analysis procedure can be employed for the design of the 3- and 9-storey BRB frames. The BRB frames are assumed to locate at the same site as the MRFs, and are considered as Risk Category II buildings with an importance factor of $I_e = 1.0$. The values of the response modification coefficient (R) and deflection amplification factor (C_d) should be 8.0 and 5.0, respectively, according to ASCE/SEI 7-10 (2010). The maximum permitted inter-storey drift is 2.0% under the DBE.

The 3- and 9-storey BRB frames considered in this chapter are illustrated in Fig. 6.4. Similar to the MRFs, basic centreline models for the BRB frames can be built in any viable numerical programme. The BRB elements can be modelled using truss elements that resist axial forces only. The material behaviour of the BRB is adequately represented using a bilinear elasto-plastic model with kinematic strain hardening. The post-yield stiffness ratio (α) for the BRBs may be assumed to be 0.05. According to the above design and modelling assumptions, the estimated maximum inter-storey drifts of the 3- and 9-storey BRB frames under the DBE are 1.60% and 1.45%, respectively, which satisfy the 2.0% inter-storey drift limit. The first mode periods of vibration of the 3- and 9-storey BRB frames are 0.64 s and 1.73 s, respectively.

It is worth mentioning that BRBs can exhibit significant isotropic hardening, especially at large deformations. The idealised bilinear elasto-plastic BRB presentation used in the current BRB frame models is mainly considered to enable a more direct comparison against flag-shaped self-centring braces, where the two cases only differ in the energy dissipation ratio β. In other words, the simplified BRB model can be deemed as a special case of $\beta = 2.0$ (Fang et al. 2018a). In fact, idealised elasto-plastic models have been adopted by many researchers for system-level analysis of BRB frames (e.g. Sabelli et al. 2003; Nguyen et al. 2010; Kari et al. 2011; Qiu and Zhu 2016), but the reader should bear in mind that the varying modelling approaches for the BRBs could have a mild, although not critical, effect on the overall structural dynamic response. Using a reversed triangle load pattern, the static lateral pushover curves of the MRFs and the BRB frames are shown in Fig. 6.5. The descending branch observed in the 9-storey structures is caused by the P-Δ effect.

One can more comprehensively evaluate the performance of the BRB frames through nonlinear time-history analysis. Earthquake records should be selected and scaled to be compatible with the design spectrum. Table 6.1 gives the basic information of the selected far-field ground motions that exhibit minimal pulsing effects. The

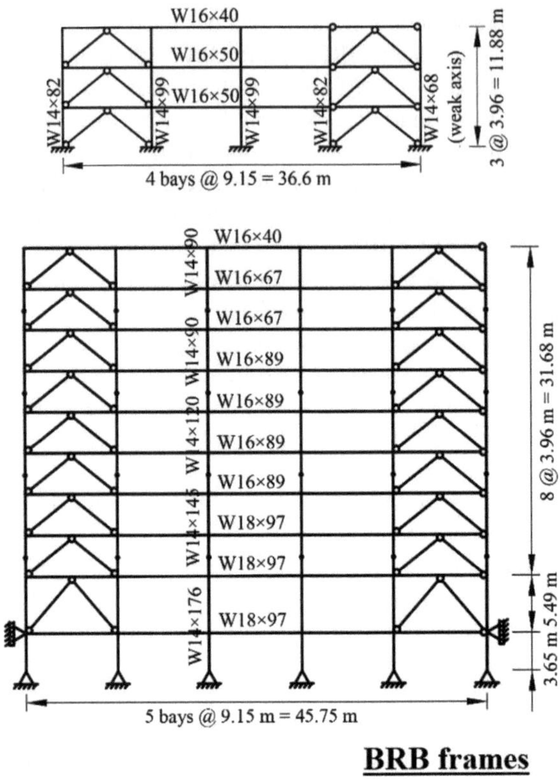

Brace information:

3-storey		
Storey	F_y (kN)	k_0 (kN/mm)
1	975.0	140.6
2	774.3	111.6
3	348.3	50.2

9-storey		
Storey	F_y (kN)	k_0 (kN/mm)
1	2918.17	356.2
2	1706.26	246.0
3	1326.22	191.2
4	1185.10	170.9
5	1157.32	166.9
6	958.03	138.1
7	804.59	116.0
8	528.34	76.2
9	320.42	46.2

Fig. 6.4 Elevation view of 3- and 9-storey prototype BRB frames

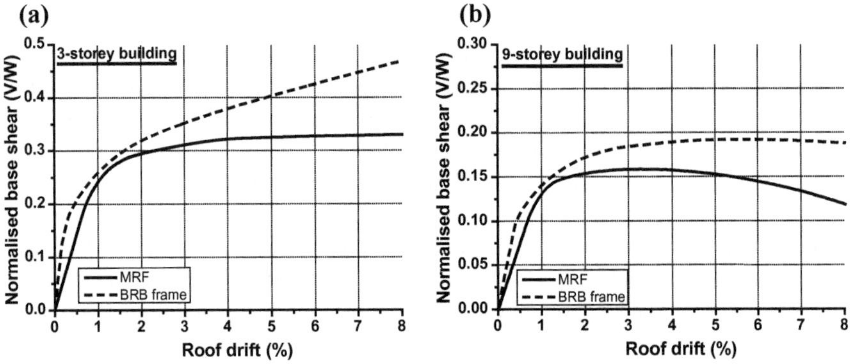

Fig. 6.5 Pushover responses of MRFs and BRB frames (V = base shear, W = seismic weight): **a** 3-storey buildings, **b** 9-storey buildings

Table 6.1 Basic information of selected far-field ground motions

Event no.	Earthquake name	Year	Station	Soil type	Magnitude (M_w)	PGA (g)	PGV (cm/s)	PGV/PGA	Scale factor	
									DBE	MCE
FF1	San Fernando	1971	LA-Hollywood Stor	Stiff soil	6.6	0.547	49.19	0.092	2.61	3.91
FF2	Imperial Valley	1979	Delta	Stiff soil	6.5	0.697	76.11	0.111	2.93	4.40
FF3	Imperial Valley	1979	Delta	Stiff soil	6.5	0.607	57.03	0.096	1.73	2.59
FF4	Imperial Valley	1979	El Centro Array #11	Stiff soil	6.5	0.318	30.03	0.096	0.87	1.31
FF5	Imperial Valley	1979	El Centro Array #11	Stiff soil	6.5	0.552	61.27	0.113	1.46	2.18
FF6	Superstition Hills	1987	El Centro Imp. Co.	Stiff soil	6.5	0.650	84.08	0.132	1.82	2.72
FF7	Superstition Hills	1987	El Centro Imp. Co.	Stiff soil	6.5	0.643	101.58	0.161	2.49	3.73
FF8	Superstition Hills	1987	Poe Road (temp)	Stiff soil	6.5	0.895	71.50	0.082	2.00	3.01
FF9	Superstition Hills	1987	Poe Road (temp)	Stiff soil	6.5	1.003	109.36	0.111	3.34	5.01
FF10	Loma Prieta	1989	Capitola	Stiff soil	6.9	0.646	42.73	0.068	1.22	1.83
FF11	Loma Prieta	1989	Gilroy Array #3	Stiff soil	6.9	0.406	26.08	0.066	0.73	1.10
FF12	Landers	1992	Cool water	Stiff soil	7.3	0.652	63.38	0.099	2.30	3.45
FF13	Northridge	1994	Beverly Hills—Mulhol	Stiff soil	6.7	0.961	136.07	0.145	2.31	3.47
FF14	Kocaeli—Turkey	1999	Duzce	Stiff soil	7.5	0.519	67.23	0.132	1.45	2.18
FF15	Duzce—Turkey	1999	Bolu	Stiff soil	7.1	0.594	46.04	0.079	0.82	1.22

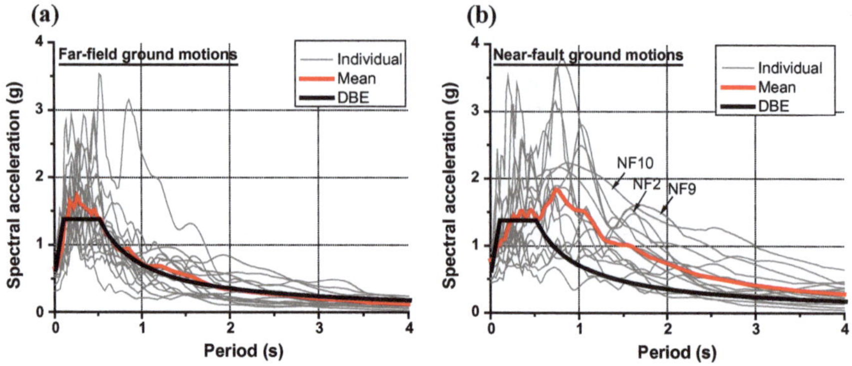

Fig. 6.6 Elastic response spectra of the selected ground motion records: **a** far-field ground motions, **b** near-fault ground motions

ground motions are further scaled by 1.5 to represent the MCE-level earthquake. The individual and mean response spectra of the considered far-field earthquake records are shown in Fig. 6.6a together with the design response spectrum.

The allowable inter-storey drift obtained from nonlinear time-history analysis can be increased by 25% for Risk Category II buildings, according to ASCE/SEI 7-10 (2010). Therefore, the limit of the maximum inter-storey drift is relaxed to 2.5% under the DBE level. No drift limitation is given in ASCE/SEI 7-10 for the MCE, as long as collapse is prevented based on engineering judgement. The nonlinear time-history analysis, employing the considered earthquake records, shows that the 3-storey BRB frame exhibits a mean maximum inter-storey drift of 1.72% and 2.24% under the DBE and MCE, respectively, and for the 9-storey BRB frame, the corresponding values are 1.47% and 2.07%, respectively. Clearly, both the BRB frames satisfy the drift requirements under the DBE, and the risk of collapse under the MCE is remote. In other words, the design is code-compliant.

Before moving on to the discussion of the self-centring braced frames, some extra comments on the design of the model structures are made herein. It is recalled that the BRB frames are designed to enable a direct comparison against the prototype MRFs. To achieve this, two different strategies may be adopted when determining the basic properties of the BRB frames:

1st strategy—the BRB frames and the MRFs have similar initial stiffness and therefore similar fundamental periods of vibration. The advantage of this strategy is that the behaviour of the different types of structures under the same suite of ground motions is minimally influenced by the varied natural periods of vibration, and as a result, can be more rationally compared, whereas the main disadvantage is that the design of the BRB frames may not be realistic.

2nd strategy—all the structures are code-compliant. This means that the different types of structures don't have to exhibit the same fundamental period of the vibration, as long as they satisfy the basic code requirements, e.g. 2% drift limit under the DBE.

This approach leads to the most practical design for the structures, but the results are inevitably affected by the period of vibration.

The 2nd strategy is adopted for the BRB frames shown in Fig. 6.4. As can be confirmed from Fig. 6.5, the initial stiffness of the BRB frames is larger than that of the MRFs, which is common in practice.

6.2.1.3 Self-centring Braced Frames

The self-centring braced frames have the same beam and column sizes as those of the BRB frames, and the two systems only differ in the energy dissipation factor β of the braces. This makes the same fundamental period of vibration of the two types of structures. It is assumed that the self-centring braces and the BRBs have the same initial stiffness, post-yield stiffness and yield strength, whereas the energy dissipation factor of the self-centring braces is half of that of the BRBs. $\beta = 1.0$ is a typical case for SMA-based members, as can be seen from Chaps. 3 and 4.

A material model that mimics the flag-shaped hysteresis is assigned to the self-centring braces. As static design procedures for self-centring systems are still lacking, e.g. neither specific response modification coefficient (R) nor deflection amplification factor (C_d) values are available, a nonlinear time-history analysis procedure can be the most efficient and reliable way to check the inter-storey drift and acceleration responses of the structures. Using the same suite of far-field ground motions given in Table 6.1, the mean MIDs obtained for the 3-storey and 9-storey self-centring braced frames under the DBE are 2.04% and 1.74%, respectively. The values under the MCE are 2.91% and 2.43%, respectively. Clearly, the codified drift limits are satisfied, according to ASCE/SEI 7-10 (2010). It is noted that due to the decreased energy dissipation of the self-centring braces, the MID responses of the self-centring frames are larger than those of the BRB frames.

For the present self-centring braced frames, the type of SMA elements (e.g. SMA cables, bars, or different types of springs) used for the braces is not particularly specified. One reason for this is that the main objective of the present section is to obtain and compare the key engineering demand parameters of the different structures in a generic way. Another reason is that the self-centring braces, even with different types of SMA elements, commonly exhibit hysteretic behaviour which is adequately represented by a typical flag-shaped hysteretic response. In practice, the strength, stiffness and energy dissipation factor of any SMA-based self-centring brace product could be readily acquired via sample tests before application.

It is also worth mentioning that the possible deteriorations or failures of the beams, columns and braces, are not considered in the building frame models. This is an acceptable assumption as long as the main frame members have compact sections and the connections are ductile (which holds true in a well-designed structure), and in addition, the MID is preferably less than 5% inter-storey drift. For simplicity, one may simply define a ‘failure criterion’ (e.g. 5% inter-storey drift), exceeding which the structure is assumed to have collapsed (Erochko et al. 2011; Özhendekci and Özhendekci 2012; Hsiao et al. 2013; Silwal et al. 2016). Nevertheless, we choose not to specify such collapse limit state in the simplified models, because (1) a definite threshold of collapse, e.g. 5%, is inaccurate itself and is subjected to uncertainty, (2) the structures generally have MIDs being less than 5% under the aforementioned DBE and MCE design earthquakes and (3) this chapter focuses more on revealing the trends caused by the varying brace characteristics rather than on collapse assessment of the structures. The latter will be discussed in the next chapter.

While the failure behaviour of the structural members is excluded from the frame model, the reader should be aware that the maximum elongation of an SMA-based self-centring brace depends on the effective deformability of the SMA elements. For typical brace configurations described in Chap. 4, it is indeed challenging, if not infeasible, to ensure a reliable and stable brace performance when the inter-storey drift exceeds 5%. The same applies to BRBs, which can hardly survive many repeated cycles of large axial deformation (e.g. inter-storey drift being larger than 2%) because of their finite low-cycle fatigue resistance (Andrews et al. 2009; NIST 2015). This raises a concern when pulse-like near-fault earthquakes are considered, as these ground motions are more damaging than far-field ones (Fang et al. 2018a). In this context, the behaviour of the MRFs, BRB frames and self-centring braced frames under near-fault earthquakes is particularly examined in the following section, before we continue to summarise the overall building behaviour in Sect. 6.2.3.

6.2.2 Consideration for Near-Fault Earthquakes

After showing that all the three types of multi-storey building frames satisfy the drift requirements specified in the current code, their performances under near-fault earthquakes need to be further confirmed. A total of 15 recorded or simulated near-fault ground motions are used for the analysis (Fang et al. 2018a). The event magnitude M_w ranges from 6.7 to 7.6, and the fault distance is less than 15 km. Most of these ground motions have also been considered in the SODF analysis described in Chap. 5. The basic information, including the peak ground acceleration (PGA), peak ground velocity (PGV), PGV/PGA ratio and pulse period (T_p) of the near-fault ground motions, is summarised in Table 6.2, and the 5% damped acceleration spectra are shown in Fig. 6.6b.

In contrast to far-field ground motions which are often uniformly scaled, scaling of near-fault ground motions should be done more cautiously. This is largely because the pulse period of near-fault ground motions is related to the seismogenic mechanism, where source parameters such as the rise time and fault dimensions generally change with magnitude. In other words, the shape of the intermediate- and long-period regions of the response spectrum changes as the magnitude increases. Therefore, the key spectral characteristics of near-fault ground records may not be sufficiently reflected by a simple monotonic scaling (Somerville 2003). In light of this, the selected near-fault ground motions are not scaled in the analysis, such that 'true' pulsing characteristics are reflected. Incidentally, the ground motions discussed in

Table 6.2 Basic information of selected near-fault ground motions

Event No.	Earthquake	Year	Station	Soil type	Magnitude	Distance (km)	PGA(g)	PGV (cm/s)	PGV/ PGA	Type	T_p (s)	$T_{p\text{-}v}$ (s)
NF1	Tabas	1978	Tabas	Stiff soil	7.4	1.20	0.978	105.81	0.110	FD	5.24	4.70
NF2	Loma Prieta	1989	Los Gatos	Stiff soil	7.0	3.50	0.718	172.84	0.246	FD	2.28	3.26
NF3	Loma Prieta	1989	Lex. Dam	Stiff soil	7.0	6.30	0.687	178.66	0.266	FD	1.56	1.08
NF4	Northridge	1994	Rinaldi	Stiff soil	6.7	7.50	0.891	174.22	0.200	FD	1.39	1.06
NF5	Kobe	1995	KJMA	Stiff soil	6.9	0.96	0.821	81.22	0.101	FD	0.84	0.90
NF6	Kobe	1995	Kobe	Stiff soil	6.9	3.40	1.089	160.17	0.150	FD	2.16	0.88
NF7	Kobe	1995	Takatori	Stiff soil	6.9	4.30	0.787	144.73	0.188	FD	1.78	1.28
NF8	Elysian Park 2	–	–	Stiff soil	7.1	10.70	0.904	96.81	0.109	FD	0.84	0.76
NF9	Elysian Park 3	–	–	Stiff soil	7.1	11.20	1.014	155.07	0.156	FD	2.64	1.92
NF10	Palos Verdes 2	–	–	Stiff soil	7.1	1.50	0.969	287.86	0.303	FD	3.28	2.60
NF11	Landers	1992	Landers	Stiff soil	7.3	1.10	0.714	136.05	0.194	FS	5.70	4.16
NF12	Chi-Chi	1999	TCU52	Stiff soil	7.6	1.84	0.448	220.67	0.503	FS	13.87	9.80
NF13	Chi-Chi	1999	TCU65	Stiff soil	7.6	2.49	0.790	132.48	0.171	FS	6.20	4.68
NF14	Chi-Chi	1999	TCU67	Stiff soil	7.6	1.11	0.499	97.47	0.199	FS	3.93	2.28
NF15	Chi-Chi	1999	TCU68	Stiff soil	7.6	3.01	0.512	280.28	0.559	FS	12.15	9.32

FD = forward directivity, FS = fling step

Chap. 5 for the SDOF analysis are also unscaled as the strength of the system has been adjusted by strength ratio R.

ASCE/SEI 7-10 (2010) requires that the mean acceleration response spectrum of the selected ground motions should not be less than the design response spectrum for periods ranging from $0.2T$ to $1.5T$, where T is the fundamental period of vibration of the structure being considered. It is confirmed from Fig. 6.6b that this condition is satisfied at the DBE level. It is also demonstrated that the near-fault ground motions tend to have increased spectral accelerations at longer periods, which, as explained in Chap. 5, is attributed to the long-period pulses.

The typical near-fault earthquake-induced structural responses of the three types of buildings are first illustrated via a case study using the Loma Prieta Lex. Dam record. Figure 6.7 gives the roof drift responses of the structures under this earthquake as a function of time. The peak roof drifts for the different structures occur at 4.5–5 s, a time which generally coincides with the occurrence of the dominant ground velocity pulse, as previously shown in Fig. 5.4. After undergoing the initial peak roof drift, the structures immediately bounce back, showing a reversed peak roof drift. The amplitude of the reversed peak is much smaller than that of the initial peak.

The fluctuation immediately after undergoing the peak roof drift seems to be more obvious in the self-centring braced frames, which results from the unique flag-shaped behaviour of the self-centring braces with reduced energy dissipation compared to the BRBs. Nevertheless, the self-centring frames tend to oscillate in the adjacent of the upright position and finally reach a negligible residual drift during the attenuation of vibration. On the other hand, the MRFs and BRB frames oscillate around certain roof drift excursions, which is directly responsible for their final residual drift. For the Loma Prieta Lex. Dam record considered in this case study, the peak roof drifts are on the order of 4–6% and 2.5–3% for the 3- and 9-storey structures, respectively. The residual roof drift is on the order of 0.5–1.5% for the MRFs and BRB frames, but almost no residual roof drift is observed for the self-centring braced frames.

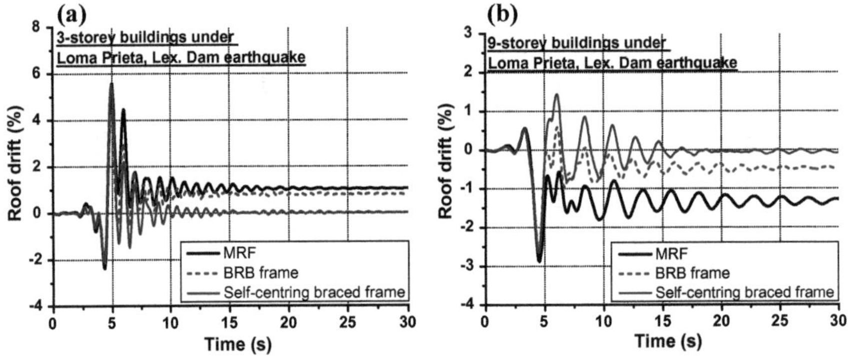

Fig. 6.7 Roof drift response of three types of buildings under Loma Prieta Lex. Dam ground motion: **a** 3-storey buildings, **b** 9-storey buildings

Figure 6.8 shows the axial force–deformation responses of the idealised BRBs and self-centring braces at selected floor levels. In contrast to the full hysteretic curves exhibited by the BRBs, flag-shaped responses accompanied by moderate energy dissipation are clearly shown for the self-centring braces. The self-centring braces and BRBs experience comparable peak deformations, but the former exhibit negligible residual deformation. This is generally in line with the roof drift responses as shown in Fig. 6.7. The peak axial deformation demand of the considered braces is on the order of 100–150 mm. Assuming a typical brace length of 6.05 m, an axial displacement of 150 mm corresponds to a relative member elongation or compression of at least 2.5%. This level of deformation demand can be easily achieved by SMA elements such as SMA cables, but may make the conventional PT steel cable/tendon impractical, noting that the effective elastic strain, i.e. that excludes the prestrain, for conventional PT tendons is typically less than 1.0%. The individual brace responses warn that a considerable deformation demand is expected for the structures subjected to near-fault earthquakes, and highlight the important role played by the SMA-based solutions which offer large deformation capability.

Figure 6.9 compares the absolute floor acceleration responses of the three different structures at selected stories. The self-centring frames experience larger floor acceleration responses than the MRFs and BRB frames. Taking the 3rd floor of the 9-storey self-centring braced frame for example, the peak floor acceleration exceeds 1.7 g, which is more than twice the PGA. This implies that endowing self-centring behaviour to a structure can potentially lead to more significant economic seismic losses associated with non-structural damages. Tremblay et al. (2008) reported a similar finding based on an analysis of structures under far-field earthquakes. The results shown in Fig. 6.9 also emphasise the fact that the observation of the acceleration response from the SDOF analysis is not applicable to the MDOF systems.

The influence of pulse characteristics can be further revealed by examining the structural responses under individual earthquake excitations, as shown in Figs. 6.10 through 6.15. The mean, median and 84th percentile values of the key responses are

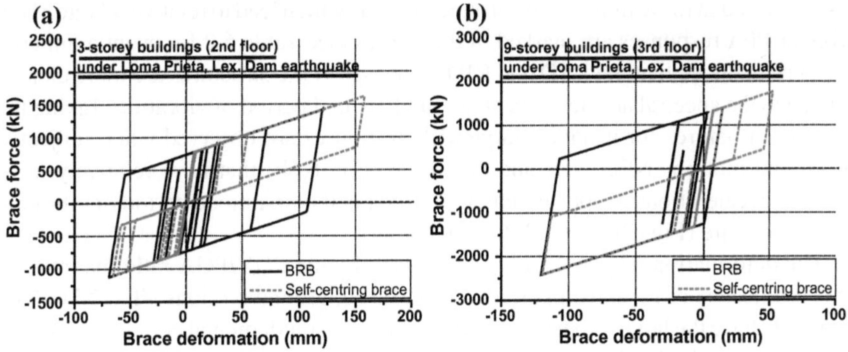

Fig. 6.8 Brace response of three types of buildings under Loma Prieta Lex. Dam ground motion: **a** 3-storey buildings, **b** 9-storey buildings

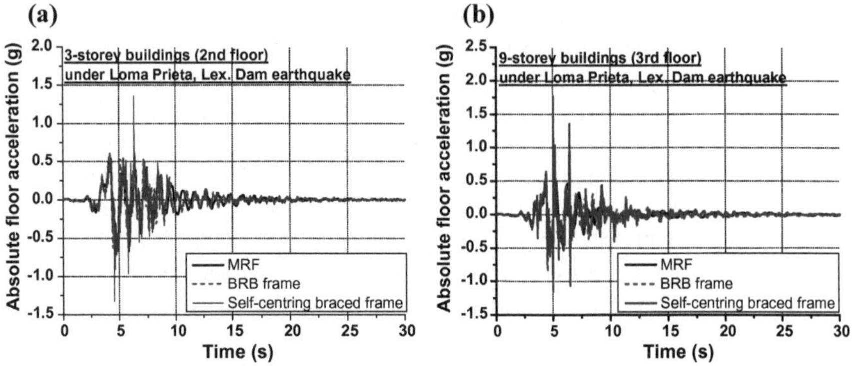

Fig. 6.9 Floor acceleration response of three types of buildings under Loma Prieta Lex. Dam ground motion: **a** 3-storey buildings, **b** 9-storey buildings

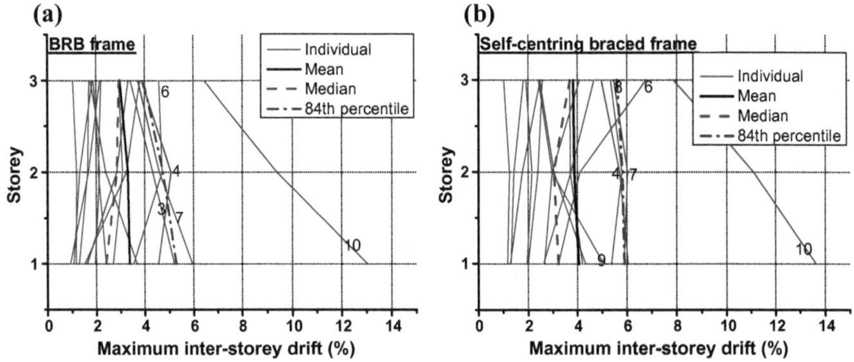

Fig. 6.10 Maximum inter-storey drifts of 3-storey buildings to individual near-fault ground motions: **a** BRB frame, **b** Self-centring braced frame

also provided in these figures. The ground motions which lead to relatively large MID, RID, or PFA responses are marked in the figures (see Table 6.2 for event numbers). It can be deduced that the MID and RID responses of the structures are more closely related to the spectral acceleration at the fundamental period of vibration. Taking the 9-storey structures for instance, large MID responses are observed when subjected to ground motions NF2, NF9 and NF10 (see Fig. 6.6b) that have relatively large spectral accelerations at approximately 1.73 s, i.e. the first mode period of vibration of the structure (Figs. 6.11, 6.12, 6.13 and 6.14).

The deformation responses are also affected by the PGV/PGA ratio that reflects the significance of the pulsing effect. For instance, ground motions NF2, NF6, NF8 generally have the largest spectral accelerations at the fundamental period of vibration for the 3-storey buildings, but the resulting MID and RID responses are much smaller than those induced by the Palos Verdes 2 earthquake (ground motion NF10). As can be seen from Table 6.2, the PGV/PGA ratio of the Palos Verdes 2 earthquake is almost

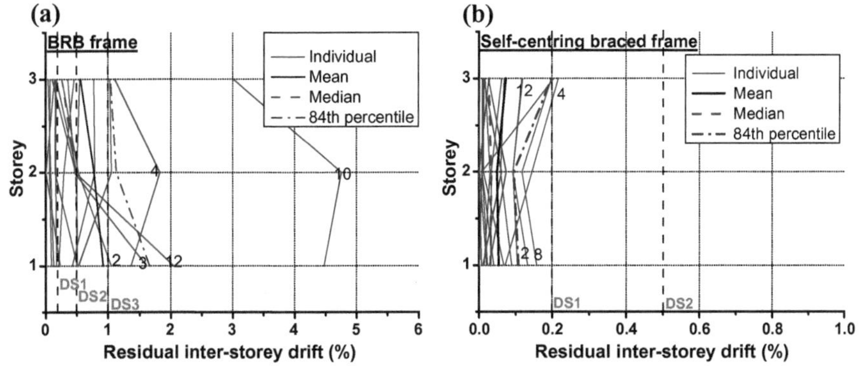

Fig. 6.11 Residual inter-storey drifts of 3-storey buildings to individual near-fault ground motions: **a** BRB frame, **b** Self-centring braced frame

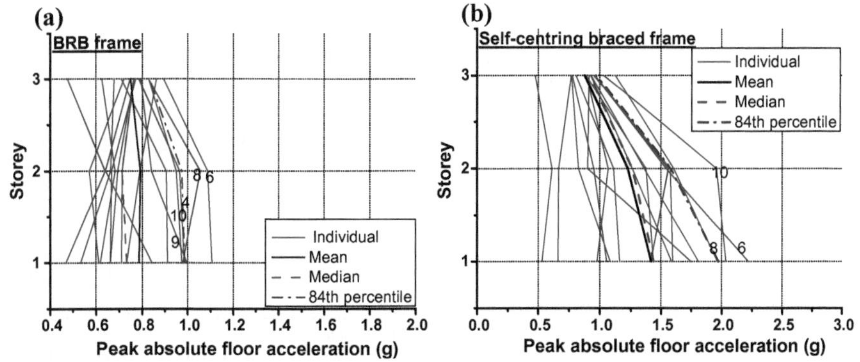

Fig. 6.12 Peak absolute floor accelerations of 3-storey buildings to individual near-fault ground motions: **a** BRB frame, **b** Self-centring braced frame

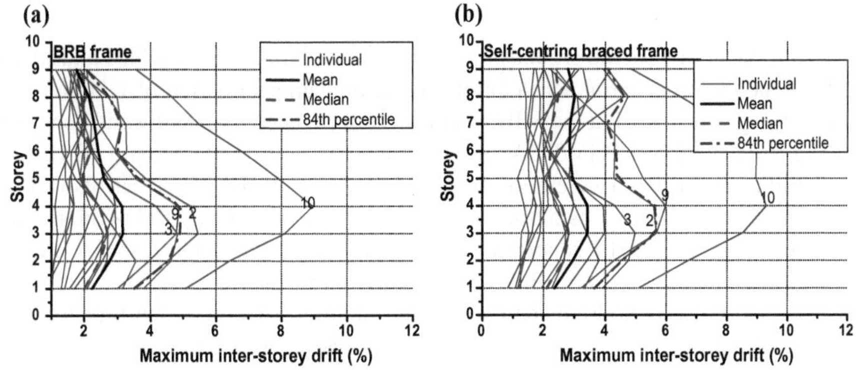

Fig. 6.13 Maximum inter-storey drifts of 9-storey buildings to individual near-fault ground motions: **a** BRB frame, **b** Self-centring braced frame

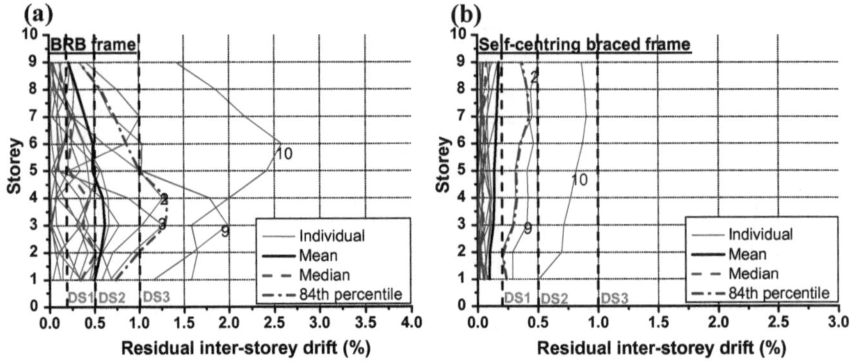

Fig. 6.14 Residual inter-storey drift s of 9-storey buildings to individual near-fault ground motions: **a** BRB frame, **b** Self-centring braced frame

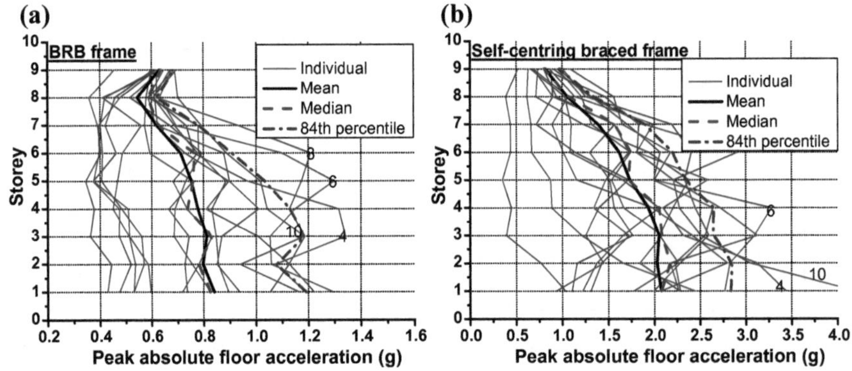

Fig. 6.15 Peak absolute floor accelerations of 9-storey buildings to individual near-fault ground motions: **a** BRB frame, **b** Self-centring braced frame

twice the mean value of the ground motions NF2, NF6, NF8. Similarly, NF9 and NF10 exhibit the largest spectral accelerations at the fundamental period of vibration for the 9-storey structures, but the MID response induced by NF9 is significantly smaller due to the much lower PGV/PGA ratio of ground motion compared with NF10.

Apart from spectral acceleration and PGV/PGA ratio, pulse period can also affect the MID and RID responses of the structures. For example, the two Chi-Chi ground motions NF12 and NF15 generally cause noncritical deformation responses, even though they have very large PGV/PGA ratios (more than 0.5). This is probably because the pulse periods of the two ground motions far exceed the fundamental periods of vibration of the considered structures, and therefore, the resonance effect decays significantly.

The PFA response of the structures is inconsistent with the MID and RID responses, as it is generally perceived that a more remarkable inelastic behaviour (e.g. a larger MID) of a structure leads to reduced floor acceleration demands (Ray-Chaudhuri and Hutchinson 2011). The floor acceleration response depends on many factors such as the period of vibration, the number of stories, modal shape and the extent of nonlinear behaviour of the structure. The frequency content of the ground motion is also highly influential.

6.2.3 Summary of Building Behaviour

A summary of the mean MID, RID and PFA responses of the MRFs, BRB frames and self-centring braced frames, under the considered near-fault and DBE/MCE-level far-field ground motions, is presented in Figs. 6.16 through 6.18 (Fang et al. 2018a).

Despite the difference in energy dissipation capability, the MRFs and self-centring braced frames have similar MID responses under the far-field earthquakes. This is because the oscillation of the self-centring frames tends to occur near the upright position, and hence permanent inter-storey drift is not significantly accumulated. Benefiting from the larger initial stiffness and sufficient energy dissipation, the BRB frames do experience smaller MID responses than the other systems. Importantly, the near-fault earthquakes tend to cause larger MID and RID responses compared with the DBE- and MCE-level far-field earthquakes. It is also noticed that the 9-storey self-centring braced frame experiences larger peak inter-storey drifts than its MRF and BRB frame counterparts at upper stories. This phenomenon is caused by the high-mode effect, which occurs more commonly in self-centring structures of which the energy dissipation is sometimes insufficient (Qiu and Zhu 2016).

As shown in Fig. 6.17, the self-centring braces reduce the RID to negligible values for the case of far-field earthquakes. Under the near-fault ground motions, the RIDs

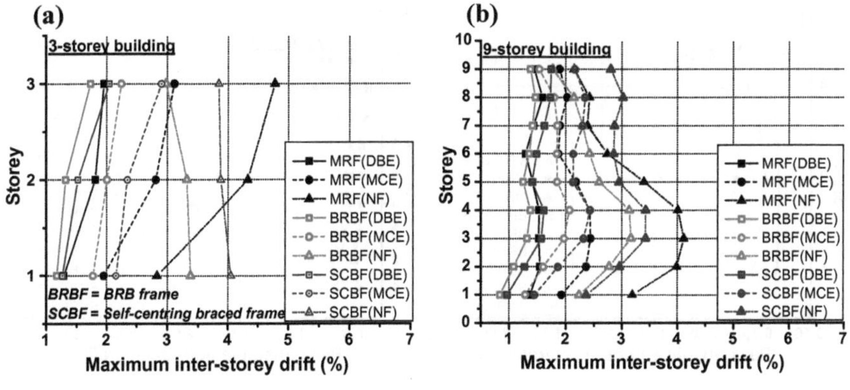

Fig. 6.16 Mean maximum inter-storey drifts of structures subjected to considered ground motions: **a** 3-storey building, **b** 9-storey building

of the self-centring braced frames are slightly larger, which is caused by larger MID responses. It is recalled that dual systems are adopted for the braced frames, so it is believed that the minor RIDs of the self-centring frames are mainly from the inelastic deformations of the boundary members. For the MRFs and BRB frames under the MCE, the mean RID is below 0.5%, but the value is significantly increased when the structures are subjected to the near-fault ground motions.

FEMA P-58 (2012) gives four RID limits (i.e. damage states, or DSs, as elaborated later in Chap. 7) with can be used to evaluate the reparability and probability of demolition of the structures after earthquakes. The first class, DS1, corresponds to RID being less than 0.2% such that 'no structural realignment is necessary for structural stability, but the building may require adjustment and repairs to non-structural and mechanical components'. The RID limit for DS2 is 0.5%, in which case realignment of structural frame and related structural repairs are economically feasible, and

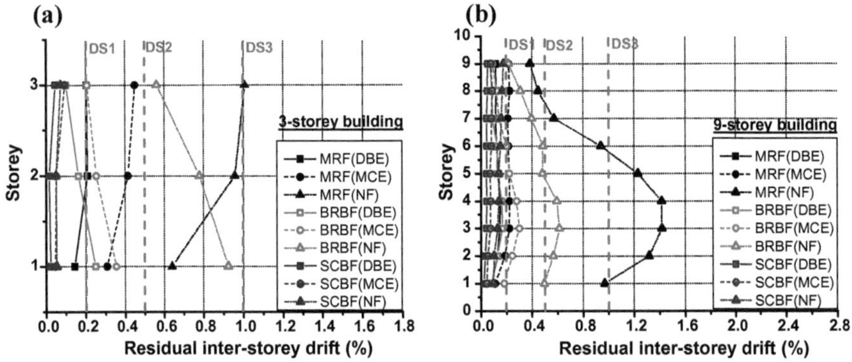

Fig. 6.17 Mean residual inter-storey drifts of structures subjected to considered ground motions: **a** 3-storey building, **b** 9-storey building

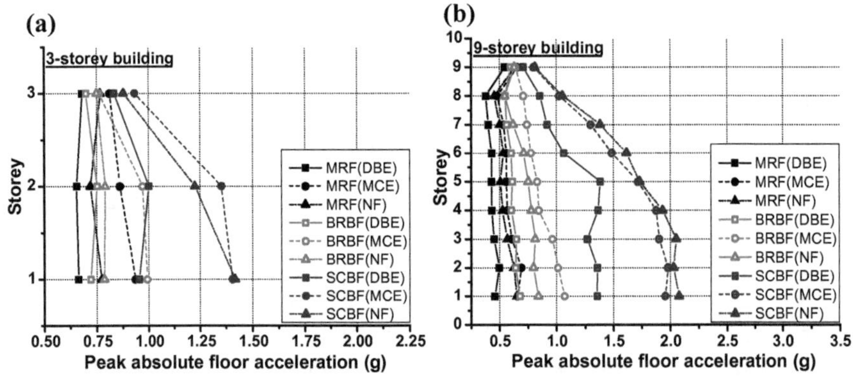

Fig. 6.18 Mean peak absolute floor accelerations of structures subjected to considered ground motions: **a** 3-storey building, **b** 9-storey building

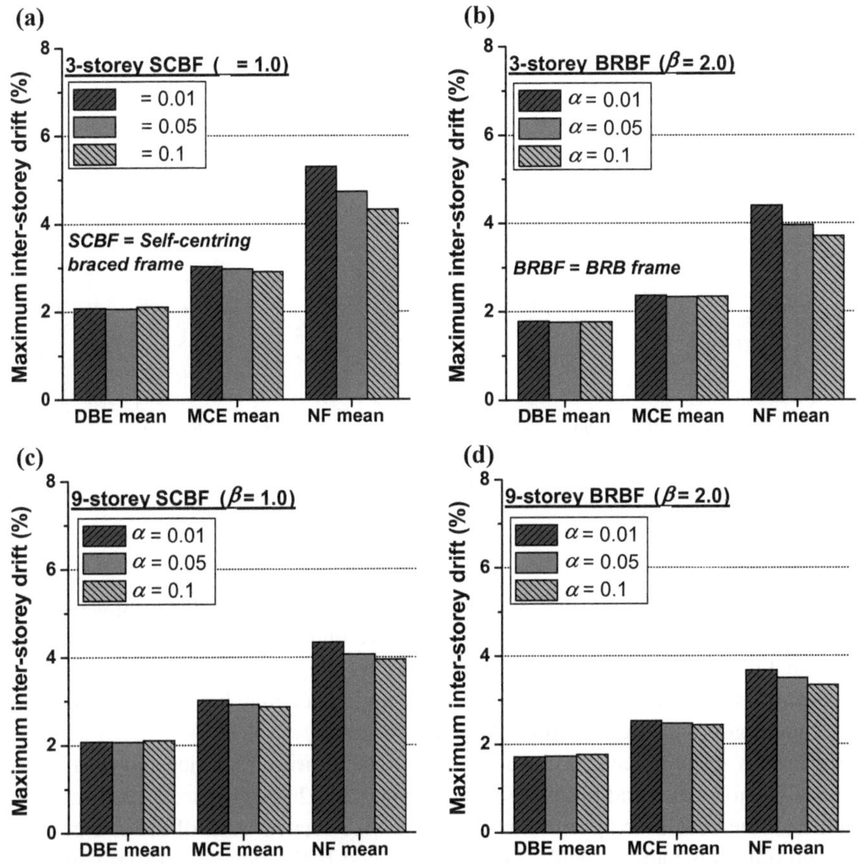

Fig. 6.19 Influence of brace post-yield stiffness on maximum inter-storey drift: **a** 3-storey self-centring braced frame, **b** 3-storey BRB frame, **c** 9-storey self-centring braced frame, **d** 9-storey BRB frame

degradation in structural stability is limited. Class DS3 corresponds to the condition that the RID exceeds 0.5% but is less than 1%, and in this case, major structural realignment, which may not be economically and practically feasible, is required to restore margin of safety for lateral stability. For Class DS4 with RID exceeding 1.0%, the residual inter-storey drift is too large that the structure is in danger of collapse from aftershocks.

According to the above thresholds, the results from Fig. 6.17 clearly indicate that the self-centring braced frames satisfy Class DS1 under both the far-field and near-fault earthquakes, and hence, the required repair work is limited. Class DS1 is also generally satisfied when the MRFs and BRB frames are under the DBE. While the MCE increases the deformation demands, Class DS2 is still satisfied. This indicates that the MRFs and BRB frames, even in the absence of self-centring devices, may

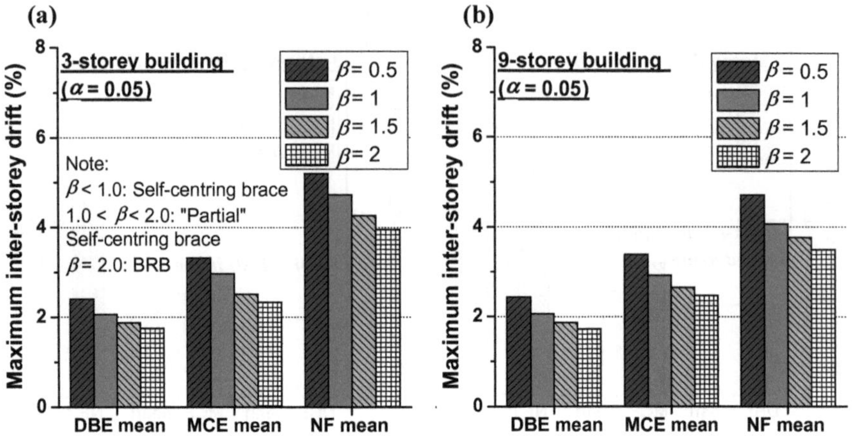

Fig. 6.20 Influence of brace energy dissipation factor on maximum inter-storey drift: **a** 3-storey building, **b** 9-storey building

be economically repairable after experiencing MCE-level far-field earthquakes. This is, however, not the case when the structures are under near-fault earthquakes. It is seen that the mean RID could fall into Class DS4, implying that these conventional steel frames are not economically repairable and hence may have to be demolished after undergoing the near-fault earthquakes.

The PFA response can be closely linked to the injuries, fatalities and repair costs caused by non-structural failure (Karavasilis and Seo 2011). Figure 6.18 reveals that under all the considered ground motions, the PFA-related issue of the self-centring braced frames is more critical than the MRFs and BRB frames, especially at lower stories. This implies that using braces with a less full hysteresis is less effective in controlling the accelerations 'transmitted' into the structures. The increase in the PFA response is attributed to the more complex transition points of the flag-shaped hysteresis compared with its elasto-plastic counterpart (Tremblay et al. 2008). Considering two adjacent stories which have inconsistent inter-storey shear forces during earthquakes (normally the shear force at one storey lags that of the other), the abrupt transitions of the unloading and reloading paths exhibited by the flag-shaped braces cause a larger difference in the shear force between the two stories. Such difference produces a very short duration, large value transient floor acceleration, which is, however, less likely to occur in MRFs and BRB frames where the transition points during the loading and unloading hysteretic paths are less abrupt (Tremblay et al. 2008). It is also observed that the PFA response is not necessarily amplified for the case of near-fault earthquakes, which indicates that floor acceleration is not sensitive to the pulsing characteristics of the ground motion.

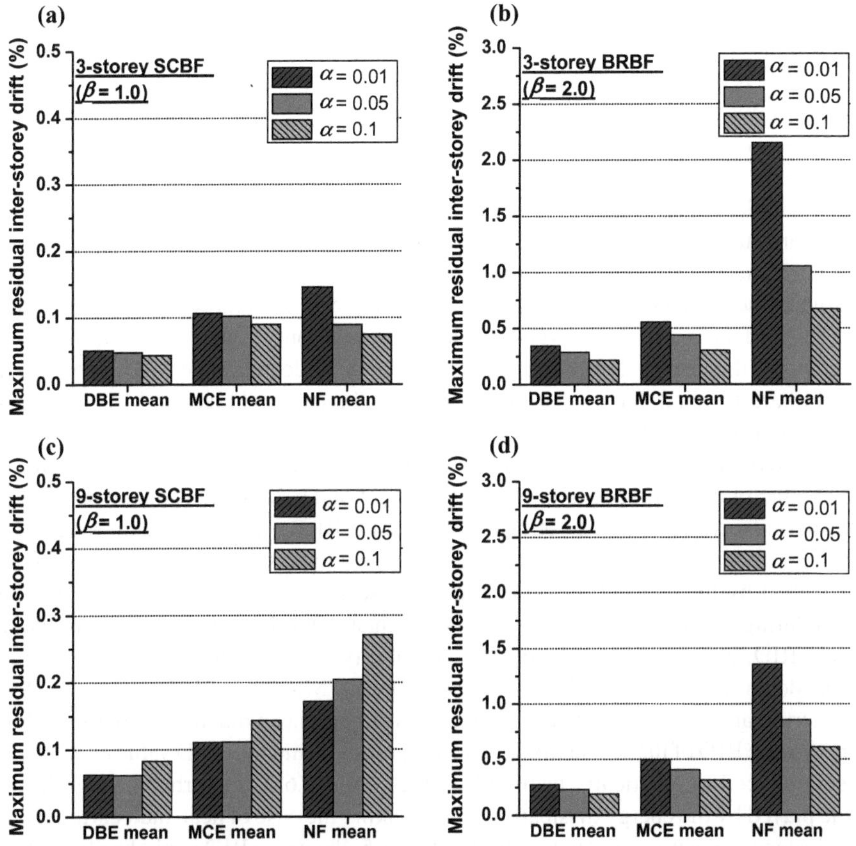

Fig. 6.21 Influence of brace post-yield stiffness on maximum residual inter-storey drift: **a** 3-storey self-centring braced frame, **b** 3-storey BRB frame, **c** 9-storey self-centring braced frame, **d** 9-storey BRB frame

6.2.4 Influence of Brace Parameters

Post-yield stiffness (α) and energy dissipation factor (β) are two brace parameters which are adjustable and of importance to engineers pursuing optimal structural and non-structural behaviour against earthquakes. In this section, a broadened brace parameter matrix is further considered. Three post-yield stiffness values, namely $\alpha = 0.01$, 0.05 (reference case) and 0.1, as well as four levels of energy dissipation factor, i.e. $\beta = 0.5$, 1.0, 1.5 and 2.0, are covered. A post-yield stiffness exceeding 0.1 is less likely in SMA-based braces. In fact, an overly large post-yield stiffness (i.e. over-strength), leading to possible damages to the adjacent steel beams, columns and connections, is undesirable. For the energy dissipation characteristic, the case of $\beta = 0.5$, as compared with the reference case where $\beta = 1.0$, represents self-centring braces with an increased self-centring tendency but reduced energy dissipation capability; $\beta = 1.5$ represents a typical 'partial' self-centring brace where an increased energy dissipation is provided but certain residual brace deformation would occur. The brace is deemed as an idealised BRB when $\beta = 2.0$.

The influences of the post-yield stiffness and energy dissipation factor on the mean MID responses of the braced frames are shown in Figs. 6.19 and 6.20, respectively (Fang et al. 2018a). The MID responses of both the BRB and the self-centring braced frames under the far-field earthquakes are not sensitive to the post-yield stiffness. For the near-fault earthquakes, however, increasing the post-yield stiffness could slightly decrease the MID, which is probably because an increased post-yield stiffness helps mitigate the concentration of inter-storey drift (i.e. weak storey), especially when the P-Δ effect is critical at large structural deformations. Figure 6.20 confirms that increasing the energy dissipation factor could indeed decrease the MID response. It is of interest to find that the partial self-centring braces could lead to MID demands close to those caused by the BRBs.

The RID responses are quite sensitive to the post-yield stiffness and energy dissipation factor of the braces, as shown in Figs. 6.21 and 6.22. Increasing the post-yield stiffness always decreases the RID response for the BRB frames, and such tendency is particularly obvious for the case of near-fault earthquakes. This phenomenon is well explained by the concept of 'probabilistic self-centring', based on the fact that the RID during dynamic shakedown is often much smaller than the maximum possible static RID, and more importantly, the probability of self-centring during dynamic shakedown is sensitive to the post-yield stiffness (MacRae and Kawashima 1997; Eatherton and Hajjar 2011). The post-yield stiffness has a much less obvious influence on the RID of the self-centring braced frames of which the RID value has already been reduced to practically negligible values. Figure 6.22 confirms the significant role played by the energy dissipation factor. Taking the case of near-fault earthquakes for instance, when β is decreased from 2.0 to 1.5, the mean RID could be effectively reduced from around 1.0% (class DS3-DS4, according to FEMA P-58) to the values below the 0.5% limit (class DS2).

As shown in Figs. 6.23 and 6.24, the PFA responses of the structures are not sensitive to the post-yield stiffness of the braces, but can be evidently influenced by the energy dissipation factor. The results particularly warn that the flag-shaped self-centring braces with $\beta \leq 1.0$ may be undesirable from the perspective of floor acceleration control. An important finding is that the PFA is remarkably suppressed when β increases from 1.0 to 1.5, whereas a further increase in β makes little further improvement. Combining the MID, RID and PFA results presented in Figs. 6.19 through 6.24, it is confident to conclude that the partial self-centring braces can lead to a harmonisation of the deformation and acceleration responses, and should be promoted for practical use.

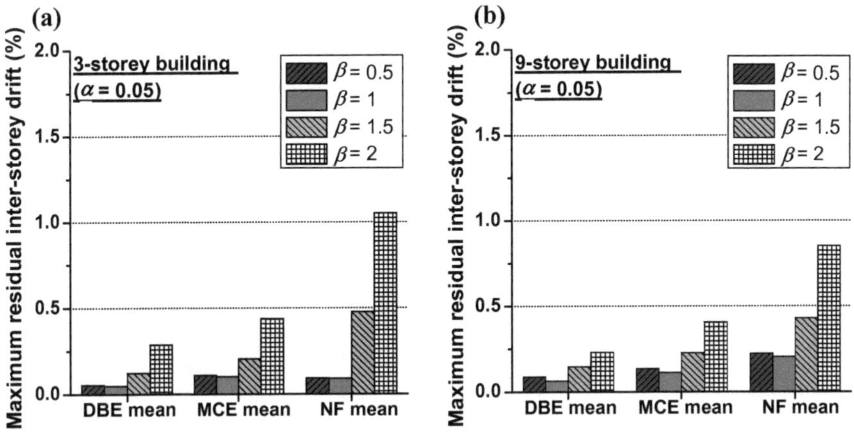

Fig. 6.22 Influence of brace energy dissipation factor on maximum residual inter-storey drift: **a** 3-storey building, **b** 9-storey building

6.2.5 *Prediction of Residual Inter-storey Drift*

Residual deformation is a critical index related to the post-earthquake resilience of a building. As mentioned in Chap. 5, engineers are often interested in knowing the relationship between the MID and RID, even from an empirical perspective. The fact is, RID is quite sensitive to the structural behaviour and modelling assumptions. An accurate statistical simulation of RID requires the use of advanced numerical models, careful attention to cyclic hysteretic response, and a sufficient number of ground motions. Many researchers have attempted to derive simplified RID prediction models for performance based design (MacRae and Kawashima 1997; Christopoulos et al. 2003; Ruiz-Garcia and Miranda 2006a, b; Erochko et al. 2011).

Similar to the definition of RDR presented in Chap. 5, a normalised RID (i.e. RDR), as expressed by Eq. (6.1), could be used as a simple index to depict the relationship between the residual and peak transient inter-storey drifts:

$$RDR = \frac{\theta_{r,max}}{\theta_{t,max} - \theta_{el}} \tag{6.1}$$

where $\theta_{r,max}$ is the maximum residual inter-storey drift of a structure among all the stories, $\theta_{t,max}$ is the maximum transient inter-storey drift of the structure among all the stories and θ_{el} is the elastic inter-storey drift which, for ease of design, may be estimated based on the roof drift pushover response. $\theta_{t,max} - \theta_{el}$ is an upper bound for the amount of RID.

Figure 6.25 shows the mean RDR values of the considered buildings with varying α and β values under the far-field DBE/MCE and near-fault earthquakes. Fitted curves using quadratic polynomials, as expressed in Eq. (6.2), are also given in the

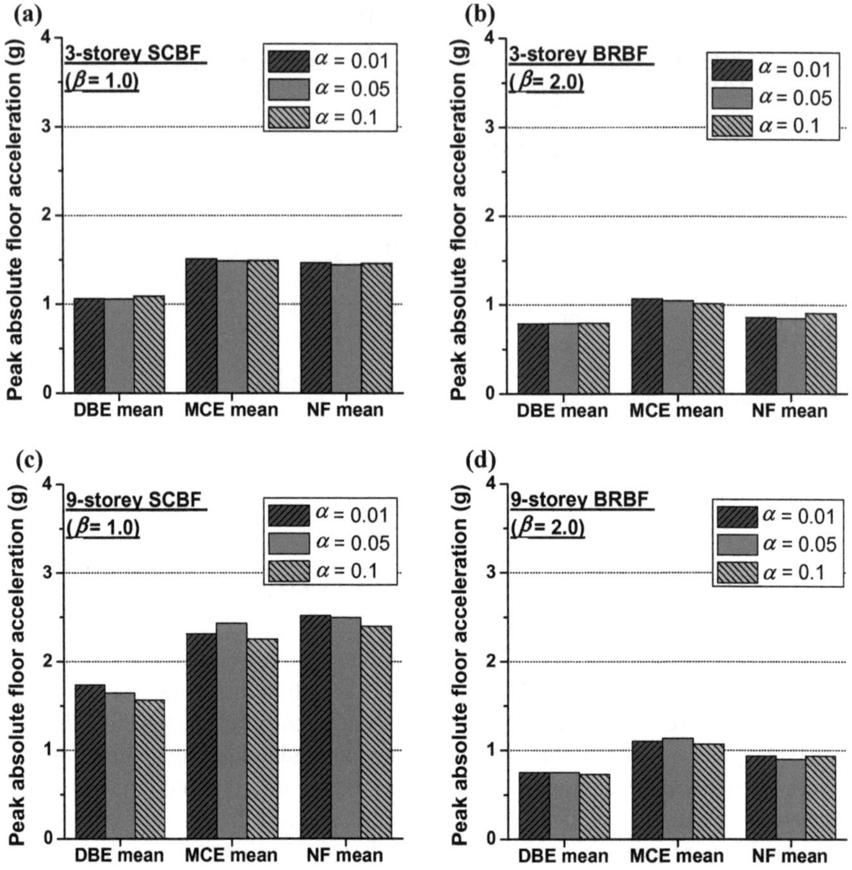

Fig. 6.23 Influence of brace post-yield stiffness on peak absolute floor acceleration: **a** 3-storey self-centring braced frame, **b** 3-storey BRB frame, **c** 9-storey self-centring braced frame, **d** 9-storey BRB frame

figure. Although some uncertainties exist, the mean RDR generally falls within the range between 0.1 and 0.4.

$$RDR = \boldsymbol{A}\beta^2 + \boldsymbol{B}\beta + \boldsymbol{C} \tag{6.2}$$

In Eq. (6.2), A, B and C are constants which vary with different cases and could be obtained via least square regression (see Table 6.3). While the constants provided in the table are derived based on special cases, the 3- and 9-storey buildings are code-compliant and represent typical dual-system multi-storey steel frames in practice. In other words, interpolated or upper limit values may be used for buildings with storey numbers and brace parameters that fall within the considered ranges, depending on the designer's judgement. Alternatively, designers may directly use an upper limit

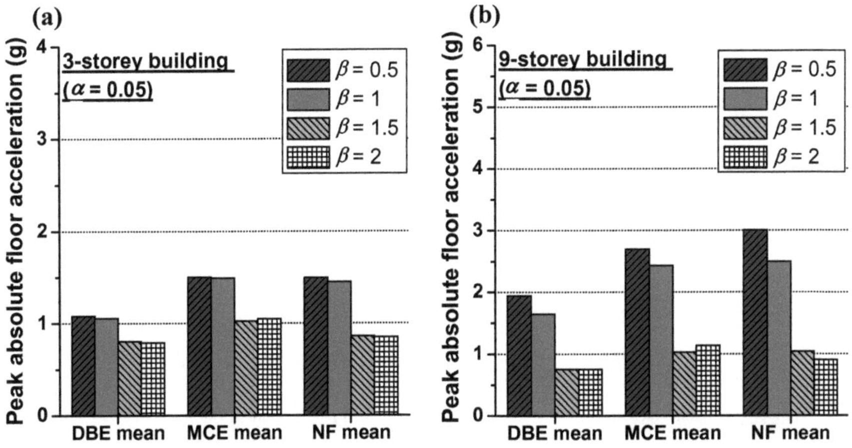

Fig. 6.24 Influence of brace energy dissipation factor on peak absolute floor acceleration: **a** 3-storey building, **b** 9-storey building

Table 6.3 Constants for RID prediction model

Case	Stiffness ratio	A	B	C
3-Storey-DBE	$\alpha = 0.01$	0.104	−0.164	0.081
	$\alpha = 0.05$	0.083	−0.127	0.062
	$\alpha = 0.10$	0.054	−0.080	0.044
3-Storey-MCE	$\alpha = 0.01$	0.110	−0.180	0.099
	$\alpha = 0.05$	0.079	−0.123	0.072
	$\alpha = 0.10$	0.046	−0.063	0.044
3-Storey-NF	$\alpha = 0.01$	0.208	−0.268	0.103
	$\alpha = 0.05$	0.138	−0.184	0.076
	$\alpha = 0.10$	0.103	−0.144	0.067
9-Storey-DBE	$\alpha = 0.01$	0.084	−0.104	0.061
	$\alpha = 0.05$	0.066	−0.078	0.053
	$\alpha = 0.10$	0.052	−0.063	0.052
9-Storey-MCE	$\alpha = 0.01$	0.100	−0.136	0.080
	$\alpha = 0.05$	0.085	−0.118	0.076
	$\alpha = 0.10$	0.061	−0.086	0.073
9-Storey-NF	$\alpha = 0.01$	0.228	−0.344	0.153
	$\alpha = 0.05$	0.146	−0.221	0.119
	$\alpha = 0.10$	0.080	−0.107	0.088

value of the RDR from Fig. 6.25 to quickly evaluate the relationship between the RID and MID, at least for preliminary design.

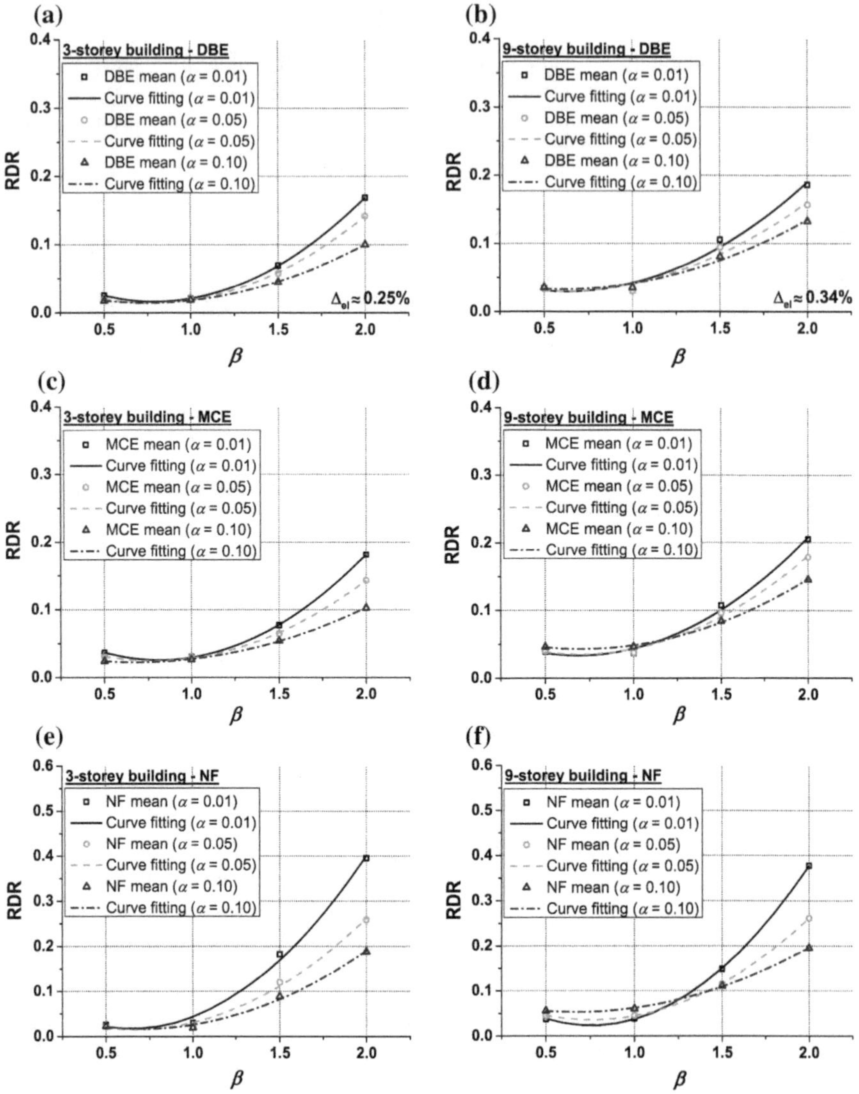

Fig. 6.25 Relationships between residual and maximum inter-storey drifts: **a** 3-storey building under DBE, **b** 9-storey building under DBE, **c** 3-storey building under MCE, **d** 9-storey building under MCE, **e** 3-storey building under NF earthquake, **f** 9-storey building under NF earthquake

6.3 Self-centring MRFs with SMA Connections

6.3.1 Building Overview

Section 6.2 has provided useful information on the behaviour of multi-storey MRFs and steel frames with different idealised braces, and shed light on the critical factors that affect the structural response. This section proceeds with the discussion of the behaviour of more comprehensively designed MRFs equipped with SMA-based beam-to-column connections. For comparison purpose, three structural types are considered, namely conventional (original) prototype MRF, self-centring MRF with fully distributed SMA connections and self-centring MRF with partially distributed SMA connections.

The conventional prototype steel frame is a 5-storey, 4 × 3 bay office building constructed on very stiff soil (CEN 2004; Dimopoulos et al. 2013). Figure 6.26 shows the plan and elevation views of the building, which consists of four identical MRFs in the longitudinal direction, and two perimeter braced frames that resist the lateral load along the transverse direction. The design yield strengths of the steel beams and columns are 300 MPa and 350 MPa, respectively. Again, the focus of the current analysis is on a typical internal 2D MRF in the longitudinal direction. Rigid and full-strength beam-to-column connections are used in this MRF, and therefore first yielding is expected to occur at the beam section near the connection region, conforming to the 'strong column-weak beam' principle. The panel zones are reinforced by doubler and continuity plates, and the shear stiffness may be assumed to be infinite. The sections selected for the beams and columns follow the European practice, as marked in Fig. 6.26 for different floor levels. The same modelling approach as that discussed in Sect. 6.2 is adopted for the structural model.

The conventional MRF is redesigned to allow the use of SMA-based self-centring connections (Fang et al. 2018b). Two modified MRFs are considered, in which the SMA connections are either fully or partially distributed within the frame, noting that the concept of the latter is similar to the idea of using partial self-centring braces, as discussed in the last section. The two self-centring MRFs, as illustrated in Fig. 6.27, have identical framing arrangement and member sizes to those of the conventional MRF, so the only difference is the connection detail. For the self-centring MRF with fully distributed SMA connections (hereafter named as MRF-FSMA), all the rigid connections used in the conventional MRF are replaced by the SMA connections. Such a strategy aims to maximise the self-centring capability of the structure and to minimise the damage to the boundary frame. The design of partially distributed SMA connections for the self-centring MRF aims to achieve a tradeoff between structural performance and cost of construction. This frame is named as MRF-PSMA, where the arrangement of the connection is illustrated in Fig. 6.27.

The three buildings are subjected to a suite of 20 earthquake ground motions recorded at locations with very stiff soil conditions, i.e. average shear wave velocity is between 360 and 800 m/s (CEN 2004). These ground motions are selected from nine events in different countries and regions, with the event magnitudes M_w ranging from

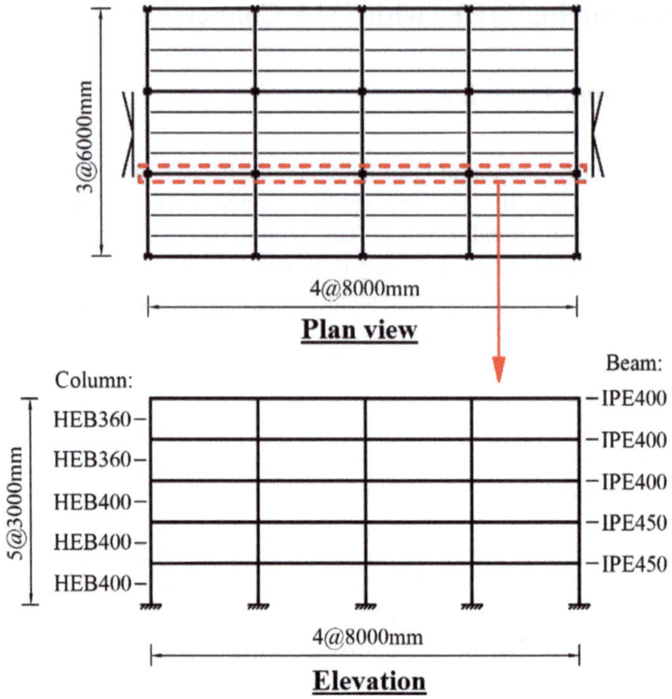

Fig. 6.26 Conventional prototype building

6.5 to 7.6. The ground motions are scaled to the DBE and MCE levels. The details of the scaled ground motions are not elaborated here, but can be found elsewhere (Dimopoulos et al. 2013).

6.3.2 Connection Design and Modelling

SMA ring spring-based connections, as discussed previously in Chap. 3, are adopted for the MRF-FSMA and MRF-PSMA. While there is no codified recommendation currently available for designing self-centring connections, the following rules need to be followed (Fang et al. 2018b):

(1) the initial stiffness of the SMA connections prior to decompression of the SMA ring springs should be comparable to the rigid connection criterion such that the conventional and modified buildings could have similar initial elastic stiffness and fundamental period of vibration;
(2) the 'yield moment' of the SMA connection should be smaller than the yield capacity ($M_{b,el}$) of the steel beam. The main concept is to ensure that the connections enter into the inelastic stage before the beam yields, and no significant

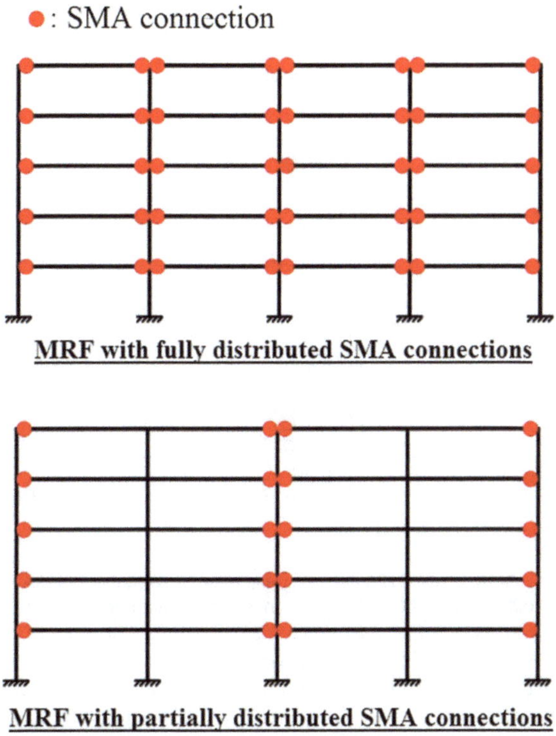

Fig. 6.27 Modified buildings with SMA connections

damage to the beam is expected after earthquakes. Due to the pronounced strain hardening of the SMA components, it is possible that the moment resistance of the connection may have exceeded $M_{b,el}$ at large connection rotations; in this case, a certain degree of beam yielding is expected. Of course, whether beam yielding is allowed or not depends on the engineers' design intention and judgement. As a preliminary attempt, the target yield moment of the SMA connection could be set as approximately 80% of the $M_{b,el}$;

(3) the SMA ring spring system should be designed to provide a sufficient concentrated connection rotation ability which satisfies the codified requirement. Keeping in mind that some pulse-like near-fault ground motions can induce much larger inter-storey drifts than far-field earthquakes do.

According to the above design recommendations, the SMA ring spring connections are designed to match the section properties of the IPE400 and IPE450 beams, as illustrated in Figs. 6.28 and 6.29, respectively. Each connection consists of six bolt rows that provide the required moment resistance, where each bolt includes one set of ring spring system including two or three SMA outer rings. Taking the SMA connection for IPE450 beams for instance, the external diameter (D_e) and thickness (T) of the SMA outer rings are 70 mm and 7 mm, respectively, and the inner high-strength alloy (HSA) rings are designed to match the geometry and strength of the

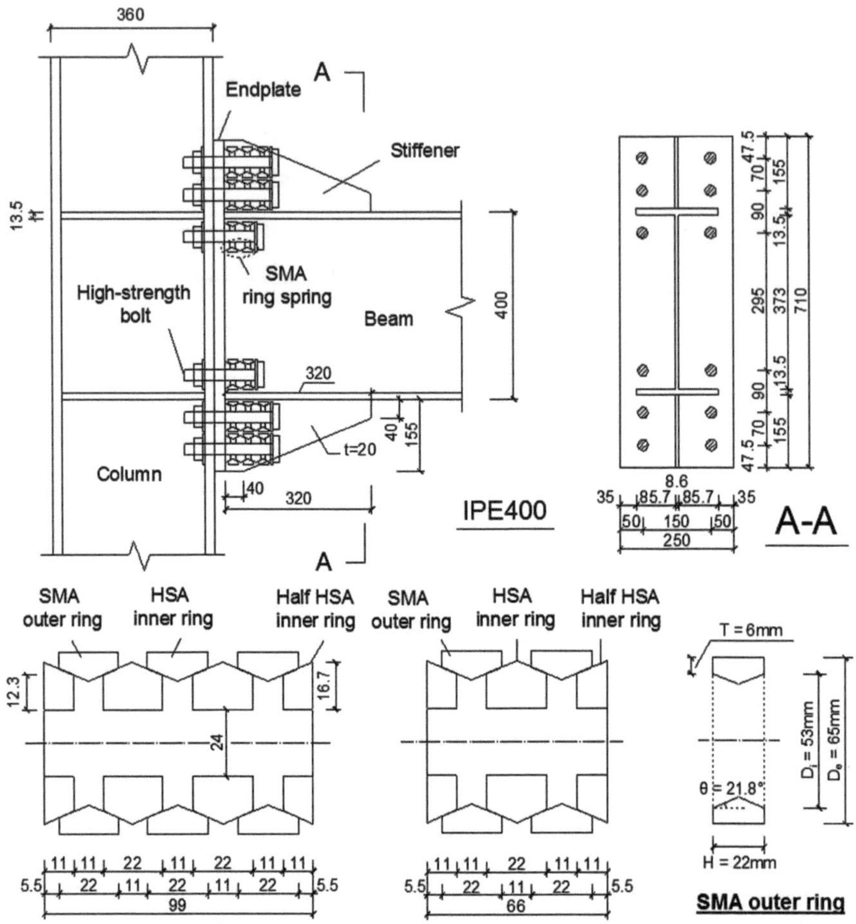

Fig. 6.28 Details of SMA connection for IPE400 steel beam

outer rings. Each SMA outer ring can offer a total deformation capacity of 12 mm, out of which 3 mm is consumed as precompression. The SMA connections are first simulated using 3D solid element models established ABAQUS (2012), with the uniaxial stress–strain relationship of the SMA material shown in Fig. 6.30 employed. The basic parameters selected for the material model are the ones typically observed in commercial NiTi SMA. The moment-concentrated rotation responses of the connections predicted by the 3D models are shown in Fig. 6.31.

To facilitate nonlinear dynamic analysis of the buildings, a simplified spring model for the SMA connections should be used, as schematically illustrated in Fig. 6.32. Rigid elements are used to represent the end-plate and the corresponding column flange, while the gap opening mechanism is simulated by applying appropriate spring sets in between. For each connection, six rows of zero-length spring element sets,

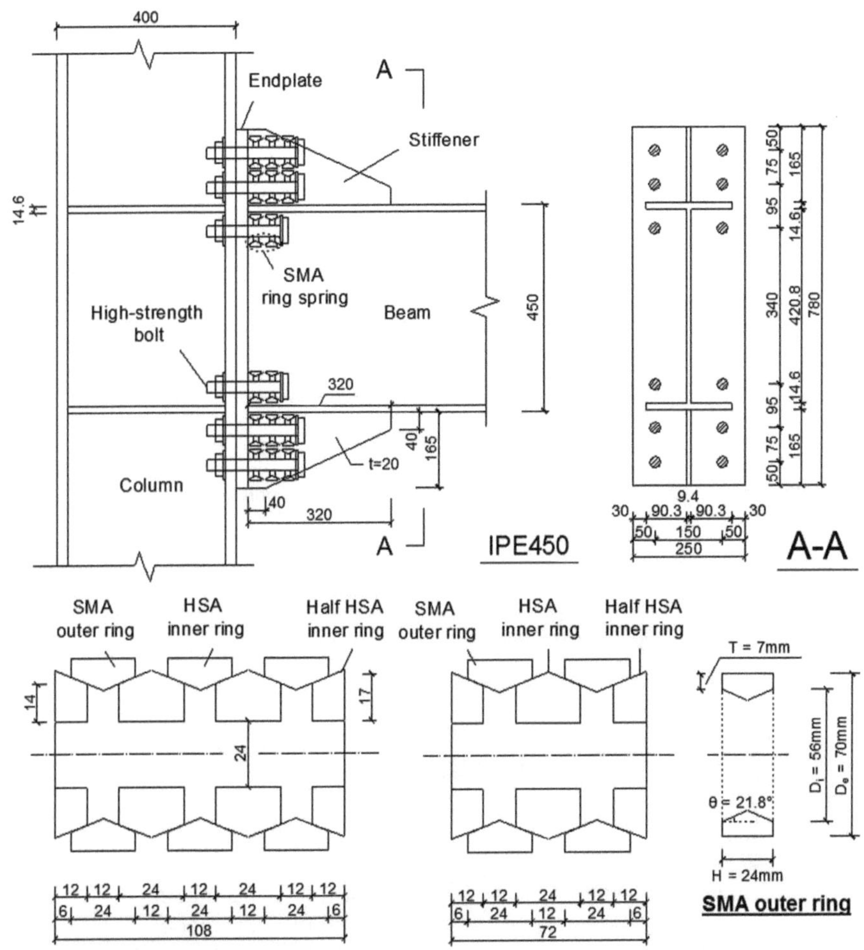

Fig. 6.29 Details of SMA connection for IPE450 steel beam

representing the associated SMA ring spring systems, are arranged along the height of the beam-to-column interface according to their positions. The springs for each bolt row may include two parallel elements with different hysteresis models, namely a flag-shaped hysteresis model and steel hysteresis model. The latter one, which is optional, can be used to mimic the possible degradation phenomenon of the SMA material (Fang et al. 2018b).

In addition, two zero-length contact spring elements must be added to the top and bottom ends of the interface to allow simulation of the gap opening mechanism. These springs can have a linear elastic response in compression with a large or infinite stiffness, but should deform freely in tension. For each side of the connection, a shear resistance spring element is also needed to provide the necessary shear resistance for the connection. If the panel zone of the column is unstiffened, this region may

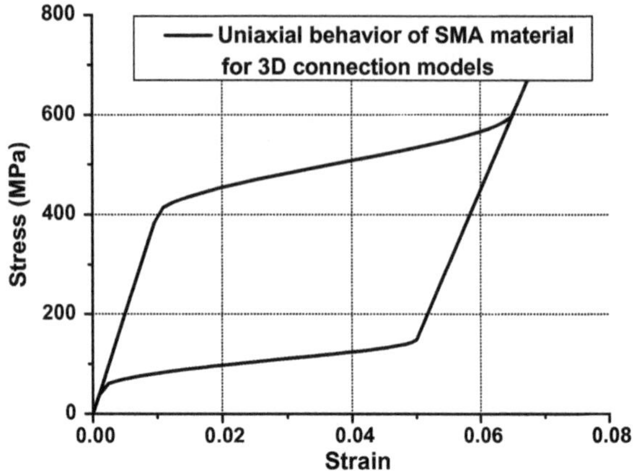

Fig. 6.30 Uniaxial stress–strain behaviour of SMA for 3D connection models

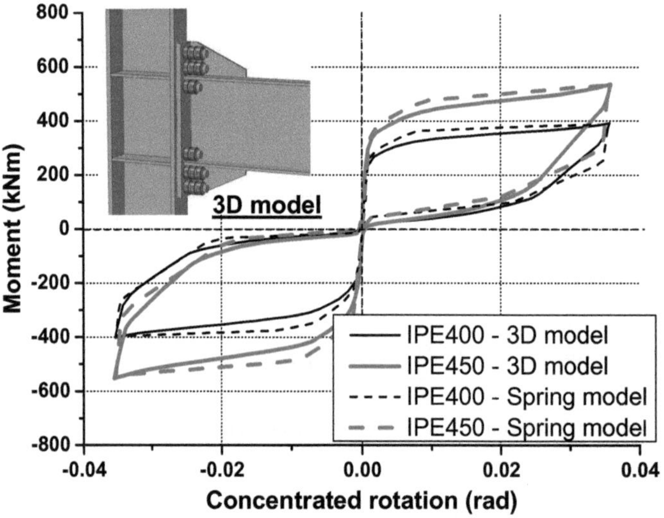

Fig. 6.31 Modelling of SMA connections using 3D and spring models

be simulated via a zero-length rotational spring element to reflect the bilinear elasto-plastic shear force–deformation panel zone behaviour. For the case of stiffened panel zone, an infinite (or very large) stiffness may be assigned to the rotational spring.

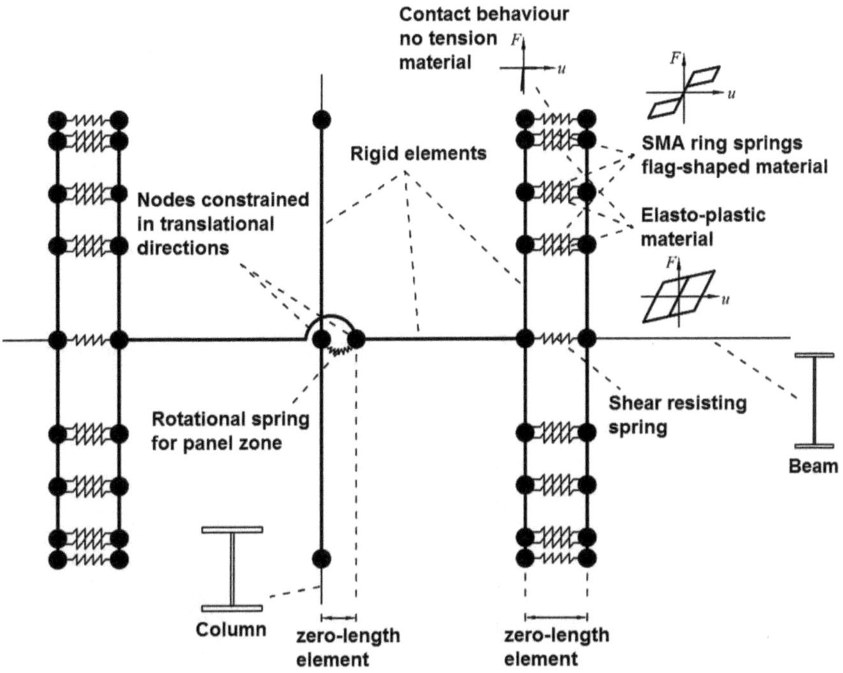

Fig. 6.32 Illustration of spring model for SMA connections

6.3.3 Seismic Responses and Comparisons

Prior to the dynamic time-history analysis, a static 'push-pull analysis' is carried out to understand the hysteretic characteristics of the frames. As shown in Fig. 6.33, the conventional MRF exhibits a full hysteretic response with an obvious residual drift induced during static unloading. The frames equipped with SMA connections tend to have less static residual drifts. The static lateral resistance of the two SMA-based frames is slightly smaller than that of the conventional frame because the yield moments of the SMA connections are designed to be smaller than those of the steel beams, as mentioned before. The three frames have similar initial stiffness, and as a result, their natural periods of vibration are similar to each other, ranging between 1.070 and 1.094 s.

The median and 84th percentile values of the MID and RID responses over the heights of the frames are shown in Figs. 6.34–6.35. The trends revealed by the analysis results are consistent with those observed in the braced frames discussed in the last section. It is confirmed again that although the SMA frames typically have less energy dissipation than the conventional MRF, all the three frames exhibit comparable MID responses. Given the 20 ground motions, the largest displacement that the bolt row springs experience is approximately 15 mm (corresponding to a maximum hoop strain of around 4.5% according to the 3D model), which is smaller

Fig. 6.33 Static push–pull response of three buildings (V = base shear, W = seismic weight): **a** conventional MRF, **b** MRF-FSMA, **c** MRF-PSMA

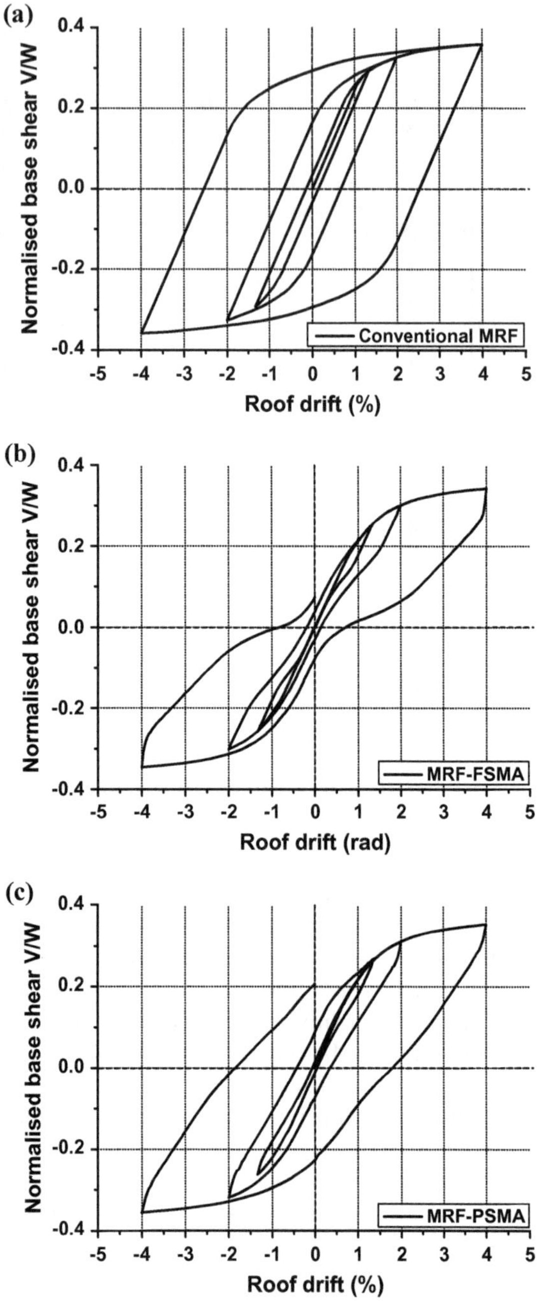

Fig. 6.34 Maximum inter-storey drift response of three buildings: **a** conventional MRF, **b** MRF-FSMA, **c** MRF-PSMA

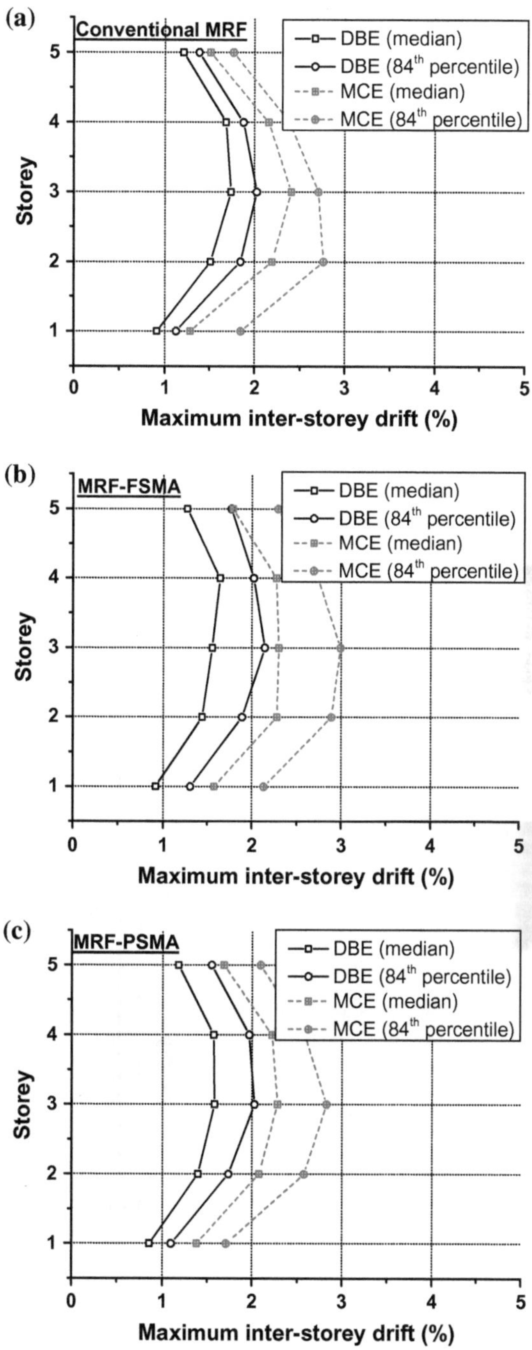

Fig. 6.35 Maximum residual inter-storey drift response of three buildings: **a** conventional MRF, **b** MRF-FSMA, **c** MRF-PSMA

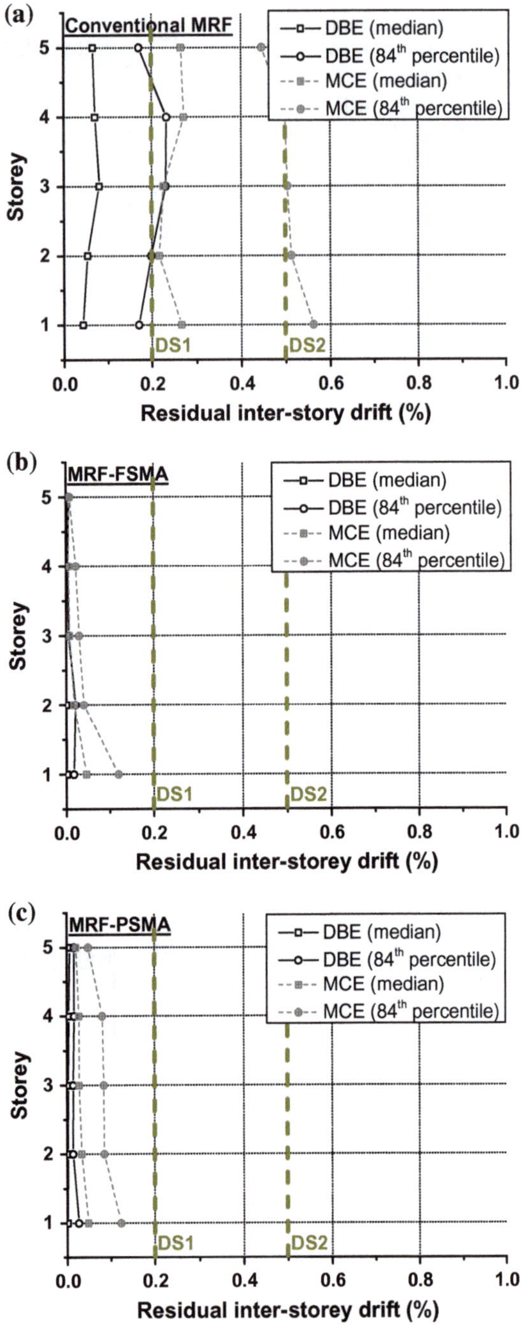

than the maximum deformation capacity. This indicates that the SMA ring springs are not fully compressed when the MID is achieved.

For the RID response, MRF-FSMA and MRF-PSMA have essentially no residual drift under the DBE, whereas the conventional MRF has median and 84th percentile RIDs of 0.07% and 0.24%, respectively. Under the MCE, the median and 84th percentile RID values for the conventional frame increase to 0.21% and 0.56%, respectively. On the other hand, both the SMA frames have small RID values under the MCE, with the median values generally smaller than 0.05%, and the 84th percentile values below 0.15%. It is of interest to note from Fig. 6.35 that MRF-FSMA and MRF-PSMA have comparable residual drift levels, although the number of the SMA connections used for the latter is half of that for the former. This phenomenon suggests that, as long as the static reversal force necessary for 'pushing back' the deformed structure is reasonably decreased (i.e. as in MRF-PSMA), the self-centring response can be significantly improved during dynamic shakedown. It is also suggested that SMA connections can be selectively used in critical regions to achieve satisfactory self-centring responses with a reduced cost.

As for codified assessment, MRF-FSMA and MRF-PSMA both satisfy class DS1 under almost all the MCE-level ground motions. For the conventional MRF, however, five out of 20 DBE-level ground motions result in RID values that exceed 0.2%. For the MCE, 35% of the ground motions result in RID values greater than 0.5%, in which case the conventional frame fails to satisfy the DS2 limit under these ground motions. The results clearly reaffirm the positive role played by the SMA connection in reducing RID for steel frames.

References

ABAQUS (2012) 6.12 analysis user's manual. Dassault Systemes Simulia Corp, Providence, RI

American Society of Civil Engineers (ASCE) (2010) Minimum design loads for buildings and other structures. ASCE/SEI 7-10, Reston, VA

Andrews BM, Fahnestock LA, Song J (2009) Ductility capacity models for buckling-restrained braces. J Constr Steel Res 65(8–9):1712–1720

Christopoulos C, Pampanin S, Priestley MJN (2003) Performance-based seismic response of frame structures including residual deformations—part I: single-degree of freedom system. J Earthq Eng 7(1):97–118

Dimopoulos AI, Karavasilis TL, Vasdravellis G, Uy B (2013) Seismic design, modelling and assessment of self-centering steel frames using post-tensioned connections with web hourglass shape pins. B Earthq Eng 11(5):1797–1816

Eatherton MR, Hajjar JF (2011) Residual drifts of self-centering systems including effects of ambient building resistance. Earthq Spectra 27(3):719–744

Erochko J, Christopoulos C, Tremblay R, Choi H (2011) Residual drift response of SMRFs and BRB frames in steel buildings designed according to ASCE 7-05. J Struct Eng-ASCE 137(5):589–599

European Committee for Standardization (CEN) (2004) EN 1998-1, Eurocode 8: design of structures for earthquake resistance—part 1: general rules, seismic actions and rules for buildings. Belgium, Brussels

Fang C, Zhong QM, Wang W, Hu SL, Qiu CX (2018a) Peak and residual responses of steel moment-resisting and braced frames under pulse-like near-fault earthquakes. Eng Struct 177:579–597

Fang C, Wang W, Ricles J, Yang X, Zhong QM, Sause R, Chen YY (2018b) Application of an innovative SMA ring spring system for self-centering steel frames subject to seismic conditions. J Struct Eng-ASCE 144(8):04018114
Federal Emergency Management Agency (FEMA) (2012) Seismic performance assessment of buildings, volume 1—methodology. FEMA P-58-1, prepared by the SAC Joint Venture for FEMA, Washington, DC
Gupta A, Krawinkler H (1999) Seismic demands for performance evaluation of steel moment resisting frame structures, Report No. 132. John A. Blume Earthquake Engineering Center, Stanford University, Stanford, CA
Hsiao PC, Lehman DE, Roeder CW (2013) Evaluation of the response modification coefficient and collapse potential of special concentrically braced frames. Earthq Eng Struct D 42(10):1547–1564
Karavasilis TL, Seo CY (2011) Seismic structural and non-structural performance evaluation of highly damped self-centering and conventional systems. Eng Struct 33(8):2248–2258
Kari A, Ghassemieh M, Abolmaali SA (2011) A new dual bracing system for improvingthe seismic behavior of steel structures. Smart Mater Struct 20(12):125020
MacRae GA, Kawashima K (1997) Post-earthquake residual displacements of bilinear oscillators. Earthq Eng Struct D 26(7):701–716
Mazzoni S, McKenna F, Scott MH, Fenves GL (2006) Open system for earthquake engineering simulation (OpenSees), OpenSees Command Language Manual. Pacific Earthquake Engineering Research Center, University of California, Berkeley
Nguyen AH, Chintanapakdee C, Hayashikawa T (2010) Assessment of current nonlinear static procedures for seismic evaluation of BRBF buildings. J Constr Steel Res 66(8–9):1118–1127
NIST (2015) Seismic design of steel buckling-restrained braced frames—a guide for practicing engineers, GCR 15-917-34, NEHRP seismic design technical brief no. 11. Produced by the Applied Technology Council and the Consortium of Universities for Research in Earthquake Engineering for the National Institute of Standards and Technology, Gaithersburg, MD
Özhendekci D, Özhendekci N (2012) Seismic performance of steel special moment resisting frames with different span arrangements. J Constr Steel Res 72:51–60
Qiu CX, Zhu SY (2016) High-mode effects on seismic performance of multi-story self-centering braced steel frames. J Constr Steel Res 119:133–143
Ray-Chaudhuri S, Hutchinson TC (2011) Effect of nonlinearity of frame buildings on peak horizontal floor acceleration. J Earthq Eng 15(1):124–142
Ruiz-Garcia J, Miranda E (2006a) Evaluation of residual drift demands in regular multi-story frames for performance-based seismic assessment. Earthq Eng Struct D 35(13):1609–1629
Ruiz-Garcia J, Miranda E (2006b) Residual displacement ratios for assessment of existing structures. Earthq Eng Struct D 35(3):315–336
Sabelli R, Mahin SA, Chang C (2003) Seismic demands on steel braced frame buildings with buckling-restrained braces. Eng Struct 25(5):655–666
Silwal B, Ozbulut OE, Michael RJ (2016) Seismic collapse evaluation of steel moment resisting frames with superelastic viscous damper. J Constr Steel Res 126:26–36
Somerville PG (2003) Magnitude scaling of the near fault rupture directivity pulse. Phys Earth Planet In 137(1–4):201–212
Tremblay R, Lacerte M, Christopoulos C (2008) Seismic response of multistory buildings with self-centering energy dissipative steel braces. J Struct Eng-ASCE 134(1):108–120

Chapter 7
Economic Seismic Loss Assessment

Abstract Economic seismic loss is one of the most important indices that quantify the seismic resilience of a structural system. This chapter provides further insight into the potential of SMA-based self-centring frames for reducing the economic seismic losses. The assessment is conducted based on three prototype steel building frames, namely, conventional moment resisting frame (MRF), buckling-restrained braced frame (BRB frame) and SMA-based self-centring braced frame. The rationale behind the seismic loss assessment framework is explained, and the FEMA P-58 methodology, a procedure which is now widely adopted in the community of seismic engineers, is particularly elaborated. The basic principles of incremental dynamic analysis (IDA) and fragility analysis are also clarified. Based on the existing methodology, the seismic losses of the three types of structures under different levels of earthquakes are obtained, and factors that affect the seismic loss performance are discussed. The assessment can offer more quantitative and rational seismic loss predictions for decision makers, especially in the feasibility analysis stage of construction projects.

7.1 Introduction

Compared with the conventional seismic design methods which are prescriptive in nature, the performance-based seismic design (PBSD) allows a more direct evaluation of the risk of collapse, possible occupancy suspension and economic loss of building structures subjected to future earthquakes. The PBSD is implemented in a probability fashion, i.e. the probability of failure is estimated. The PBSD involves the determination of the desired performance level and the selection of earthquake hazard level which leads to this performance level. Engineers then conduct performance-based assessment and make a judgment whether the selected performance level is satisfied and whether the risk/probability of loss is acceptable by the owners and other stakeholders.

Under the PBSD framework, economic seismic loss (sometimes referred to as seismic loss or repair cost) is one of the most important indices that quantify the seismic resilience of a structural system. Since 2000, basic methodologies and pro-

C. Fang and W. Wang, *Shape Memory Alloys for Seismic Resilience*, https://doi.org/10.1007/978-981-13-7040-3_7

cedures for estimating the building repair cost after earthquakes became available (Porter et al. 2001; Cornell and Krawinkler 2000), and the feasibility of the proposed approaches has been further examined by many researchers in various ways (Porter et al. 2002; Aslani and Miranda 2005; Ramirez and Miranda 2012; Gunay and Mosalam 2013). Meanwhile, computer programs have been developed to facilitate damage measurement and seismic loss estimation. Representative computer programs include PACT, short for Performance Assessment Calculation Tool documented in FEMA P-58 (2012), SLAT, short for Seismic Loss Assessment Tool (Bradley 2011) and SP3, short for Seismic Performance Prediction Program (Haselton and Baker 2017). Among them, PACT has a relatively large structural and non-structural component fragility database and is able to consider residual drifts and replacement cost thresholds in the seismic loss assessment (Yeow et al. 2018). The methodology for PACT is considered in this chapter, and this methodology is often called the FEMA P-58 procedure (FEMA 2012).

It has been confirmed from the previous chapter that the self-centring technology can decrease the RID of building frames to practically negligible levels. This definitely helps reduce the structural and non-structural damages and significantly lowers the repair cost and the probability of demolition. However, the peak floor acceleration (PFA) response may not be well controlled in self-centring structures, and as a result, an increase in the seismic loss may occur due to the more significant damage to the acceleration-sensitive contents within the building. In light of the above uncertainties, it is quite helpful to quantify, using the available PBSD tools, the potential seismic loss of emerging structural systems subjected to different earthquake hazard levels, and to provide judgment if these new structural systems are superior to the conventional ones from the perspective of seismic loss control. The assessment also helps the owners better understand whether the benefit from the possible reduction in seismic loss surpasses the initial increased cost associated with the novel structural members such as SMA-based braces or connections.

This chapter presents the results from a seismic loss assessment on both conventional and SMA-based self-centring seismic-resistant frames. The methodology of seismic loss estimation is briefly described, which is followed by an introduction of the design of the prototype buildings. By using the FEMA P-58 procedure, the different percentile values of seismic loss calculated for the various structural systems are provided, and the reasons behind the difference in the seismic loss performance are explained. Factors that can critically affect the seismic loss performance of the buildings are also highlighted.

7.2 Methodology of Seismic Loss Assessment

7.2.1 Seismic Loss Assessment Framework

The seismic loss assessment framework provided by the Pacific Earthquake Engineering Research (PEER) centre is one of the most widely used approaches among the community of seismic engineers and researchers. In this framework, the annual rate of exceeding a given value of a decision variable is expressed by

$$G(DV|IM) = \int\limits_{DM}\int\limits_{EDP} G(DV|DM)|\,dG(DM|EDP)||\,dG(EDP|IM)| \quad (7.1)$$

where the decision variable, i.e. DV, is essentially the seismic loss in the context of this chapter. IM is the ground shaking intensity measure, which is normally presented by $S_a(T_1)$, i.e. 5% damped elastic response spectral acceleration at the fundamental period of vibration of a structural system of interest. The use of $S_a(T_1)$ as the IM for far-field earthquakes is recommended by FEMA P-58, where this measure is found to produce a lower dispersion compared with other possible IM candidates such as the PGA (Vamvatsikos and Cornell 2002). However, the applicability of using $S_a(T_1)$ as IM to pulse-like near-fault earthquakes has been questioned (Luco and Cornell 2007; Sehhati et al. 2011), because the structural responses under pulse-like excitations could be more relevant to the velocity characteristics of the ground motion.

The DM in Eq. (7.1) is the damage measure, which is in practice (e.g. in the FEMA P-58 procedure) categorised into one or several discrete damage states (e.g. DSs) for each component of concern to simplify the assignment of the associated repair costs (Dimopoulos et al. 2016). Taking a reduced beam section (RBS) beam-to-column connection in a special MRF for example, three DSs are defined in FEMA P-58: DS1 corresponds to the condition that local buckling occurs, and heating is required to straightening the beam flange and web of the buckled region; DS2 is associated with lateral-torsional distortion of the beam in the hinge region requiring partial replacement of the beam flange and web, and corresponding work on adjacent structural and non-structural components; DS3 corresponds to the occurrence of low-cycle fatigue fracture of the plastic hinge, which requires significant repair work for the failure region as well as the adjacent structural and non-structural components (FEMA 2012).

In the present version of FEMA P-58, no definition of DS is available for emerging structural members such as SMA-based members/devices. Nevertheless, a similar DS classification can be proposed based on the designers' own experience and judgment, supported by experimental data. Taking a self-centring brace member equipped with SMA cables as kernel elements for instance, DS1 may be defined as the stage when minor performance degradation occurs due to the functional fatigue of the SMA, a case which may need re-tightening of the SMA cables in order to resume the necessary initial stiffness and 'yield' strength against future earthquakes. DS2 may correspond to the condition that the SMA cables undergo a significant forward transformation strain, and as a result, further degradation of the stiffness and strength of the brace is induced, and residual deformation could occur. This requires re-tightening or replacement of the SMA cables, but no other repair work is needed as the remaining parts of the brace are intact. DS3 may correspond to the occurrence of cable fracture, and due to the significant hardening effect prior to fracture, damage to the connected components such as gusset plates occurs. This requires repair work for these affected neighbouring components, in addition to a complete replacement of the SMA cables. On the other hand, if insufficient test data are available, only one DS, corresponding to the final failure of the component, is also acceptable for SMA-based braces. This single-DS strategy has also been adopted for BRBs in the FEMA P-58 procedure.

In order to establish a fragility function (or fragility curve) for each DS, an appropriate engineering demand parameter (EDP) needs to be defined. The typical fragility curves of a particular structural component are illustrated in Fig. 7.1. The probability of the violation of any DS depends on the development of the EDP, and for each component, a single EDP is often used which can be: inter-storey drift ratio, floor acceleration, floor velocity or any other responses that best estimate the potential occurrence of the DS of the component with the lowest level of uncertainty. For most displacement-sensitive structural components, such as beams, columns, shear walls and braces, and for many non-structural components such as pipelines and façades, inter-storey drift ratio is commonly selected as the EDP which best indicates their damage level. For some velocity-sensitive or acceleration-sensitive non-structural components (e.g. bookcase, ceiling and medical equipment), the use of floor velocity or floor acceleration as the EDP is recommended.

Also, in Eq. (7.1), the symbol $G(A|B)$ means the probability of the exceedance of the parameter A given a condition of B, and $dG(A|B)$ is the differentiation (increment) of the probability of exceedance. Therefore, $G(DV|IM)$ is the probability of exceeding a given value of decision variable (e.g. seismic loss) at a certain ground shaking intensity measure [e.g. $S_a(T_1)$]. What we essentially need is the result of $G(DV|IM)$,

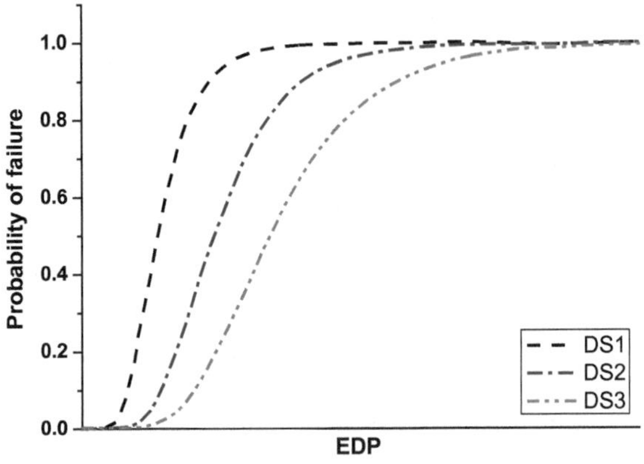

Fig. 7.1 Typical fragility curves for different damage states

i.e. the vulnerability function that represents the variation of the potential seismic loss with an increase in the IM level. The vulnerability function is often demonstrated as a continuous curve of any desired percentile statistical distributions (e.g. 16th, 50th and 84th percentile values).

Similarly, *G(DV|DM)* is the probability of exceedance of a seismic loss given a damage measure, *G(DM|EDP)* is the probability of exceedance of a damage measure given an engineering demand parameter (e.g. inter-storey drift) and finally, *G(EDP|IM)* is the probability of exceedance of an engineering demand parameter given a ground shaking intensity measure.

Provided that the seismic hazard of the site is not particularly researched, the seismic loss assessment can be carried out at the three levels associated with *G(EDP|IM)*, *G(DM|EDP)* and *G(DV|DM)*. In particular, structural analysis, i.e. incremental dynamic analysis (IDA), is required to derive *G(EDP|IM)*; fragility functions are used to acquire *G(DM|EDP)*; and finally, predefined loss distributions are employed to obtain *G(DV|DM)*, and then seismic loss (total repair cost) can be estimated. The calculation of *G(DM|EDP)* and *G(DV|DM)* can be aided by the computer program PACT, where Monte Carlo Simulation is required in such analysis. The assessment procedures are elaborated in the text below.

7.2.2 Seismic Loss Assessment Procedures

Towards the estimation of the seismic loss (repair cost) at any IM level, the relevant EDPs gleaned from the structural analysis are converted into the repair cost of the structural system. By 'linking' the repair cost calculated from each earthquake intensity, a vulnerability function is finally generated. The FEMA P-58 procedure typically involves the following seven major steps (Dimopoulos et al. 2016).

- *Step 1: Determination of EDP*. IDA is employed to obtain the distributions of the EDPs at each level of IM. IDA involves subjecting a structural model to a suite of ground motion records, each scaled to multiple levels of IM, thus producing a group of EDP vs. IM curves, i.e. IDA curves. FEMA P695 (2009) provides 22 pairs of far-field ground motions which can be readily adopted for the IDA. These ground motions are recorded from historical earthquakes with magnitudes ranging from M6.5 to M7.6. According to FEMA P-58, $S_a(T_1)$ should be chosen as the IM for these records. Denoting the number of ground motions and the number of IM levels selected in the IDA as N_{GM} and N_{IM}, respectively, a total of $N_{GM} \times N_{IM}$ non-linear dynamic analyses need to be conducted. For self-centring building frames and other conventional lateral load resisting systems such as MRFs and BRB frames, the relevant EDPs include the MID at each storey, the maximum transient inter-storey drift among all the stories in a structural system, denoted as $\theta_{t,max}$, the maximum residual inter-storey drift among all the stories in a structural system, denoted as $\theta_{r,max}$, the peak floor velocity (PFV) at each floor level, and the PFA at each floor level. These quantities are extracted from each round of IDA. It is noted that a sufficient free vibration decay time (e.g. 10 s) after each earthquake excitation is needed to allow an accurate determination of the RID response.

- *Step 2: Determination of total replacement cost of a building.* The total replacement cost of a building is not equal to its total construction cost. The former actually refers to the total costs of demolishing the original building and those of building a new structure with the same level of function as the original one at the current price level. The representative replacement costs, involving structural and non-structural components, can be obtained from various sources, e.g. https://www.steelconstruction.info for steel construction (Dimopoulos et al. 2016). Apart from that, FEMA P-58 provides the repair cost functions for a variety of building components and contents at their most severe DS, i.e. replacement is needed at this DS. While there is a distribution (i.e. uncertainty) of the total replacement cost, the mean total replacement cost of a building can be determined based on the mean repair cost corresponding to the most severe DS of each component category. In this way, the proportion of the cost associated with each component group (e.g. structural, non-structural and content) can be evaluated.
- *Step 3: Determination of a 3D repair cost table.* Imagine a 3D table that consists of three axes, as shown in Fig. 7.2. Along the x-axis, there are N_{IM} IM values increasing with an appropriate interval, the same as the incremental IMs considered for the IDA. The y-axis of the 3D table gives the number of ground motions, N_{GM}. For example, if a total of 44 ground motions are used in the IDA, then $N_{GM} = 44$. The z-axis defines the number of cost samples per distribution, N_S, which is often taken as a sufficiently large number, e.g. 1000. Therefore, this 3D table, for now, has $N_{IM} \times N_{GM} \times N_S$ 'empty cells' in total to be filled.
  For a given structural model, the EDPs (e.g. inter-storey drift) obtained from a single non-linear dynamic analysis using the ith ground motion scaled to the jth value of IM can be converted into the repair cost of the system. The repair cost

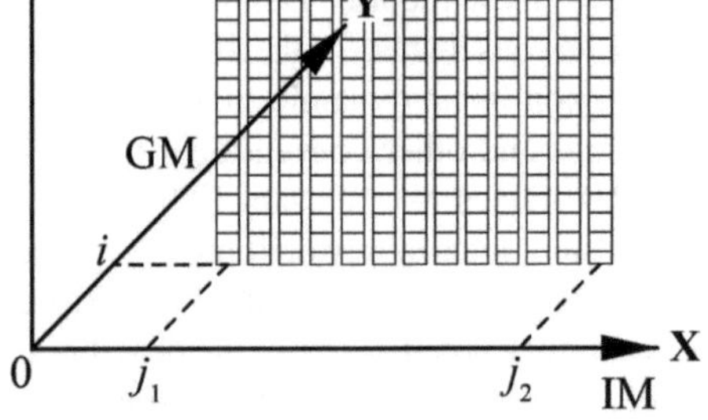

Fig. 7.2 Illustration of 3D repair cost table

is not a definite value but a random variable. An event repair cost vector, $\mathbf{C}^{i,j}$, which is part of the 3D table, can be established to consider the uncertainty of the repair cost. This vector consists of N_S (e.g. 1000) samples of the repair cost for the structure subjected to the ith ground motion scaled to the jth IM:

$$\mathbf{C}^{i,j} = \left\{ C_1^{i,j} \ldots C_k^{i,j} \ldots C_{N_S}^{i,j} \right\}_{1 \times N_S} \tag{7.2}$$

How to determine the, say, 1000 values of $\mathbf{C}^{i,j}$ is explained later in the next steps.

- *Step 4: Estimation of replacement cost if collapse occurs.* When the non-linear dynamic analysis suggests that collapse has occurred in a structural model subjected to the ith ground motion scaled to the jth IM, i.e. the MID exceeds the predetermined threshold, say, 5% inter-storey drift, then the N_S cells in the event repair cost vector are all sampled based on the probabilistic distribution of repair cost associated the worst DS. In other words, $\mathbf{C}^{i,j}(1{:}N_S)$ is filled by randomly permuting the total building replacement costs from Step 2.
- *Step 5: Estimation of demolition cost if no collapse occurs.* When the non-linear dynamic analysis suggests that collapse has not occurred in the structural model subjected to the ith ground motion scaled to the jth IM, the probability of the structure having to be demolished, $P(Dem|\theta_{r,max})$, should be calculated based on the maximum RID ($\theta_{r,max}$) among all the stories. $P(Dem|\theta_{r,max})$ can be assumed to follow a log-normal distribution. As suggested by McCormick et al. (2008), the median value of $\theta_{r,max}$ that leads to demolition should be taken as 0.5%, although some researchers think that this threshold may need to be relaxed. The log-standard deviation of the log-normal distribution of $P(Dem|\theta_{r,max})$ is often taken as 0.3. Accordingly, N_{dem} [i.e. $N_S \times P(Dem|\theta_{r,max})$] out of the N_S cells in $\mathbf{C}^{i,j}(1{:}N_S)$ are filled by randomly permuting the total building replacement costs from Step 2.
- *Step 6: Estimation of repair cost if neither collapse nor demolition occurs.* Before explaining how to fill the remaining N_S-N_{dem} empty cells in this step, the probability of a particular component to suffer damage associated with a certain DS should be first elaborated. For the kth DS (assuming two or more DSs are defined), denoted as DS(k), the corresponding hypothetical fragility curve, such as those shown in Fig. 7.1, is available in FEMA P-58 and can be directly used. The fragility function for each DS is defined by a median demand value at which there is a 50% chance that the damage state will initiate, and a dispersion which indicates uncertainty that the damage state will initiate at this value of demand. For a specific value of EDP associated with the ith ground motion scaled to the jth IM, the fragility curve essentially reports the probability that the component suffers damage equal or higher than DS(k). Subtracting the probabilities of DS(k) and DS($k+1$) gives the probability of the component to experience damage equal to DS(k) under a certain EDP, i.e. P(DS(k)|EDP), as illustrated in Fig. 7.3.
  Subsequently, P(DS(k)|EDP) is multiplied by the remaining $N_R = N_S$ - N_{dem} cells in $\mathbf{C}^{i,j}(1{:}N_S)$ to obtain the number of repair cost cells associated with each DS. The number of cells corresponding to no damage is equal to $N_R \times (1 - \sum P(\text{DS}(k)|\text{EDP}))$, and each cell is set as zero. Component repair cost values can be

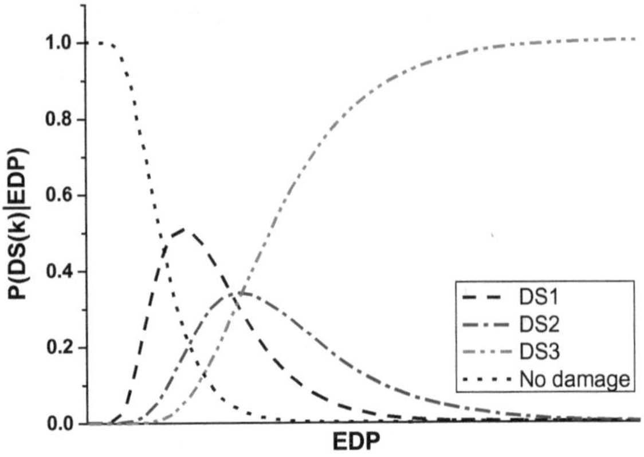

Fig. 7.3 Typical fragility curves for different damage states

obtained from the cost function provided for each specific component experiencing DS(k). Repeating this procedure for all DSs, N_R component repair cost values are obtained based on the probabilistic distribution of repair cost. For each cell, the total repair costs for all the considered components are added in a random fashion, in which way the remaining N_R empty cells in $\mathbf{C}^{\mathrm{i,j}}(1{:}N_S)$ are successfully filled.

- *Step 7: Determination of vulnerability function.* Assembling the vectors of different ground motions, the system repair cost vector, $\mathbf{C}^{\mathrm{j}}$, associated with the entire ground motion set scaled to the jth value of IM, can be expressed by

$$\mathbf{C}^{\mathrm{j}} = \left\{\mathbf{C}^{1,\mathrm{j}} \ldots \mathbf{C}^{2,\mathrm{j}} \ldots \mathbf{C}^{N_{\mathrm{GM}},\mathrm{j}}\right\}_{1\times(N_{\mathrm{S}}\times N_{\mathrm{GM}})} \tag{7.3}$$

A dataset is formed consisting of the $N_{GM} \times N_S$ cells in $\mathbf{C}^{\mathrm{j}}$. The percentile values (e.g. 16th, 50th and 84th percentile values) can be obtained from the dataset. These are indeed the percentile total repair cost values for the jth IM. Repeating this procedure using various IMs, the percentile total repair cost (seismic loss) versus IM curves, i.e. the vulnerability functions, are generated for each considered structural model.

7.3 Prototype Buildings for Seismic Loss Assessment

7.3.1 Prototype Conventional Steel Frames

Two conventional structural systems, namely a 9-storey MRF and a 9-storey BRB frame, are considered for seismic loss assessment (Hu et al. 2019). Again, owing to regular structural layout and symmetry, only 2D lateral-resisting frames are considered to numerically replicate the seismic responses of the entire prototype building along the direction of interest.

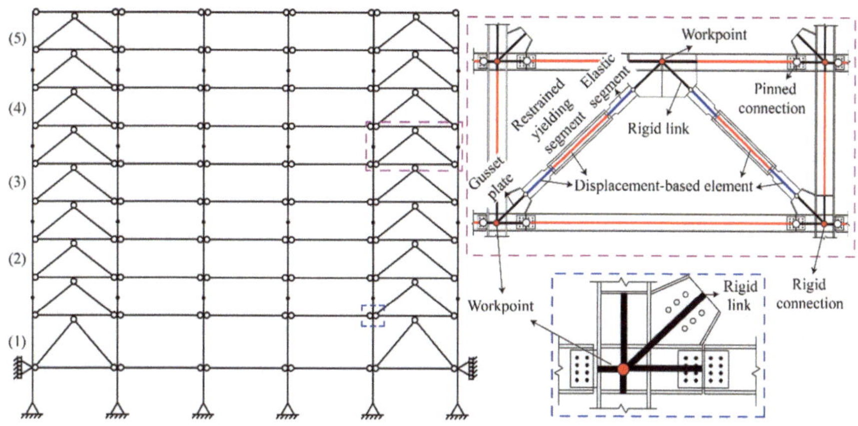

Fig. 7.4 Layout and connection details of BRB frame

For seismic loss assessment, the design of the structural systems should be as close as possible to the real practice. The considered MRF is the one designed for the SAC steel project (Gupta and Krawinkler 1999), where the structural information has been discussed in Chap. 6.

The details of the BRB frame are shown in Fig. 7.4 and Table 7.1. The fundamental period of the frame is 1.75 s. The frame member sizes and the connection details are different from the idealised 9-storey BRB frame discussed in Chap. 6. In particular, pinned beam-to-column connections are used here in the BRB frame for seismic loss assessment, instead of the rigid connections used in the previous version. This is because according to ASCE 7-10 (2010), the beam-to-column connections are allowed to be designed as pinned in BRB frames that are not taller than 49 m. As the total height of the prototype BRB frame is 37.2 m, engineers would choose to design pinned connections instead of rigid ones as the former is much more cost effective. Similar to the design principle discussed in Chap. 6, the BRBs are designed to meet the strength requirement and to ensure that the MID is not larger than 2%, following the standard design procedure such as the equivalent lateral force procedure or the modal response spectrum analysis. The beams and columns are designed based on the capacity design principle according to ANSI/AISC 341-10 (2010). Again, the BRB frames can be simulated via OpenSees (Mazzoni et al. 2006). In order to more realistically capture the fundamental hysteretic behaviour of the BRBs towards a more accurate assessment of the seismic loss, a built-in *SteelBRB* material model, which allows the consideration of cyclic hardening and unequal tension and compression properties, is used here (Zona and Dall'Asta 2012; Gu et al. 2014), instead of the idealised elasto-plastic material model used in the BRB frame considered in Chap. 6.

Table 7.1 Summary of design of BRB frames

Column sizes from bottom to top	Floor level	Beam size	Brace core cross-sectional area (mm^2)	Brace yield strength (kN)	Brace initial stiffness (kN/mm)
W14 × 176	Basement	W16 × 89	N.A.	N.A.	N.A.
	1	W16 × 89	6693	1572.86	393.74
	2	W16 × 77	5564	1307.54	386.59
W14 × 176	3	W16 × 77	5359	1259.37	372.34
	4	W16 × 77	5036	1183.46	349.90
W14 × 159	5	W16 × 77	4584	1077.24	318.50
	6	W16 × 77	3993	938.36	277.43
W14 × 120	7	W16 × 77	3256	765.16	226.23
	8	W16 × 77	2366	556.01	164.39
W14 × 120	9	W16 × 77	2015	473.53	140.00

7.3.2 *Prototype Self-centring Braced Frames*

The considered self-centring braced frame has exactly the same spatial distribution of the braces as that of the BRB frame. From the perspective of seismic loss comparison, it is reasonable to assume that the initial stiffness and the 'yield' resistance of the two types of braces at each storey are similar, and in addition, the boundary frame layout and connection design for the two systems are identical. As a result, the self-centring and BRB frames essentially have the same fundamental period, but vary in the inelastic hysteretic behaviour. The details of the prototype self-centring steel frame with SMA-based braces are shown in Fig. 7.5.

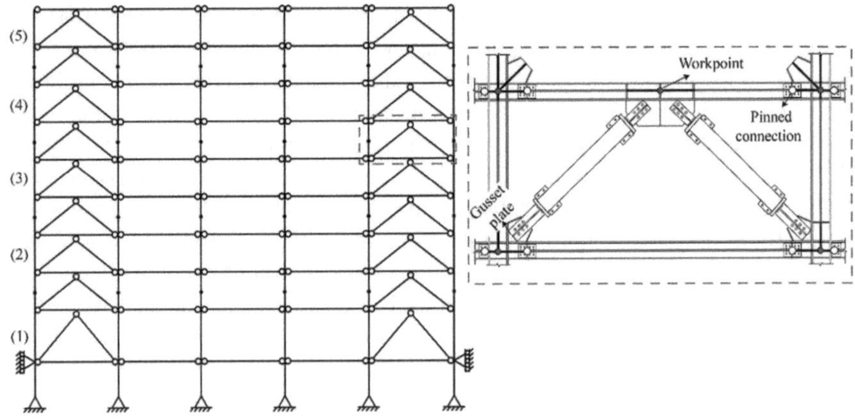

Fig. 7.5 Layout of self-centring frame

The axial force–deformation response of a typical self-centring brace at the 6th floor is shown in Fig. 7.6. The kernel part of the brace consists of an outer tube and an inner tube, together with 4 preloaded SMA cables and a pair of friction pads that provide load resistance and energy dissipation, as shown in Fig. 7.7. The brace is capable of providing both tensile and compressive deformations, and in either condition, the SMA cables are always stretched. The working principle of the self-centring brace is similar to that described in Chap. 4. The idealised material properties of the SMA cables and the basic design information of the brace are given in Tables 7.2 and 7.3, respectively.

Figure 7.8 shows the non-linear static pushover curves (i.e. normalised base shear versus roof drift curves) of the two braced frame models. Being in line with the design intention, the two models have similar initial stiffness and yield strengths. The post-yield stiffness of the BRB frame seems to be higher than that of the self-centring frame, which is primarily due to the increase in the compressive resistance of the BRBs while the difference of the tensile and compressive behaviours does not exist in the examined SMA-based self-centring braces.

It is worth noting that a detailed equivalent lateral force or modal response spectrum analysis procedure is not readily available in the current design standards for self-centring frames. The same stiffness and strength properties assumed for the self-centring and BRB frames implicitly indicate that the same response modification factor (i.e. R) of 8 is considered. To make sure that this simple assumption leads to safe design, a supplementary non-linear time-history analysis is conducted for the self-centring frame. Here, the 15 far-field ground motions adopted in Chap. 6 are used again. These ground motions have the mean spectral accelerations compatible with the DBE design spectrum and can be scaled by a factor of 1.5 to represent the MCE level earthquake.

The non-linear time-history results show that the mean MIDs of the critical storey for the self-centring and BRB frames under the DBE are 2.3% and 1.3% respectively, both satisfying the 2.5% limit stipulated in ASCE 7-10 (2010). Incidentally,

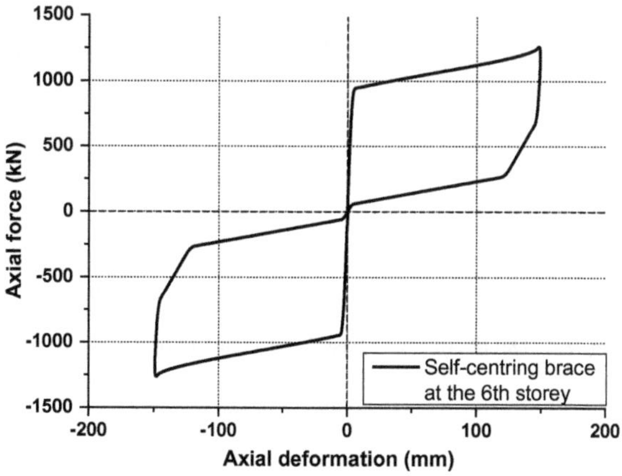

Fig. 7.6 Typical axial force–deformation response of SMA-based braces

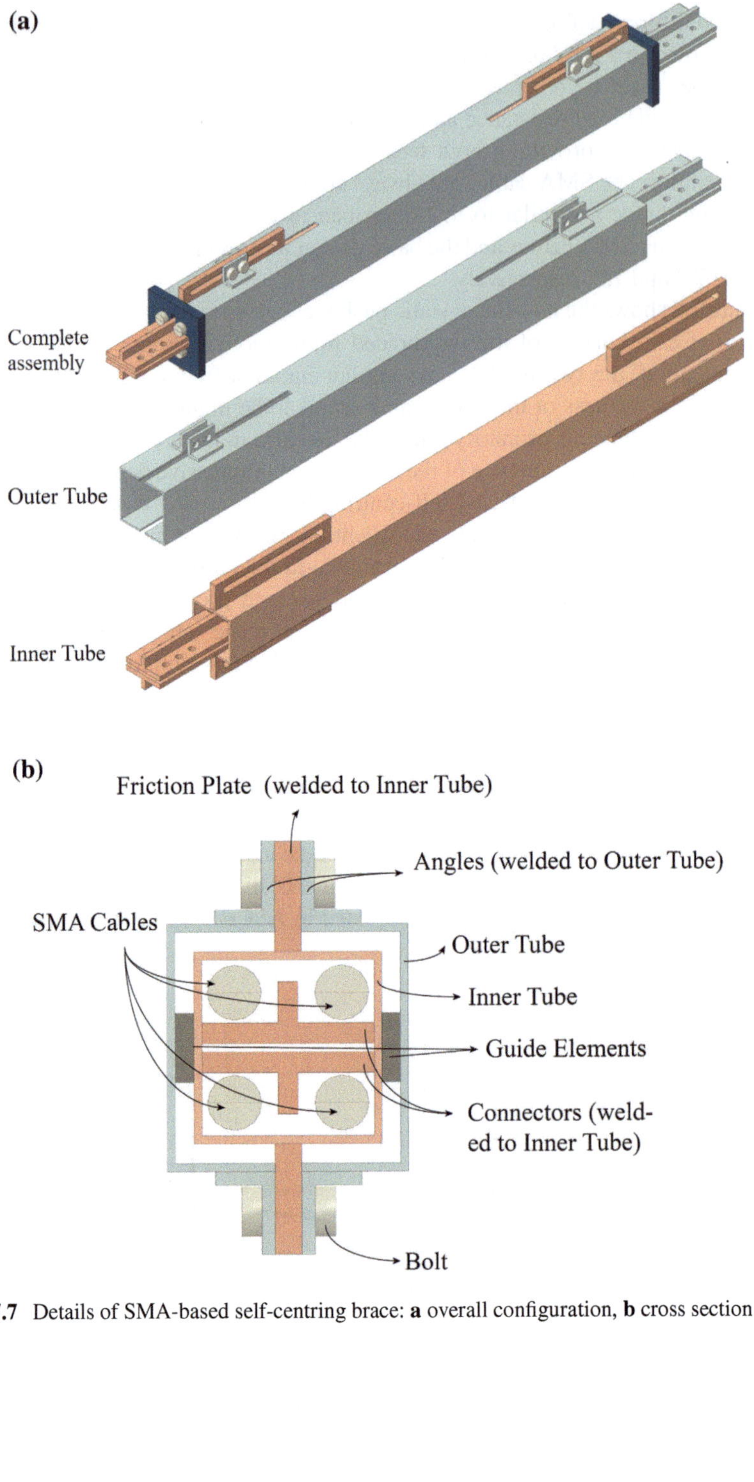

Fig. 7.7 Details of SMA-based self-centring brace: **a** overall configuration, **b** cross section

Table 7.2 Idealised material properties of SMA cables

Material properties	Value
Forward transformation (start) stress σ_{Ms}	400 MPa
Forward transformation (finish) stress σ_{Mf}	700 MPa
Reverse transformation (start) stress σ_{As}	350 MPa
Reverse transformation (finish) stress σ_{Af}	200 MPa
Young's modulus (austenite) E_A	40 GPa
Young's modulus (martensite) E_M	30 GPa
Maximum transformation strain ε_L	7.5%

Table 7.3 Design information of typical SMA-based self-centring brace at the 6th storey

Design information	Value
Yield resistance of each brace	938.4 kN
Length of SMA cables	2015 mm
Diameter of SMA cables	22.8 mm
Number of SMA cables per brace	4
Preload for each SMA cable	167.5 kN
Total extra force from frictional pads per brace	268.7 kN

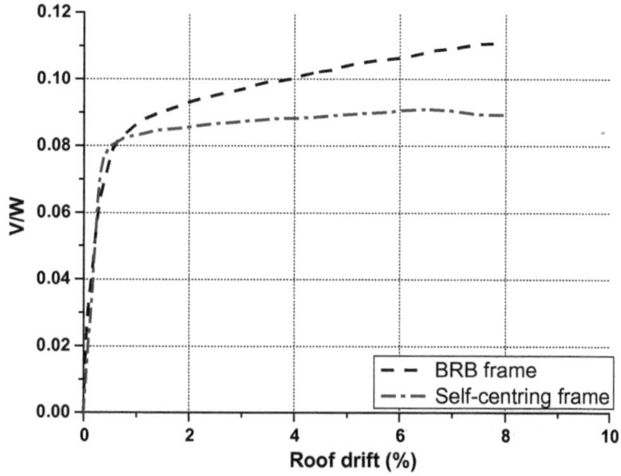

Fig. 7.8 Static pushover responses of braced frames

the mean MIDs of the critical storey of the two frames under the MCE are 3.1% and 2.6%, respectively, indicating that the risk of collapse is remote. The enlarged MID response of the self-centring frame is due to the decreased energy dissipation capability provided by the SMA-based braces. More energy dissipation can be promoted by increasing the frictional force of the self-centring brace, but no such effort is made in this chapter. In any case, the 9-storey self-centring braced frame in its present form is code compliant. In order to avoid a biased assessment, the self-centring system is adopted as it is without making any further changes.

7.3.3 Components and Component Fragility

For the considered MRF, BRB frame and SMA-based self-centring frame, Tables 7.4 and 7.5 list the tributary structural and non-structural components, respectively. Also presented in the tables are the number of component units, the corresponding EDP, and the ID number of each component given in FEMA P-58 (2012). Summing up the mean repair cost associated with the most severe DS of each component, the mean total repair costs tributary to the three frames are 7.795, 7.554 and 8.194 million US dollars, respectively. The total repair cost for the self-centring frame is the highest, which is due to the more expensive SMA-based self-centring braces used in the system. Figure 7.9 further shows the percentages of the repair costs from different component categories. It is seen that structural components take up more than 60% of the total repair cost, where the braces (if any) make up a large part, especially for the self-centring frame.

For each component, the extent of damage subjected to a particular EDP is uncertain. Component fragility functions typically follow log-normal statistical distributions, which indicate the conditional probability of damage at a given EDP. For a considered DS, the median EDP is the value that corresponds to a 50% probability that this DS is exceeded. The value of dispersion, β, reflects the uncertainties related to manufacturing tolerance, construction process, material quality, imperfections, the

Table 7.4 Summary of tributary structural components per storey

MRF	BRB frame	Self-centring frame	ID in FEMA P-58	Units	EDP
Steel column base plate	Steel column base plate	Steel column base plate	B1031.011a/B1031.011b/B1031.011c	6	MID
Post-Northridge welded steel moment connection, not RBS, beam one side	–	–	B1035.021/B1035.022	1	MID
Post-Northridge welded steel moment connection, not RBS, beam two sides	–	–	B1035.031/B1035.032	4	MID

(continued)

Table 7.4 (continued)

MRF	BRB frame	Self-centring frame	ID in FEMA P-58	Units	EDP
Bolted shear tab gravity connections	Bolted shear tab gravity connections	Bolted shear tab gravity connections	B1031.001	1/10	MID
–	Steel buckling-restrained brace (BRB), chevron brace	Self-centring braces, chevron brace	B1033.101b	4	MID

Table 7.5 Summary of tributary non-structural components and contents per storey

MRF	BRB frame	Self-centring frame	ID in FEMA P-58	Units	EDP	Category
Curtain walls	Curtain walls	Curtain walls	B2022.001	90/65	MID	Non-structural component
Suspended ceiling	Suspended ceiling	Suspended ceiling	C3032.003a	15	PFA	Non-structural component
Sanitary waste piping	Sanitary waste piping	Sanitary waste piping	D2031.011b	1	PFA	Non-structural component
HVAC	HVAC	HVAC	D3041.001a	5	PFA	Non-structural component
Modular office work stations	Modular office work stations	Modular office work stations	E2022.001	45	PFA	Contents
Unsecured fragile objects on shelves	Unsecured fragile objects on shelves	Unsecured fragile objects on shelves	E2022.010	45	PFA	Contents
Fragile contents on shelves in storage cabinets with latches	Fragile contents on shelves in storage cabinets with latches	Fragile contents on shelves in storage cabinets with latches	E2022.012	45	PFA	Contents

(continued)

Table 7.5 (continued)

MRF	BRB frame	Self-centring frame	ID in FEMA P-58	Units	EDP	Category
Electronic equipment on wall	Electronic equipment on wall	Electronic equipment on wall	E2022.021	1	PFA	Contents
Desktop electronics	Desktop electronics	Desktop electronics	E2022.022	45	PFA	Contents
Bookcase	Bookcase	Bookcase	E2022.103a	45	PFV	Contents

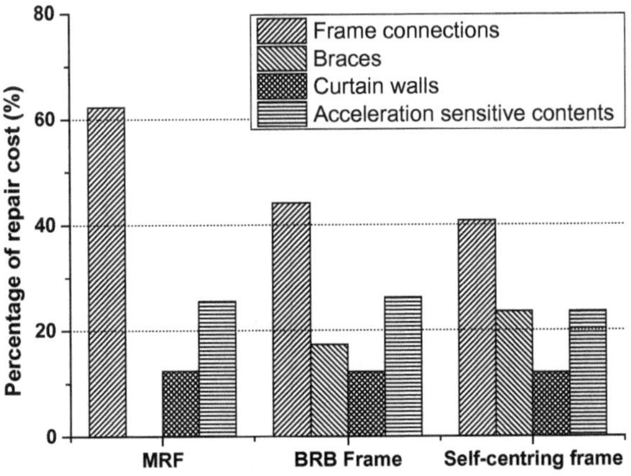

Fig. 7.9 Percentages of repair costs for three different structural systems

level of knowledge regarding the likely behaviour of a component under a specified value of EDP, and the extent to which the occurrence of damage can be predicted by a single EDP (FEMA 2012).

Component fragility functions can be determined through various ways such as physical tests, numerical/analytical investigations, data collected from post-earthquake survey, engineering judgment, or a combination of these approaches. A fragility function database, covering more than 700 typical structural and non-structural components and contents, is now available in FEMA P-58 (2012). However, for the SMA-based self-centring braces, information on the DSs and the corresponding probabilistic distribution of repair cost associated with each DS, is currently lacking. A fairly reasonable assumption is that each self-centring brace has a similar definition of DS and probabilistic distribution of repair cost to the conventional BRB with the same loading capacity, with the exception that the total replacement cost of the former is 50% higher than that of the latter. In FEMA P-58, only one DS is defined for BRBs, where the median demand is 2% inter-storey drift.

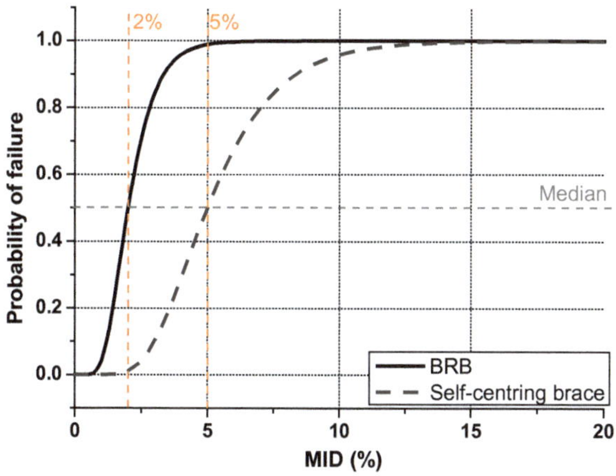

Fig. 7.10 Component fragility functions of BRB and self-centring brace

For the self-centring braces, a 5% inter-storey drift is recommended as the median demand for the DS, which is larger than that used for BRBs. The lower MID threshold considered for the BRBs is attributed to their higher risk of low-cycle fatigue, whereas for the SMA-based self-centring braces, the final failure mode is governed by cable fracture which typically occurs at larger inter-storey drifts. The component fragility curves associated with the DS defined for both the BRBs and self-centring braces are illustrated in Fig. 7.10.

7.4 Incremental Dynamic Analysis and Collapse Fragility

Incremental dynamic analysis (IDA) is performed up to the predefined collapse threshold on the structures subjected to 22 far-field earthquake time histories selected from FEMA P695 (2009). The relevant EDPs recorded from the IDA are: $\theta_{t,max}$, $\theta_{r,max}$, PFA and PFV. The typical IDA curves associated with $\theta_{t,max}$ of the three frames are shown in Fig. 7.11. The RTR dispersion of the structural responses at a certain IM can be clearly observed. Taking the MRF for example, the $\theta_{t,max}$ ranges from 1.7 to 3.7% when $S_a(T_1)/S_a(T_1,DBE) = 1.0$, which indicates that the MRF does not collapse under the DBE. For the BRB frame, however, the $\theta_{t,max}$ exceeds 2% inter-storey drift under some of the ground motions when $S_a(T_1)/S_a(T_1,DBE) = 1.0$, implying that the risk of collapse exists when the BRB frame is under the DBE. It is recalled that the collapse thresholds for the MRF and BRB frame are defined as 5% and 2% inter-storey drifts, respectively. For the self-centring frame, the $\theta_{t,max}$ exceeds 5% inter-storey drift under 5 out of the 22 ground motions, indicating that the risk of collapse also exists when the system is under the DBE.

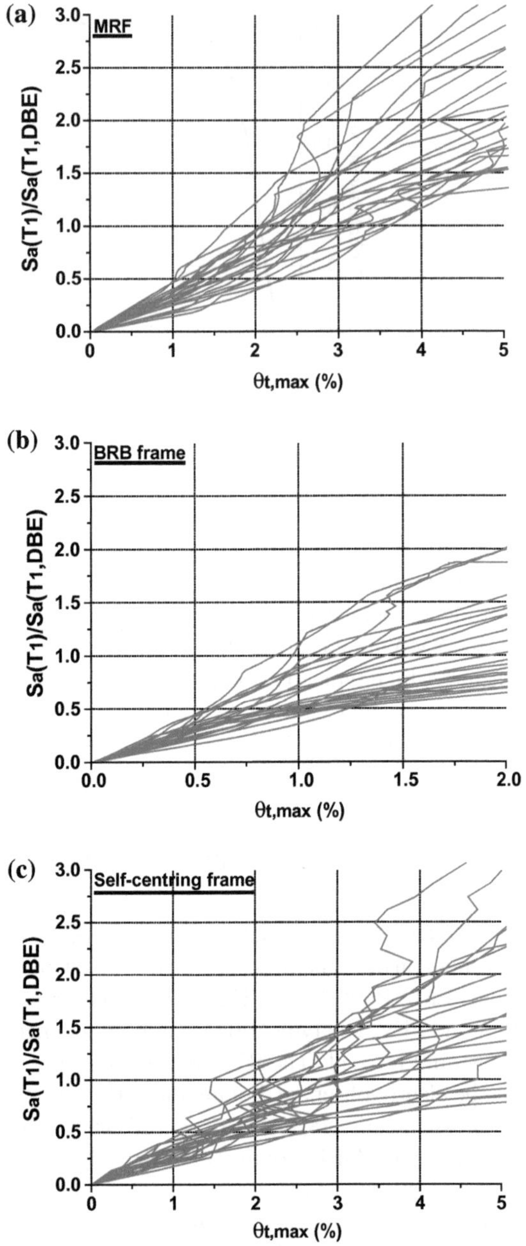

Fig. 7.11 IDA curves associated with $\theta_{t,max}$ of considered frames at the most critical storey: **a** MRF, **b** BRB frame, **c** self-centring frame

While the IDA curves are most commonly presented via IM-$\theta_{t,max}$ relationships, the variations of $\theta_{r,max}$, PFA and PFV at the most critical storey with incremental IMs can also be monitored, as shown in Fig. 7.12 for the self-centring frame. The results indicate that the residual inter-storey drift responses, as expected, are insignificant for the self-centring frame under the IM up to $S_a(T_1)/S_a(T_1,DBE) = 3.0$. The PFA of the self-centring frame at the critical storey under the DBE generally ranges from 1.0 to 2.0 g, and the PFV ranges from 3.5 to 20 m/s. These quantities generally increase with an increase in the IM. Figure 7.13 further shows the variation of $\theta_{r,max}$ with an increase in IM of the MRF and BRB frame. In contrast to the low $\theta_{r,max}$ values exhibited by the self-centring frame, the residual inter-storey drift responses of the two conventional frames are much more significant.

Considering the most critical MID responses among all the stories (i.e. $\theta_{t,max}$), the IDA results can be used for constructing the collapse fragility curves of the structures. The use of the IDA responses for seismic loss assessment will be discussed later in Sect. 7.5. Employing a commonly considered relationship between EDP and IM, i.e. $EDP = \alpha(IM)^{\beta}$, and assuming that the $\theta_{t,max}$ follows a log-normal distribution, the probabilistic demand model for $\theta_{t,max}$, which relates the median value $\bar{\theta}_{t,max}$ to the IM, $S_a(T_1)$, is given by

$$\ln \bar{\theta}_{t,max} = \ln \lambda + \kappa \ln[S_a(T_1)] \tag{7.4}$$

where λ and κ are constants which are determined via least square regression. The probability of the structure failing to meet a target performance objective such as collapse prevention at a given IM is expressed by

$$P(\text{failure}|IM) = P[\theta_{t.max} > \theta_c | S_a(T_1)] \tag{7.5}$$

where θ_c is the MID associated with the performance objective of collapse prevention, and the median values $\bar{\theta}_c$ are 5% inter-storey drift for the MRF and self-centring frame. A reduced $\bar{\theta}_c$ value of 2% inter-storey drift is considered for the BRB frame. Assuming log-normal distributions for both $\theta_{t,max}$ and θ_c, Eq. (7.5) can be written as

$$\begin{aligned} P[\theta_{t.max} > \theta_c | S_a(T_1)] &= 1 - \Phi\left[\frac{\ln(\bar{\theta}_{t\cdot max}) - \ln(\bar{\theta}_c)}{\sqrt{\beta_c^2 + \beta_d^2}}\right] \\ &= 1 - \Phi\left[\frac{\ln(\bar{\theta}_{t\cdot max}) - \ln(\bar{\theta}_c)}{0.4}\right] \end{aligned} \tag{7.6}$$

In Eq. (7.6), $\Phi[x]$ is the cumulative normal distribution function; β_d and β_c are the dispersions of $\theta_{t,max}$ and θ_c, respectively. The combined dispersion may be taken as 0.4 according to HAZUS99 (1999).

Figure 7.14 shows the collapse fragility curves of the three considered structural systems. The risk of collapse is generally low when these structures are subjected to the DBE, where the probabilities of failure are 5.8, 11.9, to 24.8% for the MRF, self-

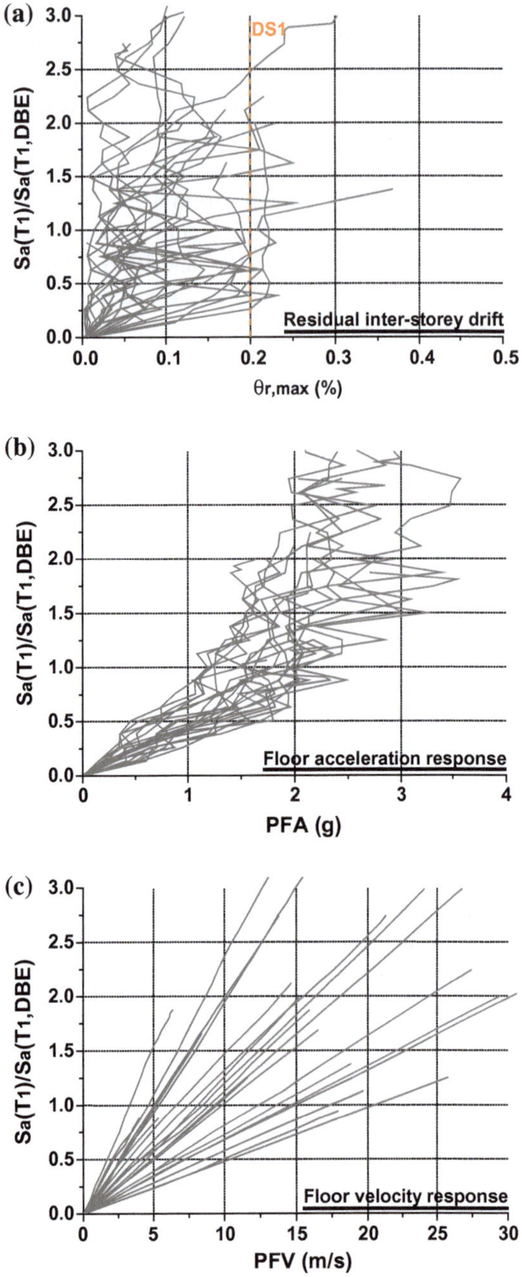

Fig. 7.12 IDA curves associated with $\theta_{r,max}$, PFA and PFV of self-centring frame at the most critical storey: **a** residual inter-storey drift, **b** floor acceleration, **c** floor velocity

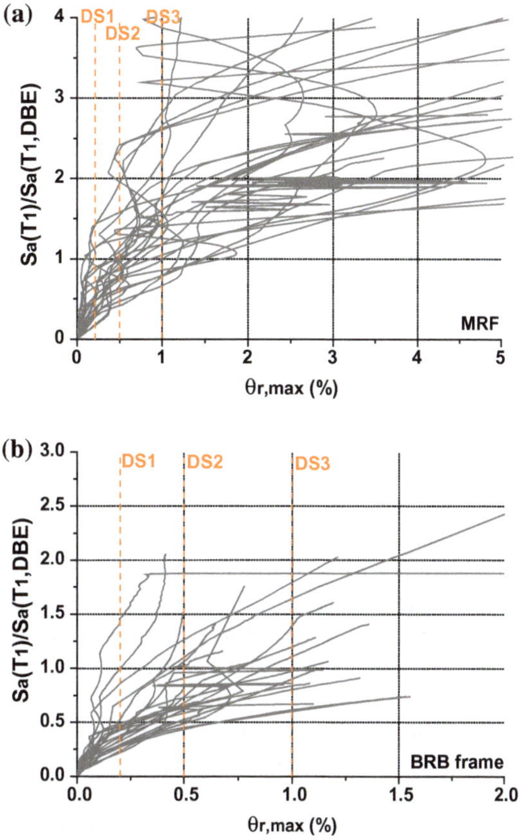

Fig. 7.13 IDA curves associated with $\theta_{r,max}$ at the most critical storey: **a** MRF, **b** BRB frame

centring frame and BRB frame, respectively. Under the MCE, i.e. 1.5 × $S_a(T_1,DBE)$, the risk of collapse increases significantly, especially for the BRB frame, where the probability of failure achieves 60%. When under 4.0 × $S_a(T_1,DBE)$, the probability of failure reaches 100% for the BRB and self-centring frames. Generally speaking, the BRB frame exhibits the highest probability of failure among the three systems, which is due to the lower $\bar{\theta}_c$ value considered. With the same $\bar{\theta}_c$ value, the probability of failure of the MRF is lower than that of the self-centring frame, which is attributed to the more energy dissipation provided by the former.

7.5 Economic Seismic Losses

According to the FEMA P-58 procedure described in Sect. 7.2, the vulnerability functions of the considered structures are obtained from the IDA results, as shown in Fig. 7.15, where the 16th, 50th and 84th percentile values are given. The vulnerability

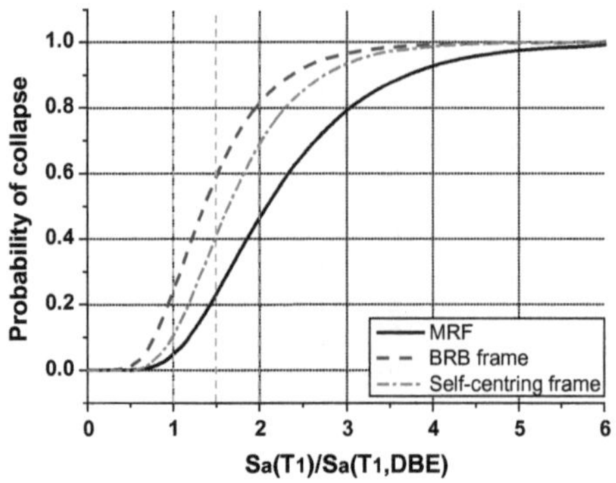

Fig. 7.14 Probability of collapse for three structural systems under varying IMs

functions are constructed based on the assumption that the median value and the standard deviation of the $P(Dem|\theta_{r,max})$ distribution are 0.5% and 0.3, respectively. The repair cost is normalised by the mean total repair cost mentioned in Sect. 7.3. It is observed that the BRB frame and MRF have comparable seismic loss performances, whereas the seismic loss is reduced evidently for the self-centring frame. It is believed that the reduced seismic loss is attributed to the low RID response, as well as the relatively high ductility of the SMA-based self-centring braces.

Concentrating on the 50th percentile values, it is found that the seismic losses of the MRF and BRB frame under the DBE are close to their mean total repair costs. The excessive economic losses are related to the pronounced residual inter-storey drifts in both frames, which leads to a high demolition potential. On the other hand, the seismic losses of the self-centring frame under the DBE and MCE are approximately 4.5% and 46% of the total repair cost, respectively. The increase in the 'budget' due to the use of the more expensive SMA-based braces is on the order of 0.5 million US dollars, which takes up around 6.25% of the mean total repair cost.

In order to understand the sensitivity of the vulnerability functions to the $\theta_{r,max}$ threshold, the median value of the log-normal distribution of $P(Dem|\theta_{r,max})$ is extended to 1.0 and 1.5% inter-storey drifts while the standard deviation is kept unchanged. Figure 7.16 shows the 50th percentile vulnerability curves of the different systems, considering the different $\theta_{r,max}$ thresholds. The MRF is found to exhibit the most significant sensitivity to the $\theta_{r,max}$ threshold. This is because the $\theta_{r,max}$ responses of the structure are distributed over a relatively wide range, and hence, an adjustment of the $\theta_{r,max}$ threshold alters the potential for demolition. The BRB frame appears to be sensitive to the $\theta_{r,max}$ threshold only when the median of $P(Dem|\theta_{r,max})$ increases from 0.5 and 1.0% inter-storey drifts. A further increase in the $\theta_{r,max}$ threshold has a limited influence on the vulnerability curve, probably because the $\theta_{r,max}$ of the BRB frame is below 1% inter-storey drift for most cases. For the same reason, the seismic loss of the self-centring frame is not sensitive to the $\theta_{r,max}$ threshold as the SMA-based braces help eliminate the RID responses to very small values being less than 0.5%.

Fig. 7.15 Vulnerability curves of structural systems under varying IMs: **a** 16th percentile, **b** 50th percentile, **c** 84th percentile

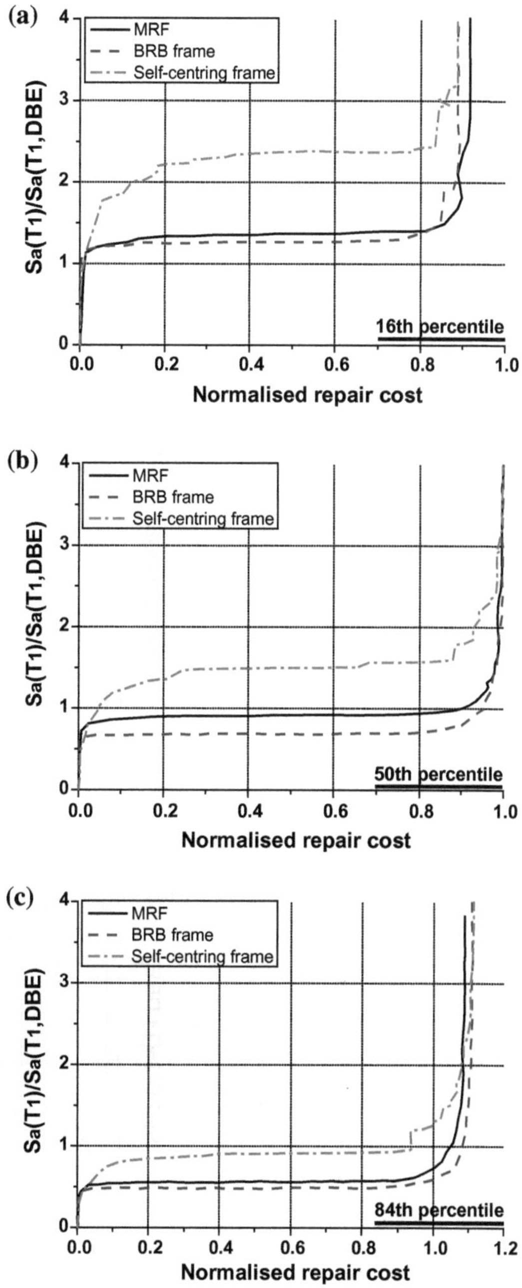

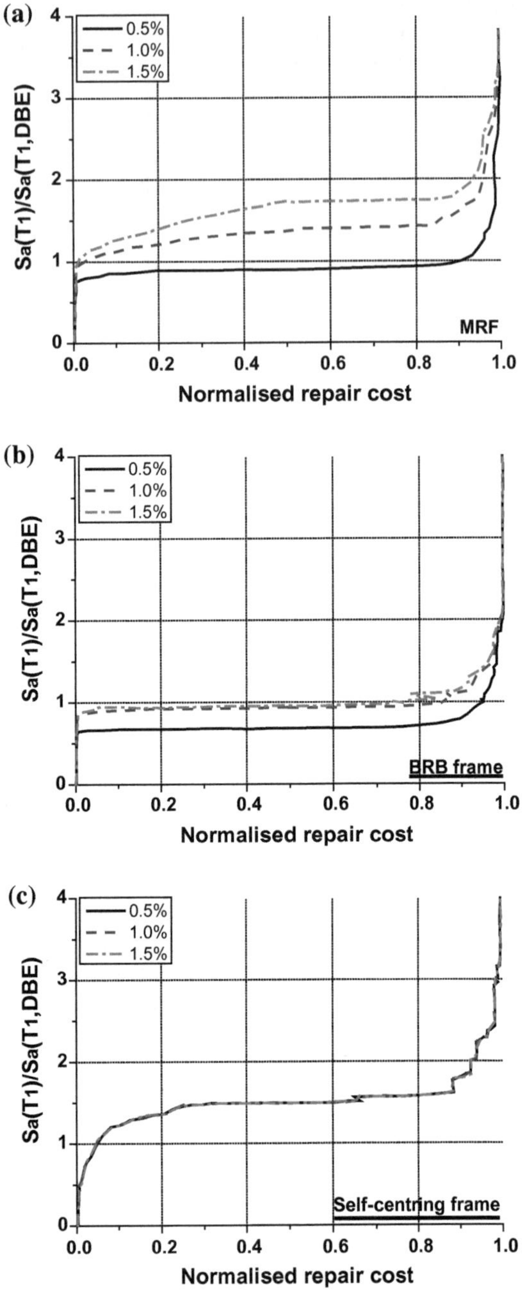

Fig. 7.16 Influence of $\theta_{r,max}$ threshold on 50th percentile vulnerability functions: **a** MRF, **b** BRB frame, **c** Self-centring frame

Fig. 7.17 Influence of θ_c threshold on vulnerability functions

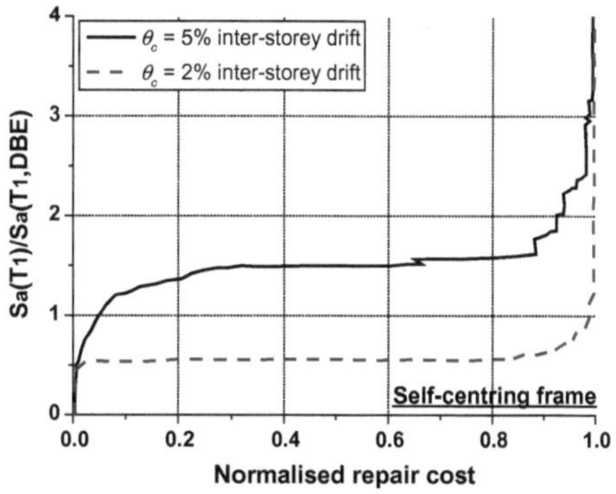

A further comparison is given in Fig. 7.17 to examine the sensitivity of the vulnerability function to the θ_c threshold which indicates the collapse resistance. Recall that the θ_c threshold is assumed to be 5% inter-storey drift for the self-centring frame. This assumption is reasonable for the case of SMA-based self-centring braces that have sufficient ductility, but is not realistic for 'conventional' steel cable- or tendon-based self-centring braces which often possess lower deformability. Considering two θ_c thresholds, which are 5% and 2% inter-storey drifts, the figure clearly shows that a decrease in the θ_c threshold significantly increases the seismic loss, as the potential for collapse increases remarkably. This highlights the necessity of accurately estimating the θ_c threshold in order to ensure a fairly reasonable estimation of the seismic loss.

References

AISC (2010) Seismic provisions for structural steel buildings, (ANSI/AISC 341-10). American Institute of Steel Construction, Chicago, IL, USA

American Society of Civil Engineers (ASCE) (2010) Minimum design loads for buildings and other structures. ASCE/SEI 7-10, Reston, VA

Aslani H, Miranda E (2005) Probabilistic earthquake loss estimation and loss disaggregation in buildings, Report No. 157. John A. Blume Earthquake Engineering Research Center, Stanford University, Stanford, CA

Bradley BA (2011) User manual for SLAT: seismic loss assessment tool, version 1.16. Department of Civil and Natural Resources Engineering, University of Canterbury, Christchurch, New Zealand

Cornell CA, Krawinkler H (2000) Progress and challenges in seismic performance assessment. PEER Center News 3(2):1–4

Dimopoulos AI, Tzimas AS, Karavasilis TL, Vamvatsikos D (2016) Probabilistic economic seismic loss estimation in steel buildings using post-tensioned moment-resisting frames and viscous dampers. Earthq Eng Struct D 45(11):1725–1741

Federal Emergency Management Agency (FEMA) (2009) Quantification of building seismic performance factors. FEMA P695, prepared by the Applied Technology Council for the FEMA, Washington, DC

Federal Emergency Management Agency (FEMA) (2012) Seismic performance assessment of buildings, volume 1—methodology. FEMA P-58-1, prepared by the SAC Joint Venture for FEMA, Washington, DC

Gu Q, Zona A, Peng Y, Dall'Asta A (2014) Effect of buckling-restrained brace model parameters on seismic structural response. J Constr Steel Res 98:100–113

Gunay S, Mosalam KM (2013) PEER performance-based earthquake engineering methodology, revisited. J Earthq Eng 17(6):829–858

Gupta A, Krawinkler H (1999) Seismic demands for performance evaluation of steel moment resisting frame structures, Report No. 132. John A. Blume Earthquake Engineering Center, Stanford University, Stanford, CA

Haselton CB, Baker JW (2017) Seismic performance prediction program, 2.6th edn. Haselton Baker Risk Group, Chico

HAZUS99 (1999) Earthquake loss estimation methodology: user's manual. Federal Emergency Management Agency, Washington DC

Hu S, Wang W, Qu B (2019) Seismic economic losses in mid-rise steel buildings with conventional and emerging lateral force resisting systems. Eng Struct Under Rev

Ramirez CM, Miranda E (2012) Significance of residual drifts in building earthquake loss estimation. Earthq Eng Struct D 41(11):1477–1493

Luco N, Cornell CA (2007) Structure-specific scalar intensity measures for near-source and ordinary earthquake ground motions. Earthq Spectra 23(2):357–392

Mazzoni S, McKenna F, Scott MH, Fenves GL (2006) Open system for earthquake engineering simulation (OpenSees), OpenSees Command Language Manual. Pacific Earthquake Engineering Research Center, University of California, Berkeley

McCormick J, Aburano H, Ikenaga M, Nakashima M (2008) Permissible residual deformation levels for building structures considering both safety and human elements. In: Proceedings of 14th world conference on earthquake engineering, Seismological Press of China, Beijing

Porter KA, Beck JL, Shaikhutdinov RV (2002) Sensitivity of building loss estimates to major uncertain variables. Earthq Spectra 18(4):719–743

Porter KA, Kiremidjian AS, LeGrue JS (2001) Assembly-based vulnerability of buildings and its use in performance evaluation. Earthq Spectra 17(2):291–312

Sehhati R, Rodriguez-Marek A, ElGawady M, Cofer WF (2011) Effects of near-fault ground motions and equivalent pulses on multi-story structures. Eng Struct 33(3):767–779

Vamvatsikos D, Cornell CA (2002) Incremental dynamic analysis. Earthq Eng Struct D 31(3):491–514

Yeow TZ, Orumiyehei A, Sullivan TJ, MacRae GA, Clifton GC, Elwood KJ (2018) Seismic performance of steel friction connections considering direct-repair costs. B Earthq Eng 16(12):5963–5993

Zona A, Dall'Asta A (2012) Elastoplastic model for steel buckling-restrained braces. J Constr Steel Res 68:118–125

MIX
Papier aus verantwortungsvollen Quellen
Paper from responsible sources
FSC® C105338

If you have any concerns about our products,
you can contact us on
ProductSafety@springernature.com

In case Publisher is established outside the EU,
the EU authorized representative is:
Springer Nature Customer Service Center GmbH
Europaplatz 3, 69115 Heidelberg, Germany

Printed by Libri Plureos GmbH
in Hamburg, Germany